HVAC Fundamentals
3rd Edition

HVAC Fundamentals

3rd Edition

Samuel C. Sugarman

River Publishers

Routledge
Taylor & Francis Group

LONDON AND NEW YORK

Published 2020 by River Publishers
River Publishers
Alsbjergvej 10, 9260 Gistrup, Denmark
www.riverpublishers.com

Distributed exclusively by Routledge
4 Park Square, Milton Park, Abingdon, Oxon OX14 4RN
605 Third Avenue, New York, NY 10017

First issued in paperback 2023

Library of Congress Cataloging-in-Publication Data

Sugarman, Samuel C., 1946-
    HVAC fundamentals / Samuel C. Sugarman. -- 3rd edition.
        pages cm
    Includes index.
        ISBN 0-88173-752-6 (Fairmont Press: alk. paper) -- ISBN 978-1-4987-5704-1
(Taylor & Francis distribution: alk. paper) -- ISBN 978-8-7702-2328-7 (electronic) 1.
Heating--Equipment and supplies. 2. Ventilation--Equipment and supplies. 3. Air
conditioning--Equipment and supplies. I. Title.

    TH7345.S795 2016
    697--dc23

                                                                    2015032347

HVAC Fundamentals/Samuel C. Sugarman. Third edition
First published by Fairmont Press in 2016.

Routledge is an imprint of the Taylor & Francis Group, an informa business

Notice:
Product or corporate names may be trademarks or registered trademarks and are
used only for identification and explanation without intent to infringe.

Publisher's Note
The publisher has gone to great lengths to ensure the quality of this reprint but
points out that some imperfections in the original copies may be apparent.

ISBN 978-87-7022-935-7 (pbk)
ISBN 978-1-4987-5704-1 (print) (hbk)
ISBN 978-8-7702-2328-7 (online)
ISBN 978-1-0031-5197-5 (ebook master)

While every effort is made to provide dependable information, the publisher,
authors, and editors cannot be held responsible for any errors or omissions.

The views expressed herein do not necessarily reflect those of the publisher.

Table of Contents

Chapter 1

Introduction to HVAC Systems

HEATING, VENTILATING,
AND AIR CONDITIONING SYSTEMS

Properly designed, installed, tested and maintained heating, ventilating and air conditioning (HVAC) systems provide conditioned air for the work process function, occupancy comfort, and good indoor air quality while keeping system costs and energy requirements to a minimum. Commercial HVAC systems provide building work areas with "conditioned air" so that occupants will have a comfortable and safe work environment. People respond to their work environment in many ways and many factors affect their health, attitude and productivity. "Air quality" and the "condition of the air" are two very important factors. "Conditioned air" and "good air quality," means that air should be clean and odor-free and the temperature, humidity, and movement of the air is within certain acceptable comfort ranges. ASHRAE, the American Society of Heating, Refrigerating and Air Conditioning Engineers, has established standards which outline indoor comfort conditions that are thermally acceptable to 80% or more of a commercial building's occupants. Generally, these comfort conditions, sometimes called the "comfort zone," are between 68F and 75F for winter and 73F to 78F during the summer. Both these ranges are for room air at approximately 50% relative humidity and moving at a slow speed (velocity) of 30 feet per minute or less.

The HVAC system is simply a group of components working together to provide heat to, or remove heat from, a conditioned space. For example the components in a typical roof-mounted package unit HVAC system (Figure 1-1) are:

1. An indoor fan (aka blower) to circulate the supply and return air.

2. Supply air ductwork in which the air flows from the fan to the conditioned space.

3. Air devices such as supply air outlets and return air inlets.

1

4. Return air ductwork in which the air flows back from the conditioned space to the unit.

5. A mixed air chamber (aka mixed air plenum) to receive the return air and mix it with outside air.

6. An outside air device such as a louver, screened opening or duct to allow for the entrance of outside air into the mixed air plenum.

7. A filter section to remove dirt, debris and dust particles from the air.

8. Heat exchangers such as a refrigerant evaporator coil and condenser coil for cooling, and a furnace for heating.

9. A compressor to compress the refrigerant vapor and pump the refrigerant around the refrigeration system.

10. An outdoor fan (aka blower) to circulate outside air across the air-cooled condenser coil.

11. Controls to start, stop and regulate the flow of air, refrigerant, and electricity.

Figure 1-1. Roof-mounted Unit (aka rooftop unit, RTU). Natural gas heating and vapor compression DX cooling.

HVAC COMPONENTS

H is for Heating using:

1. Boilers (Figure 1-2)
    Types of Boilers: steam or water
    Boiler Pressures: Various pressures from low to high
    Boiler Fuels*: coal, natural gas, oil, electricity
    Boiler Configurations: fire tube or water tube

2. Furnaces (Figure 1-3)
    Furnace Fuels: coal, liquid petroleum gas, natural gas, oil, electricity*

3. Heating Coils (Figure 1-4)
    Types of Heating Coils**: water, steam, electrical

---

*Type of fuel used will vary by region, codes, customs, availability and cost, etc.
**Cooling and heating coils are heat transfer devices or heat exchangers. They come in a variety of types and sizes. Water coils are used for heating, cooling, humidifying or de-humidifying air and are most often made of copper headers and tubes with aluminum or copper fins and galvanized steel frames.

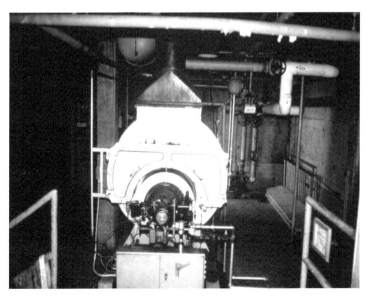

Figure 1-2. Fire Tube Boiler. Boiler exhaust stack is at the front of the boiler indicating a 2-pass or 4-pass boiler (2-pass for this boiler). Water expansion tank at top left rear of boiler.

Figure 1-3. Natural Gas Furnace. Heat exchanger shown.

Figure 1-4. Hot water heating coil on right. Left is opening to supply air fan. This a draw-though system.

V is for Ventilating for:

1. Outside air (OA). Approximately 20 cubic feet per minute (cfm) of air volume per person for ventilating non-smoking areas.

2. Make-up air (MUA) for exhaust systems, e.g., kitchen hoods, fume hoods, toilets and etc.

3. Room (aka conditioned space or occupied space) pressurization. Typically, +0.03 to +0.05 inches of water gauge for commercial buildings.

AC is for Air Conditioning

For most people air conditioning means comfort cooling using either chilled water or refrigerant DX systems. Both of these systems include cooling coils to remove heat from the air. For chilled water systems the refrigeration side is either an absorption system or a vapor-compression system. Refrigeration DX (aka direct expansion or dry expansion) systems are vapor-compression. The cooling coils** are water or refrigerant.

AC (Air Conditioning) also means conditioning the air by:

1. Heating (adding heat) or cooling (removing heat)
2. Humidifying (adding moisture) or dehumidifying (removing moisture)
3. Volume of airflow measured in cubic feet per minute (cfm)
4. Velocity (speed) of airflow measured in feet per minute (fpm)
5. Filtering and cleaning
6. Pattern or direction of airflow (horizontal and vertical)

HOW THE HVAC SYSTEM WORKS

An HVAC system is designed to provide conditioned air to the occupied space, also called the "conditioned" space, to maintain the desired level of comfort. To begin to explain how the HVAC system works let's set some design conditions. First, we will need to determine the ventilation requirements. We know that in the respiratory process the contaminate carbon dioxide is exhaled. In buildings with a large number of people carbon dioxide and other contaminants such as smoke from cigarettes and odors from machinery must be continuously removed or unhealthy conditions will result. The process of supplying "fresh" air (now most often called outside air) to buildings in the proper amount to offset the contaminants produced by people and equipment is known as "ventilation." Not only does the outside air that is introduced into the conditioned space offset the contaminants in the air but because of its larger ion content, outside air has a "fresh air" smell in contrast to the "stale" or "dead air" smell noticed in overcrowded rooms that do not have proper ventilation. National guidelines and local building codes stipulate the amount of ventilation required for buildings and work environments.

Let's say that an HVAC system supplies air to a suite in an office complex and the code requirement is for 20 cubic feet per minute (abbreviated cfm) of outside air for each building occupant. If the suite is designed for 10 people then the total outside air requirement for the people in the suite is 200 cfm. An additional amount of outside ventilation air may be required if there are exhaust hoods such as laboratory fume hoods, kitchen hoods, etc., or there are other areas where the air needs to be exhausted or vented to the outside such as bathrooms and restrooms. This ventilation air is called make-up air (MUA).

If more air is brought into a room (conditioned space) than is taken out of a room the room becomes positively pressurized. If more air is taken out of a room than is brought into a room the room becomes negatively pressurized. These air pressures, whether positive or negative are measured in inches of water gauge (in. wg) also known as inches of water column (in. wc).

Commercial office buildings are typically positively pressurized to about 0.05 inches of water pressure. This is done to keep outside air from "infiltrating" into the conditioned space through openings in or around doorways, windows, etc. Other areas that need positive pressurization are hospital operating rooms and cleanrooms. Examples of negative rooms are commercial kitchens, hospital intensive care units and fume hood laboratories.

Air Volume

The volume of air required to heat, cool, ventilate and provide good indoor air quality is calculated based on the heating, cooling and ventilating loads. The air volumes are in units of cubic feet per minute (cfm). Let's say the total volume of air is calculated to be 5250 cfm. Constant volume supply air and return air fans circulate the conditioned air to and from the occupied conditioned space.

The total volume of return air back to the air handling unit is 4200 cfm. The difference between the amount of supply air (5250 cfm) and the return air (4200 cfm) is 1050 cfm. This is the ventilation air. It is used in the conditioned space for make-up air for toilet exhaust and other exhaust systems. Ventilation air is also used for positive pressurization of the conditioned space, and for "fresh" outside air to maintain good indoor air quality for the occupants. The return air, 4200 cfm, goes into the mixed air chamber (aka mixed air plenum). The return air is then mixed with 1050 cfm, which is brought in through the outside air dampers into the mixed air plenum. This 1050 cfm of outside air is the minimum outside air required for this system. It is 20% of the supply air (1050/5250). It mixes with the 4200 cfm of return air (80%, 4200/5250) to give mixed air (100%). Next, the 5250 cfm of mixed air then travels through the filters and into the coil sections. If more outside air than the minimum is brought into the system, perhaps for air-side economizer operation, any excess air is exhausted through exhaust air dampers to maintain the proper space pressurization. For example, if 2050 cfm is brought into the system through the outside air dampers and 4200 cfm comes back through the return duct into the unit then 1000 cfm is exhausted through the exhaust air dampers. This maintains the total supply cfm (5250) into the space and maintains the proper space pressurization.

The airflow diagram looks like this

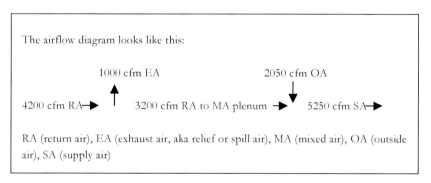

The airflow diagram looks like this:

1000 cfm EA   2050 cfm OA

4200 cfm RA→   3200 cfm RA to MA plenum →   5250 cfm SA→

RA (return air), EA (exhaust air, aka relief or spill air), MA (mixed air), OA (outside air), SA (supply air)

Heating

The heating load requirement is based on design indoor and outdoor winter conditions. The design conditioned space heating load is 198,450 Btu per hour (aka Btuh or Btu/hr). This is the amount of heat lost in the winter (mainly by conduction) through the walls, windows, doors, roofs, etc. An additional amount of heat is required to heat the outside ventilation air based on design conditions. This additional amount of heat is 45,360 Btuh. To maintain the temperature and humidity in the comfort zone for the conditioned space the heating cycle is: the supply air leaves the heating coil carrying 198,450 Btuh of heat. The air goes through the supply air fan, down the insulated supply duct, past the manual volume dampers which have been set for the correct amount of air for each diffuser, and into the conditioned space. The supply air gives up all its 198,450 Btuh of heat to the conditioned space to replace the 198,450 Btuh that is leaving the space through the walls, doors, windows, ceiling, roof, etc. As the air gives up its heat it makes its way through the room and into the return air inlets, then into the return air duct and back to the air handling unit.

The return air goes into the air handling unit through the return air fan (if there is one), through the return air automatic temperature control dampers into the mixed air chamber and mixes with the outside air. The mixed air flows through the filters into the heating coil. Now depending on the system design the mixed may also go through a preheat coil and/or a cooling coil (both coils in this example are off or de-energized meaning no heating or cooling is happening) prior to entering the heating coil.

The mixed air travels through the heating coil where it picks up heat via conduction through the hot water tubes in the coil. In addition to the tubes, the heating coil also has fins attached to the tubes to facilitate the heat transfer. 243,810 Btuh of heat is transferred from the coil into the air. From this amount of heat, 45,360 Btuh heats the outside air to bring it up to the design room air temperature. The remainder, 198,450 Btuh of heat, leaves the coil in the supply air and goes into the space and the air cycle repeats. The heating water, after giving up heat to the air, leaves the coil and goes back to the oil-fired boiler through the heating water return pipe and into the boiler where it picks up the same amount of heat that it has just given up in the coil. The water leaves the boiler, flows through the heating water pump and is pumped through the heating water supply (aka heating hot water supply) piping into the heating coil to give up its heat into the mixed air and the water cycle repeats.

Ventilating

The ventilation requirement is 1050 cfm. 1050 cfm of outside air is brought in through the outside air (OA) dampers into the mixed air plenum. This 1050 cfm of outside air mixes with the 4200 cfm of return air to form 5250 cfm of mixed air, which goes through the coil(s) and becomes supply air.

Cooling

For this system, in the summer the total heat given off by the people, lights and equipment in the conditioned space plus the heat entering the space through the outside walls, windows, doors, roof, etc., and the heat contained in the outside ventilation air will be approximately 154,000 Btu/hr. A ton of refrigeration is equivalent to 12,000 Btu/hr of heat. Therefore, this HVAC system requires a chiller that can provide approximately 13 tons of cooling (154,000 Btu/hr ÷ 12000 Btu/hr/ton = 12.83 tons).

To maintain the proper temperature and humidity in the conditioned space the cooling cycle is: the supply air (which is 20F cooler than the air in the conditioned space) leaves the cooling coil and goes through the heating coil (which is off), through the supply air fan, down the duct and into the conditioned space. The cool supply air picks up heat in the conditioned space. The warmed air makes its way into the return air inlets, then into the return air duct and back to the air handling unit. The return air goes through the return air fan, if there is one, into the mixed air chamber and mixes with the outside air. The mixed air goes through the filters and into the cooling coil. Once again, depending on the system design the mixed may also go through a preheat coil and/or a heating coil. In this example if these heating coils were in the unit both would be off or de-energized meaning no preheating or heating is happening.

The mixed air flows through the cooling coil where it gives up its heat into the chilled water tubes in the coil. This coil also has fins attached to the tubes to facilitate heat transfer. The cooled supply air leaves the cooling coil and the air cycle repeats. The water, after picking up heat from the mixed air, leaves the cooling coil and goes through the chilled water return pipe to the water chiller's evaporator. The "warmed" water flows into the chiller's evaporator—sometimes called the water cooler—where it gives up the heat from the mixed air into the refrigeration system. The newly "chilled" water leaves the evaporator, goes through the chilled water pump and is pumped through the chilled water supply piping into the cooling coil to pick up heat from the mixed air and the water cycle repeats. The evaporator is a heat exchanger that allows heat from the chilled water return to flow by conduction into the refrigerant tubes. The liquid refrigerant in the tubes "boils off" to a vapor removing heat from the water and conveying the heat to the compressor and then to the condenser. The heat from the condenser is conveyed to the cooling tower through the condenser water in the condenser return pipe. As the condenser water cascades down through the tower, outside air is drawn across the cooling tower removing heat from the water through the process of evaporation. The "cooled" condenser water falls to the bottom of the tower basin and is pumped from the tower through the condenser water pump and back to the condenser in the condenser water supply piping and the cycle repeats.

CENTRAL HVAC SYSTEM

Let's take a closer look at a central HVAC system and some of the major components. We'll start with the system as a whole (Figure 1-5) and then do individual components. Below are common names and abbreviations for components. The numbering for the terms correspond to numbers on figures 1-5 through 1-14. Terms for components can vary regionally and internationally. Location and types of components in the drawings are for illustration and general concept and may vary based on system design and intent.

1. Cooling Tower, CT
2. Condenser Water Pump, CWP
3. Condenser Water Supply, CWS
4. Condenser Water Return, CWR
5. Condenser, Cond., C
6. Evaporator Evap., E, (aka "Water Chiller" or "Water Cooler"),
7. Compressor, Comp. Note: Components 5, 6 and 7 as a group are commonly called a Chiller, CH
8. Chilled Water Pump, CHWP
9. Chilled Water Supply, CHWS
10. Chilled Water Return, CHWR
11. Chilled Water Coil or Cooling Coil, CC
12. Hot Water Coil or Heating Coil, HC
13. Hot Water Supply, HWS (also HHWS for Heating Hot Water Supply)

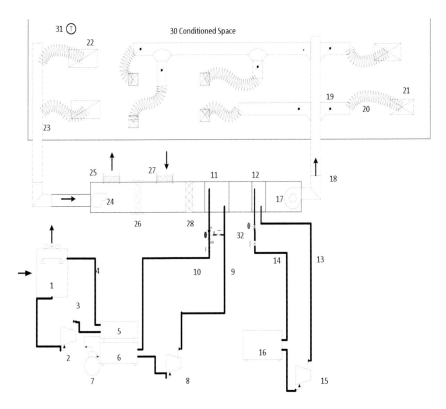

Figure 1-5. Central HVAC System: Major Components and Common Terminology.

14. Hot Water Return, HWR (also HHWR Heating Hot Water Return)
15. Hot Water Pump, HWP
16. Boiler, B
17. Supply Air Fan, SAF
18. Supply Air Duct, SA
19. Manual Volume Damper, MVD
20. Flex Duct, Flex
21. Ceiling Diffuser, CD
22. Return Air Inlet, RA
23. Return Air Duct, RA
24. Return Air Fan, RAF
25. Exhaust Air (EA) Damper, (ATCD* or MVD)
26. Return Air (RA) Damper, (ATCD or MVD)

27. Outside Air (OA) Damper, (ATCD or MVD)
28. Filters, F
29. Air Handling Unit, AHU Note: Components 17 & 24 (fans), 11 & 12 (coils), 28 (filters), and 25, 26, & 27 (dampers) are contained in a single housing and as a group are known as an Air Handling Unit (AHU). Other descriptive terms depending on size and system are Fan-Coil Unit (FCU), Roof-Top Unit (RTU), etc.
30. Conditioned Space
31. Room or conditioned space thermostat, T
32. Water Valves, Example shown: 3-way ATCV** in chilled water return pipe and 2-way ATCV in heating water return followed by MBV*** and sensor, or CBV****

---

*Automatic Temperature Control Damper
**Automatic Temperature Control Valve
***Manual Balancing Valve
****Calibrated Balancing Valve

Cooling Coil—Cooling Mode

Air temperature entering the coil, 78F

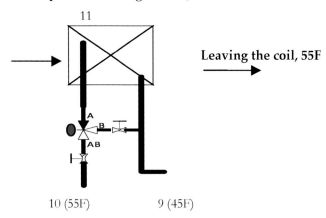

Figure 1-6. Cooling Coil (11), Chilled Water Supply (9), CHWR (10)

The cooling coil in this example is a chilled water coil. It is component number 11, the same number on the drawings in Figures 1-5 and 1-6. It is not a refrigerant evaporator, which is also known as a cooling coil. The volume of airflow through the cooling coil is 25,000 cfm (cubic feet per minute). The temperature of the water in the chilled water return (CHWR) pipe (10) is 55F. So the temperature of the water leaving the cooling coil (leaving water temperature, LWT) is 55F which is the same as the temperature entering the water chiller (entering water temperature, EWT, Figure 1-7). There has been no loss in temperature because the chilled water return pipe is insulated. The temperature of the water in the chilled water supply (CHWS) pipe (9) is 45F which is the temperature of the water leaving the water chiller and the temperature entering the cooling coil. This pipe is also insulated. The coil is piped supply at the bottom and return at the top.

The dry bulb temperature of air entering the cooling coil (entering air temperature, EAT) is 78F. The dry bulb temperature of air leaving the cooling coil (leaving air temperature, LAT) is 55F. As I said, the temperature of the chilled water supply is 45F. Therefore, this chilled water coil is piped counter flow, meaning the CHWS is piped into the coil on the side the air is leaving. So the coldest water (45F) is in contact with the coldest air (55F). Heat in the air flows by conduction to the water in the tubes in the coil. The counter flow piping configuration provides the greatest amount of heat transfer.

The amount of heat in the air entering the chilled water coil is the total heat (sensible and latent) from the conditioned space (heat from occupants, lights, equipment, and heat transferred though the building envelope—walls, doors, windows, roof, etc.) plus heat brought into the air handling unit though the outside air duct. This system is working at maximum cooling with the return air dampers full open and the outside air dampers at 20% open, the required minimum for this system.

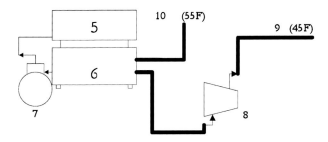

Figure 1-7. Water Chiller, CHWS (9), CHWR (10), Condenser (5), Evaporator (6), Compressor (7), Chilled Water Pump (8)

As the air flows through the coil, over the water tubes and through the fins attached to the tubes, it releases (gives up) 742,500 Btuh (heat power in British thermal units per hour). In units of tons of refrigeration (TR) the total heat removed from the air is 61.875 tons. The air is cooled from 78F to 55F, a 23F drop in dry bulb (sensible) temperature. 742,500 Btuh is now in the chilled water return and is conveyed to the water chiller. The volume of water flow required to transport this heat load is 148.5 gpm (gallons per minute). In Chapter 2, Heat Flow, you'll find the heat transfer equations for Btuh, cfm and gpm. Chapter 3, Psychrometrics, will provide more in depth information regarding the cooling process.

Water Chiller

The temperature of the water in the chilled water return pipe (10) is 55F. The water enters the water chiller and is cooled to 45F. The volume of water flow through the water chiller (6) is 148.5 gpm (gallons per minute). The chilled water pump (8, CHWP) has an electric motor and is a direct-drive, centrifugal pump*. As the 148.5 gpm of supply chilled water leaves the pump it has a head pressure of 100 feet of water. The units of water pressure are in feet of water (ft. of water or ft. H_2O). The CHWP electric motor is rated at 7.5 nameplate horsepower (nhp) and is operating at 6.2 brake horsepower (bhp).

The supply water is pumped from the water chiller through the chilled water supply pipe (9) to the cooling coil located in the air handling unit (AHU). The chilled water pipe system from the water chiller to the chilled water coil and back is a closed loop, i.e., not open to the atmosphere.

The amount of heat in the chilled water return is 742,500 Btuh, or 61.875 tons of refrigeration (TR). The water will be cooled to 45F as heat is released into the liquid refrigerant via conduction between fluid and tubes in the water chiller**. As the heat flows into the liquid refrigerant it will boil off (evaporate) the liquid into a gas (vapor). Evaporation is a cooling process. The water is cooled sensibly and the refrigerant is heated latently as the change in state from liquid to vapor takes place.

The refrigerant vapor conveying a heat load of 742,500 Btuh will flow into the compressor (7)***. The compressor will compress the vapor adding pressure and heat. This is the process in the compression phase of the refrigeration cycle. The amount of heat in the vapor leaving the compressor is now 928,125 Btuh or 77.344 TR. The higher pressure and

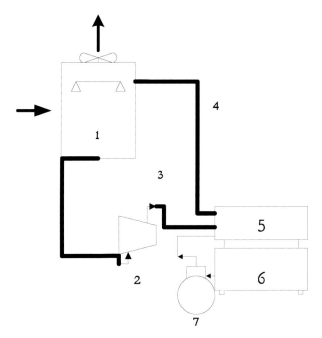

Figure 1-8. Cooling Tower (1), Condenser Water Pump (2) CWS (3), CWR (5)

higher temperature vapor (as compared to the pressure and temperature of the refrigerant vapor in the water chiller) will flow to the condenser (5) carrying a heat load of 928,125 Btuh.

---

*HVAC water pumps are centrifugal pumps, not positive displacement pumps. Almost all are direct drive, occasionally belt-driven.
**A water chiller can have a DX (dry-expansion, aka, direct-expansion) evaporator or flooded evaporator. DX evaporators have refrigerant in the tubes and water around the tubes. Flooded evaporators have water in the tubes and refrigerant around the tubes.
***A compressor may be centrifugal, rotary or reciprocal.

Cooling Tower

The refrigerant vapor enters the un-insulated water-cooled condenser* (5) and comes in contact with condenser tubes carrying water from the condenser water supply pipe (3). The water is pumped into the condenser by the electric motor, direct-drive, centrifugal condenser water pump (2, CWP). The pump using centrifugal force increases the water pressure and the water flows through the condenser water supply (CWS) pipe into the condenser. The CWS enters the condenser at 85F.

As the refrigerant vapor flows around the condenser tubes it releases heat into the water and condenses back to a liquid to be pumped by the compressor back to the water chiller where it again removes heat from the CHWR and the refrigeration cycle continues. The temperature of the water in the condenser increases to 95F. The change in water temperature through the condenser is the "condenser rise." The water flows from the condenser through the condenser water return pipe (4, (CWR) and enters the cooling tower (1) at 95F. Condenser water pipes are not insulated.

The condenser water enters into the top of the tower (also known as "on the tower") and the CWR is sprayed or cascades down the fill or slats in the tower to break the water into small droplets for more efficient evaporation. This tower is a forced air**tower because it has an electric motor-driven fan*** at the top which draws outside air through the tower. The air is drawn in through the sides of the tower through the fill and up and out at the top. As the air comes in contact with the condenser water it evaporates some of the water. As mentioned previously evaporation is a cooling process and some of the heat in the CWR is released into the air and then the warmed air is exhausted out the top of the tower. The water flowing into the tower basin at the bottom of the tower is 85F. The leaving water is also called "off the tower."

The amount of heat picked up (removed) by the water in the condenser is 928,125 Btuh (77.344 TR). The amount of water flowing through this open pipe (open to the atmosphere) condenser loop system is 185.625 gallons per minute (gpm). The amount of heat released from the water into the outside air is 928,125 Btuh.

The dry bulb temperature of the outside air entering the tower is 90F with a 75F wet-bulb temperature. The air leaves the tower at a higher temperature and humidity (a higher total heat content). The water temperature change through the tower is the "tower range." In this system both

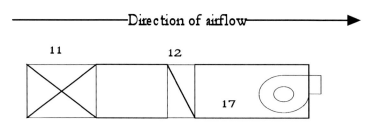

Figure 1-9. Cooling Coil (11), Heating Coil (12), Supply Air Fan (17)

the tower range and the condenser rise are 10F, 85F to 95F in the condenser and 95F to 85F in the tower. The tower approach is also 10F****.

---

*Condensers are also air-cooled.
**A natural draft tower does not have a fan.
*** Fans may be belt-driven or direct drive (directly coupled to the motor).
**** Tower approach is the difference between the temperature of the water off the tower (85F) and the wet bulb temperature of the outside air (75F). For this system the approach is 10F (85-75). Tower water cannot be cooled lower than the entering air wet bulb temperature.

Supply Air Fan

The sensible temperature of the air entering (EAT) the cooling coil (11) is 78F. The air volume totals 100% and it is a mixture of outside air (20%) and return air (80%). The mixture is called mixed air (MA) and it is always 100%. In cubic feet per minute, the volume of mixed air is 25.000 cfm; the outside air (OA) is 5000 cfm, and the return air (RA) is 20,000 cfm. The sensible temperature of the air leaving (LAT) the cooling coil (CC) is 55F. The air is now called supply air (aka conditioned air). The air enters and leaves the heating coil (HC, 12)*. The heating coil is not energized so the LAT, now designated supply air temperature (SAT), remains 55F.

This system has an electric motor, belt-driven**, centrifugal, airfoil supply air fan (SAF, 17). The fan is located after the cooling and heating coils so this fan is a draw-through fan***. It has a top horizontal discharge and the fan wheel rotates clockwise. The supply air enters the inlet (suction) side of the supply air fan where centrifugal forces increase the pressure of the air and moves the air out the discharge (outlet) side of the fan and down the distribution air ducts. The airflow is 25,000 cfm (cubic feet per minute) through the fan. As the supply air leaves the fan it has a positive static pressure of 3 inches water pressure, which is in the medium pressure range for HVAC systems. The units of air pressure are in inches water column, aka "inches water gauge" (in. wc, aka in. wg).

---

*The coils in this system are arranged with a heating coil after the cooling coil so it is a reheat system. When a heating coil is before the cooling coil it is a preheat system. A steam or hot water preheat coil is typically used as a safety measure to protect downstream coils from damage by cold (freezing) outside air temperatures in less temperate climates than San Diego, California with 1000-1500 heating degree days. A reheat coil is used for temperature and humidity control in the conditioned space. Fans may be belt-driven or direct drive (directly coupled to the motor).
**A supply fan placed before the coils is a blow-through fan.

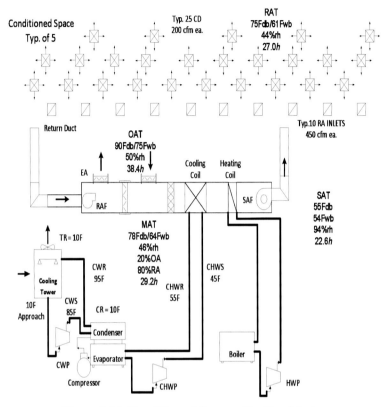

Figure 1-10. Conditioned Space—Cooling Mode

Conditioned Space—Cooling Mode

The airflow volume is 25,000 cfm and the average velocity in feet per minute in the main supply air duct is 1585 fpm. The duct is "60 x 38"(15.8 sq. ft.) and it is insulated on the outside with vapor-barrier wrap. There are 25 supply air ceiling diffusers (CD) in each of the 5 conditioned spaces and 10 return air inlets. The outlets are designed for 200 cfm each and the inlets are 450 cfm each. Each conditioned space is positive by volume of airflow, 5,000 cfm into the space and 4500 cfm out.

The supply air temperatures (SAT) are 55F dry bulb (db) and 54F wet bulb (wb). The relative humidity (rh) of the air is 94% and the total heat content (enthalpy, h) is 22.6 Btu/lb. As the supply air flows into the conditioned space it mixes with the warmer room air which has heat from occupants, equipment and heat entering through the building envelope. The 55F supply has absorbed 621,000 Btuh. The mixture of room air (aka secondary air) and supply air (aka primary air) now called return air (RA) flows into the return air vents. The return air temperatures (RAT) are 75Fdb and 61Fwb with 44%rh and 27.0 Btu/lb heat content.

Return Air, Outside Air, Mixed Air—Cooling Mode

The return air exits the five conditioned spaces via the return air inlets and the return air ducts. The total return airflow volume is 22,500 cfm and the average velocity in feet per minute in the main return air duct is 1585 fpm. The duct is 60" x 34"(14.2 sq. ft.). The return duct is not insulated. This system has a return air fan* (RAF, 24) designed for 22,500 cfm. The pressure on the suction side of the return fan is 1 in wg (this is a negative or less than atmosphere pressure)**. 22,500 cfm of return air comes through the fan into the return air plenum where 2,500 cfm is exhausted out the exhaust dampers (EA, 25) into the outside air***. The remaining 20,000 cfm return air goes through the return air dampers (RA, 26) into the mixed air plenum and mixes with 5000 cfm of outside air entering through the outside air dampers (OA, 27). Maintaining the correct amount of outside air is important for indoor air quality, energy costs, and conditioned

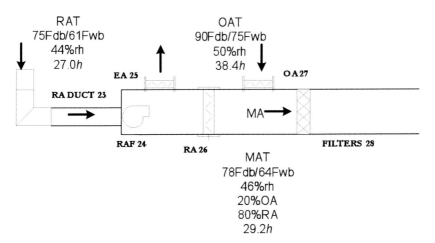

Figure 1-11. Return Air duct (23), Return Air fan (24), Exhaust Air dampers(25) Return Air dampers (26), Outside Air dampers (27), Mixed Air plenum, Filters (28). System is in cooling mode. The EA, RA and OA dampers are automatic temperature control dampers (ATCD) and are the damper components of the air-side economizer.

space pressurization.

The mixed air flows through the filters (28) and into the cooling coil. The difference between the 742,500 Btuh of heat removed by the cooling coil and the heat absorbed in the conditioned space (621,000 Btuh) is the amount of heat in the outside air that must also be removed (121,500 Btuh)*.

---

*Most HVAC systems have one supply fan to move the air to and from the conditioned space. Some systems will be like the example with a supply fan going to the conditioned space and a return fan bringing the air back to the air handling unit. Some systems have multiple fans arranged in parallel or series fashion for air movement.
**Fans must always have a negative pressure on the suction and a positive pressure on the discharge.
***Exhaust air is also called relief air or spill.

Heating Coil—Heating Mode

The heating coil in this example is a hot water coil (12), aka heating water coil or heating hot water coil. Other types of heating coils are steam coil and electric coil. This reheat coil is piped counter flow for maximum heat transfer. Water coils are typically piped supply at the bottom and return at the top to allow any entrained air to be vented out the top of the system. Steam coils are piped supply (steam) at the top and return (liquid condensate) at the bottom.

The temperature of the water in the hot water return (HWR) pipe (14) is 160F which is the coil leaving water temperature, LWT. This pipe is insulated and it returns water from the heating coil to the boiler. The

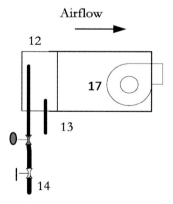

Figure 1-12. Heating Coil (12), HWS (13), HWR (14), SAF (17)

temperature of the water entering the boiler (EWT) is 160F. The temperature of the water in the hot water supply (HWS) pipe (13) is 180F which is the temperature of the water leaving the boiler (LWT) and the temperature entering the heating coil (EWT). This pipe is also insulated. The volume of airflow through the heating coil and supply air fan (17) is 25,000 cfm (cubic feet per minute). As the air flows through the coil, over the water tubes and through the fins attached to the tubes, it absorbs (picks up) 675,000 Btuh sensible heat raising the supply air temperature 25F. The SAF and duct distributes the heated air to the conditioned spaces.

In this example 675,000 Btuh has been released from the HWS reducing the water temperature 20F (180-160). The volume of water flow required to transport this heat load from the boiler to the heating coil is 67.5 gpm (gallons per minute). See Chapter 4, figure 4-1 for more information on the heating mode of a different central system.

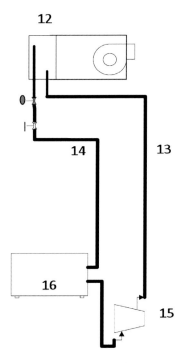

Figure 1-13. Heating Coil (12), HWS Pipe(13), HWR Pipe (14), Heating Water Pump (15) Boiler (16)

Boiler

The temperature of the water in the hot water return pipe (14, HWR) is 160F. The water enters the fire-tube boiler and is heated to 180F absorbing 675,000 Btuh from the hot gases in the boiler tubes. The return water flows around the fire-tubes and heat is transferred through conduction into the water. The volume of water flow through the boiler (16, B) and hot water pump (15, HWP) is 67.5 gpm (gallons per minute).The HWP has an electric motor and is a direct-drive, centrifugal pump. The supply water is pumped from the boiler through the hot water supply pipe (13, HWS) to the heating coil (12, HC) which is located in the air handling unit (AHU). The hot water pipe system from the boiler to the hot water coil and back is a closed loop.

Chapter 2

Heat Flow

HEAT AND TEMPERATURE

Heat is energy in the form of molecules in motion. As a substance becomes warmer, its molecular motion and energy level (temperature) increases. Temperature describes the level of heat (energy) with reference to no heat. Heat is a positive value relative to no heat. Because all heat is a positive value in relation to no heat, cold is not a true value. It is really an expression of comparison. Cold has no number value and is used by most people as a basis of comparison only. Therefore, warm and hot are comparative terms used to describe higher temperature levels. Cool and cold are comparative terms used to describe lower temperature levels. The Fahrenheit scale is the standard system of temperature measurement used in the United States. The U.S. is one of the few countries in the world still using this system. Most countries use the metric temperature measurement system, which is the Celsius scale. The Fahrenheit and Celsius scales are currently used interchangeably in the U.S. to describe equipment and fundamentals in the heating, ventilating and air conditioning industry.

The temperatures on the Fahrenheit and Celsius scales for (pure) water are: freezing point 32 degrees Fahrenheit (32F) and zero degrees Celsius (0C). Boiling point is 212 degrees Fahrenheit (212F) and 100 degrees Celsius (100C).

Converting degrees Celsius and degrees Fahrenheit
$$C = (F-32) \div 1.8$$
$$F = 1.8C + 32$$
Examples: Celsius conversion to Fahrenheit

25C	77F
20C	68F
5C	41F
0C	32F
-5C	23F
-40C	- 40F

Celsius and Fahrenheit conversion for everyday temperatures
    0C is 32F
    16C is approximately 61F
    28C is approximately 82F
    37C is 98.6F
    40C is 104F

HVAC Fahrenheit temperatures

The temperatures associated with most HVAC system components range from 0F to 250F. The indoor comfort range for most people is between 68F and 78F.

HEAT TRANSFER

Heat naturally flows from a higher energy level to a lower energy level. In other words, heat travels from a warmer substance to a cooler substance. Therefore, heat energy flows downhill and the expression "heat rises" is incorrect, it's heated air that rises and cooled air falls.

When there is a temperature difference between two substances, heat transfer will occur. In fact, temperature difference is the driving force for heat transfer. The greater the temperature difference, the greater the heat transfer. The three types of heat transfer are conduction, convection and radiation.

Conduction

Heat transfer by conduction is heat energy traveling from one molecule to another. A heat exchanger, for example a water coil, in a commercial HVAC system or home furnace uses conduction to transfer heat. Your hand touching a cold wall is an example of heat transfer by conduction. However, heat does not conduct at the same rate in all materials. For example, metals are good conductors, but copper, a very good conductor, conducts at a different rate from iron. Glass, wood and air on the other hand are poor conductors of heat. Very poor conductors are called insulators.

Convection

Heat transfer by convection is when some substance that is readily movable such as air, water, steam, or refrigerant moves heat from one location to another. Compare the words "convection" (the action of conveying) and "convey" (to take or carry from one place to another). An HVAC system uses air, water, steam and refrigerants in ducts and piping to convey heat energy to various parts of the system. As stated, when air is heated, it rises; this is heat transfer by "natural" convection. "Forced" convection is when a fan or pump is used to convey heat in fluids such as air and water. For example, many large buildings have a central heating plant where water is heated and pumped throughout the building to terminals, such as heating coils (aka heat exchangers). Fans move heated air into the conditioned space.

Radiation

Heat transferred by radiation travels through space without heating the space. Radiation (aka radiant heat) does not transfer the actual temperature value. A portable electric space heater is example of heat transfer by radiation. As the electric heater coil glows red-hot it radiates heat into the room. The space heater does not heat the air (the space) instead it warms the solid objects that the radiant heat encounters. However, radiant heat diminishes by the distance traveled. An example of diminishing heat is the sun. The earth does not experience the total heat of the sun—approximately 27,000,000F at the sun's core but only about 10,000F where the sun's radiation is detected as sunlight to the earth—because the sun is some 93 million miles from the earth.

UNITS OF ENERGY AND POWER

Heat Energy and Power

A British thermal unit (Btu) describes the quantity of heat contained in a substance. Because the Fahrenheit scale is the standard system of temperature measurement used in the United States, a Btu is defined as the amount of heat required to raise the temperature of 1 pound of water 1 degree Fahrenheit. Btu is the unit of heat energy.

The rate of heat power (consumption) is energy consumed over some period of time. For example, 1 Btu consumed in an hour is 1 Btu per hour of heat power, written Btu/hr or Btuh. Another unit of heat power is Btu per minute (Btu/m or Btum). Other expressions of heat power are: MBh (1000 Btuh) and MMBh (1,000,000 Btuh). The letter "M" is the Roman numeral for 1000. See Roman Numerals Chapter 14.

Electrical Energy and Power

Watt-hour (Whr) and kilowatt-hour (kWh) are the units of electrical energy. Mathematically, kWh is also expressed kW/hr. Kilo (k) is Greek for 1000.

The rate of electrical power is energy consumed over some period of time, e.g., 1 kiloWatt-hour consumed in an hour equals 1 kW. Watt (W) and kiloWatt (kW) are units of electrical power. A kilowatt is equal to 1000 watts.

Summary: Heat and Electrical Energy and Power

To sum up this section on power and energy the equation for power is: power = energy x time.

Heat: 1 Btu/hr (power) = 1 Btu (energy) x 1 hour (time) and

Electrical: 1 kW (power) = 1 hr/hr (energy) x 1 hour (time).

As you can see it is easy to misuse the units of energy and power. Btu (British thermal unit) and kWh (kilowatt-hour) are units of energy. Btuh (British thermal unit per hour) and kW (kilowatt) are units of power.

ELECTRIC AND HEAT POWER EQUIVALENTS

1 watt = 3.412 Btuh
1 kW = 3412 Btuh

746 w = 1 hp (motor horsepower) or 2545 Btuh
0.746 kW = 1 hp or 2545 Btuh

HEAT CONTENT AND SPECIFIC HEAT OF A SUBSTANCE

The heat content of a substance is called enthalpy. The symbol is "h." The units of enthalpy are Btu/lb or Btu/lbF (Btu per pound or Btu per pound degrees Fahrenheit). Specific heat is the amount of heat necessary to raise the temperature of 1 pound of a substance 1 degree Fahrenheit. Every substance has a different specific heat. The specific heat of water is 1.0 Btu/lb. The specific heat of air is 0.24 Btu/lb and the specific heat of steam or ice is approximately 0.5 Btu/lb.

HEAT MEASUREMENTS IN AIR SYSTEMS

In HVAC, heat in air is defined and measured as sensible, latent and total. Heat measurements include heat level, heat content, and moisture content.

Sensible Heat

The definition of sensible heat (Hs) is heat that is readily measured with an ordinary analog or digital thermometer. Sensible heat is measured in degrees Fahrenheit (F) and is indicated as dry bulb (db) temperature. Enthalpy of sensible heat is in Btu/lbF. People, lights, motors and computers are examples of substances that give off sensible heat. Depending on the person and the level of activity, a seated office worker gives off (or loses) sensible heat if the air temperature surrounding that person is below the skin temperature. If the air temperature is above the skin temperature then that person is gaining sensible heat from the air. When the surrounding, or ambient, air is about 73F most people lose sensible heat at a rate that keeps them comfortable. For estimation purposes we'll say that this office person loses approximately 225 Btuh sensible heat. Sensible heat measurements are taken in the outside air, return air, supply air, room air, and across heating and cooling coils, etc. Sensible temperatures are written degrees Fahrenheit dry bulb, °Fdb. For example, 55°Fdb. I shorten that to 55Fdb.

Latent Heat

Latent heat (HL) aka "hidden heat" is when heat is known to be added or removed but no temperature change is recorded. The heat released by boiling water is an example of latent heat. Once water is brought to the boiling point, adding more heat only makes it boil faster; it does not raise the temperature (heat level). Latent heat is measured in degrees Fahrenheit (F) and is indicated as dew point (dp) temperature (for example 60Fdp). Its enthalpy is in Btu/lbF. People and water sources are examples of substances that give off latent heat. For estimation purposes a seated office worker gives off approximately 225 Btuh of latent heat. Latent heat measurements are taken in mixed air plenums, outside air, return air, supply air, room air, etc.

Total Heat

The sum of sensible heat and latent heat is total heat (HT). Total heat is measured in degrees Fahrenheit and is indicated as wet bulb (wb) temperature (example, 54Fwb). Enthalpy of total heat is in Btu/lbF. For estimation purposes a seated worker in an office gives off approximately 450 Btuh of total heat. Total heat is measured with an ordinary analog dry bulb thermometer; however, the thermometer tip is covered with a sock made from a water-absorbing material. The thermometer is now a wet bulb thermometer. The sock is wetted with distilled water and the thermometer is placed in the airstream in the air handling unit or duct or manually twirled around as in a room. As air moves across the wet sock some of the water is evaporated. Evaporation cools the remaining water in the sock and cools the thermometer. The decrease in the temperature of the wet bulb thermometer is called "wet bulb depression." The instrument that measures dry bulb and wet bulb temperatures in a room or other location is a psychrometer. Psychrometers may be digital, fan-equipped or sling.

HEAT TRANSFER AND HEAT LOAD EQUATIONS

The following air and water equations are typical for calculating heat transfer across or through heating and cooling coils, through the building envelope (walls, doors, windows, roofs, etc.), and heat load in conditioned spaces, etc.

Air Systems Sensible Heat Transfer
Btuhs = cfm x 1.08 x TD or Btuhs = cfm x 1.08 x ΔT
Terminology:

Btuhs Btu per hour sensible heat

cfm   volume of airflow, cubic feet per minute

1.08   constant: 0.24 Btu/lbF (specific heat of air) x 0.075 lb/cf (density at standard conditions) x 60 min/hr

TD   dry bulb temperature difference of the air entering and leaving a cooling or heating coil. Entering Air Temperature and Leaving Air Temperature (EAT-LAT or LAT-EAT). TD (temperature difference) can also be written as delta T or ΔT.

Air Systems Latent Heat Transfer
Btuhl = cfm x 4.5 x Δhl
Terminology:
Btuhl Btu per hour latent heat

cfm   volume of airflow, cubic feet per minute

4.5   constant: 0.075 lb/cf x 60 min/hr

Δhl   Btu/lb change in latent heat content (enthalpy) of the supply air. (Latent enthalpy (dew point) is found using a psychrometric chart, table or software program). The symbol Δ (delta), meaning difference, in this case difference in latent heat enthalpies

Or

Btuhl = Btuht – Btuhs

Terminology:

Btuhl Btu per hour latent heat

Btuht Btu per hour total heat

Btuhs Btu per hour sensible heat

Or

Btuhl = cfm x 0.68 x ΔHR or Btuhl = cfm x 0.68 x Δgr

Terminology:

Btuhl Btu per hour latent heat

cfm   volume of airflow, cubic feet per minute

0.68   constant: 0.075 lb/cf x 60 min/hr x 1060 Btu/lb (latent heat of vaporization of water) ÷ 7000 grains per pound of air

ΔHR difference in humidity ratios in grains per pound between entering and leaving air

Δgr difference in grains per pound between entering and leaving air

Air Systems Total Heat Transfer

Btuht = cfm x 4.5 x Δh

Terminology:

Btuht Btu per hour total heat

cfm   volume of airflow, cubic feet per minute

4.5   constant: 0.075 lb/cf x 60 min/hr

Δh   Btu/lb change in total heat content (enthalpy) of the supply air (Enthalpy is found from wet bulb and dry bulb temperatures using a psychrometric chart or table or software program). The symbol Δ (delta), meaning difference, in total heat enthalpies.

Water Systems Heat Transfer

Btuh = gpm x 500 x TD

Terminology:

Btuh   Btu per hour

gpm   volume of water flow, gallons per minute

500   constant: 1 Btu/lbF (specific heat at std cond.) x 8.33 lb/gal
       (weight of water) x 60 min/hr

TD   temperature difference of the water entering and leaving the coil.
       Temperature difference can also be written as delta T or ΔT.

CONDITIONED SPACE HEATING AND COOLING LOADS

Sensible heat is added to the conditioned space by conduction, convection, and radiation. Heat in the conditioned comes from conduction through exterior walls, interior walls, doors, floors, ceilings and roofs. There is also the heat released by people, lights, computers, and other equipment and heat from the outside air through ventilation and infiltration and by solar radiation through windows and other transparent surfaces. Because heat always flows from a warm temperature to a colder temperature the term heat loss refers to heating loads and the term heat gain refers to cooling loads.

Heating Load

The conditioned space heating load is the rate at which heat must be added to the space to maintain a constant air temperature. To determine heating load for a winter condition we would need to know the average normal coldest outdoor temperature during the winter and at what temperature we want to maintain the space. We also need to know how much heat flows from the inside of the building to the outside through walls, doors, roofs, etc. This information is based on the heat transfer factors for each of the building materials and installation in walls, roof, etc. Then we would calculate how much heat is added to the conditioned space from people, lights, computers, and so forth. Subtracting the heat in the building from the loss through the building envelope (walls, etc.) would give us how much heat needs to be added by the heating system. Example, estimation: There are 80 people in a building with lights and computers for each person. Multiplying 225 Btuh (sensible heat) per person and adding the Btuh for the lights and electrical equipment (total watts times 3.412 Btuh/watt) gives us 110,000 Btuh sensible heat added to the space. The roof surface is 11,000 sf and 3360 sf of wall and window surface. For estimation purposes we'll say the average coefficient of heat transfer for all materials is 0.38 Btu/sf-hr-F. The calculation for heat loss is Btuh = A x U x TD (A is surface area, U is coefficient of heat transfer, TD is temperature difference between indoor design temperature and outdoor design temperature). Therefore, Btuh = (14,360 x 0.38 x 73 – 25) or 261,900 Btuh is lost through the building envelope. Subtracting heat gain in the space from heat leaving the space gives us 151,900 Btuh sensible heat that must be added to the room from the heating system. Additionally, the outside air being brought in for ventilation to maintain good indoor air quality requires about 100,000 Btuh sensible heat. So, the heating coil is operating at 252,000 Btuh. It is very important to maintain outside air at the correct minimum otherwise extra OA requires more heat from the heating system and therefore extra energy costs.

Cooling Load

The conditioned space cooling load is the rate at which heat must be removed from the space to maintain a constant air temperature. To determine cooling load for a summer condition we would need to know the average normal hottest outdoor temperature during the summer and at what temperature we want to maintain the space. We also need to know how much heat flows from the outside of the building to the inside through the building envelope (walls, doors, roofs, etc.) This information is based on the heat transfer factor for each on the building materials and installation in walls, etc. Then we would calculate how much heat is added to the conditioned space from people, lights, computers, and so forth. Adding the heat in the building (from people, etc.) to the gain through the building envelope would give us how much heat needs to be removed by the cooling system. Example, estimation: There are 80 people in a building with lights and computers for each person. Multiplying 225 Btuh (sensible heat) per person and adding the Btuh for the lights and electrical equipment (total watts times 3.413 Btuh/watt) gives us 110,000 Btuh sensible heat added to the space. The roof surface is 11,000 sf and 3360 sf of wall and window surface. For estimation purposes we'll say the average coefficient of heat transfer for all materials is 0.38 Btu/sf-hr-F. The calculation for heat transfer is Btuh = A x U x TD (A is surface area, U is coefficient of heat transfer, TD is temperature difference between outdoor design temperature and indoor design tem-

perature). Therefore, Btuh = (14,360 x 0.38 x 90 – 75) or 81,852 Btuh added to the space through the building envelope. Now adding the 110,000
Btuh from lights, computers and people the total sensible heat is 191,852
Btuh. But wait! If you order today you can get a cooling system that
also takes care of the latent heat at no extra charge, except of course for
shipping and handling. The 80 people are also adding 225 Btuh of latent
heat, or another 18,000 Btuh. And the outside air being brought in for
ventilation to maintain good indoor air quality is adding both sensible
and latent heat. So the total heat (cooling load) that must be removed by
this cooling system is 265,652 Btuh total or about 23 tons of refrigeration.
Energy conservation opportunity: Maintain OA at the correct minimum
otherwise the extra OA requires that the cooling system provides more
cooling, and therefore extra energy costs.

HEAT AND TONS OF REFRIGERATION

A ton of refrigeration (12,000 Btu/hr) is equivalent to the amount
of refrigeration produced by melting 1 ton of ice at 32F in 24 hours.
Here's the calculation: One ton of ice weights 2000 pounds and it takes
144 Btu of heat to change 1 pound of ice at 32F to 1 pound of water at
32F. Therefore, 12,000 Btuh/ton = 2000 lb/ton of ice x 144 Btu/lb heat of
fusion ÷ 24 hr. See Chapter 4, Figure 4-3 (Btu Change from One Pound
Ice to Water to Steam to Superheated Steam).

Chapter 3

Psychrometrics

PSYCHROMETRIC CHART

Psychrometrics is derived from Greek, "psychro" meaning "cold" and "metrics" meaning "measure of." However, HVAC psychrometrics is more than just the measure of cold (a comparative term discussed in Chapter 2); it is the science of atmospheric air and water vapor thermodynamics applied to HVAC systems. With an understanding of psychrometrics, one can use psychrometric charts or "psych" charts to illustrate the changes that occur to air as it goes through the air conditioning processes of heating, cooling, humidifying, and dehumidifying. This chapter is an introduction to psychrometric charts and describes some of their uses in analyzing HVAC systems. The psychrometric chart is a valuable tool for HVAC engineers and technicians alike.

The psychrometric charts in this chapter are for sea level pressures (the standard chart) but psych charts for other elevations are available. Printed or digital psychrometric charts, software programs and Smartphone apps provide the following psychrometric information:

Dry bulb, wet bulb and dew point temperatures in degrees Fahrenheit. Specific volume cubic feet per pound (cf/lb). Density in pounds per cubic foot (lb/cf) is not shown directly on the psychrometric chart but can be calculated as the reciprocal of specific volume.

Enthalpy in Btu per pound of dry air.

Sensible Heat Ratio which is calculated: Btu/hr sensible heat divided by Btu/hr total heat.

Moisture Content (Specific Humidity) which is grains of moisture per pound of dry air or pounds of moisture per pound of dry air.

Relative humidity as a percentage.

Vapor pressure in inches of mercury (in. Hg).

Standard air dot aka standard air reference point.

Dry Bulb (DB) Temperature

Dry bulb temperature is the temperature of the air measured by a standard thermometer. On the psychrometric chart, the straight vertical lines are dry bulb temperatures (example: 60Fdb shown as #1 in Figure 3-1). The dry bulb temperature scale in degrees Fahrenheit is along the bottom of the chart (scale shown from 40Fdb to 110°Fdb). Follow the #1 dry bulb line down to intersect the scale at 60F. Psychrometric charts are available in the following approximate dry bulb temperature ranges: low temperatures: -20Fdb to 50Fdb, normal temperatures: 20Fdb to 100Fdb, and high temperatures from 60Fdb to 250Fdb. All charts used in this text are normal temperatures.

Wet Bulb (WB) Temperature

To get wet bulb temperatures above 32F use a standard thermometer. Cover the sensing bulb with a wet (distilled water) wick. Expose the thermometer wick to air moving at a velocity greater than 900 feet per minute. The wet bulb temperature scale on the psychrometric chart is along the 100 percent saturation line on the chart (the curved left line indicated by #3, wet bulb temperatures indicated from 40F to 85F). The wet bulb lines run at about a 30 degree angle across the chart as shown by #2 in Figure 3-1 (40Fwb).

Dew Point (DP) Temperature

Dew point temperature is the temperature at which moisture will start to condense from the air. The scale (shown from 0F to 85F) for dew point temperatures is along the right side of the chart (#7, Figure 3-1). The dew point temperature lines run horizontally, at right angles to the dry bulb temperature lines (#4 in Figure 3-1); in fact, they are the same lines as the specific humidity (also known as humidity ratio) lines (#6, Figure 3-1). For some charts, dew point temperature is along the curved line on the left side of the chart (#8, Figure 3-1). This curved line is the 100 percent saturation line. The dew point is where the humidity ratio line intersects the 100 percent saturation curve.

Specific Volume and Density

Specific volume is the volume of a substance per unit weight. For air, the units of specific volume are cubic feet per pound. Specific volume lines run across the chart at about a 70 degree angle (#10, Figure 3-1). Density and specific volume are reciprocals. Once specific volume is found on the psych chart, density is calculated using the equation: density = 1/specific volume. For example, if the specific volume is 13.33 cubic feet per pound then the density is 0.075 pounds per cubic foot (0.075 lb/cf = 1/13.33 cf/lb).

Enthalpy

Enthalpy (h) is a thermodynamic property, which serves as a measure of heat energy. Chapter 2 discussed heat in terms of temperature and Btu. To review: temperature is an indication of heat intensity and Btu is the quantity of heat in a substance. Enthalpy is the measurement of the heat content of a substance used in psychrometrics to find the amount of heat necessary for various processes. On the psych chart, enthalpy represents the energy in one pound of dry air and the grains of moisture associated with it at a given condition. Enthalpy is used in HVAC calculations to find the total heat (typically in Btu per hour) of the air at a given condition. For example, the total heat of the air entering a cooling coil less the total heat of the air leaving the cooling coil would give the total heat removed by the cooling coil. The enthalpy scale is on the far left side of the psych chart (represented as #9 in Figure 3-1). To find enthalpy on some psych charts extend the wet bulb line from the condition point past the 100 percent saturation line to the enthalpy scale. On other charts (such as represented in Figure 3-1) use a straight edge and draw a line from the left hand enthalpy scale to another enthalpy scale at the bottom or right side of the chart (also shown as #9). The enthalpy line pivots on the condition point. The enthalpy is correct when the line passes through condition point and the enthalpy is the same on the left hand scale and the right or bottom scale. The units of enthalpy are Btu per pound of dry air.

Sensible Heat Ratio

The sensible heat ratio (SHR), also known as sensible heat factor (SHF), is the ratio of sensible heat to total heat (Qs ÷ Qt, where Q is quantity). It is also expressed as Btu per hour sensible heat divided by Btu per hour total heat (Btuhs ÷ Btuht). The sensible heat ratio scale is shown on the right side of the chart (#11, Figure 3-1), and is used in conjunction with the SHR reference dot near the center of the chart (#12, Figure 3-1). The SHR reference dot is located at 50% relative humidity and 78Fdb on some charts and 50% relative humidity and 80Fdb on other charts. The sensible heat ratio line is drawn from the SHR dot and extended through the SHR scale.

Moisture Content

Moisture content (w) is the weight of water vapor in each pound of dry air. Moisture content is also known as humidity ratio or specific humidity. The moisture content scale is on the right side of the chart from 0 to 200 grains of moisture per pound of dry air (#5, Figure 3-1). Some psychrometric charts also have a scale that gives the moisture content in pounds of moisture per pound of dry air from 0.0 to 0.026 pounds. Specific humidity is the actual amount of moisture in a pound of air. On the psych chart, the specific humidity lines (#6, Figure 3-1) run horizontally at right angles to the dry bulb lines. The units of specific humidity are grains of moisture per pound of dry air or pounds of moisture per pound of dry air. 7000 grains of moisture equals one pound of moisture.

Relative Humidity

Relative humidity (rh) is the ratio of the amount of water vapor, aka moisture, present in the air to the total amount of moisture that the air can hold at a given temperature. Relative humidity is expressed as a percentage. The percent relative humidity lines are the curved lines that start on the left side of the chart at the 100 percent saturation line and decrease going to the right side of the chart. On most psych charts, the designated relative humidity lines are in 10% increments starting at the saturation line. The next line to the right of the saturation line is 90%, then 80%, 70%, etc., with the last line at 10%. In Figure 3-1: #8, indicates relative humidity lines shown at 100% rh, 50% rh and 10% rh. Table 3-1 shows the changes in relative humidity as air temperature changes with a constant amount of moisture content (specific humidity) in the air.

Table 3-1. Change in relative humidity with change in air temperature and constant specific humidity.

Specific Humidity	Air Temperature	Relative Humidity
60 grains per pound	100F	21%
60 grains per pound	70F	54%
60 grains per pound	55F	94%

Vapor Pressure

Vapor pressure (Pw) is the pressure exerted by water vapor in the air (#13, Figure 3-1). The units of vapor pressure are inches of mercury (in. Hg).

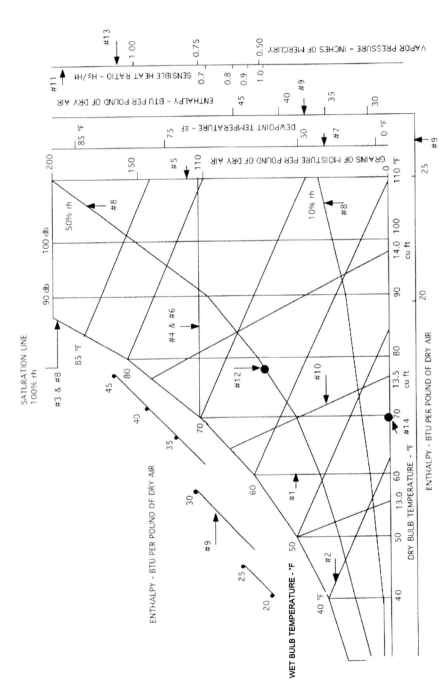

Figure 3-1. Representation of a Psychrometric Chart—Not to Scale

Standard Air Dot

Some psych charts have a dot indicating standard air (#14, Figure 3-1). Standard air is dry air with the following properties: Temperature @ 70F.

Barometric Pressure @ 29.92 inches mercury or 14.7 pounds per square inch

Volume (specific volume) @ 13.33 cubic feet per pound

Weight (density) @ 0.075 pounds per cubic foot

HVAC PROCESSES ON THE PSYCHROMETRIC CHART

Sensible Heating

Sensible heating is heat added to the air, which causes the air temperature to increase. Sensible heat is measured using a standard thermometer. A horizontal line on the psych chart represents any heating process, such as a typical residential or commercial office heating process, that adds sensible heat only. Figure 3-2: Air enters the heating coil at 50Fdb and 40Fwb (condition point A) and leaves the heating coil at 100Fdb and 62.5Fwb (condition point B). As the air is heated, the dry bulb and wet bulb temperatures increase but the dew point temperature and the moisture content (specific humidity) remain constant. The relative humidity also changes from 40% to less than 10%. This change in relative humidity is because as the air gets warmer, the air can hold more moisture (per pound of air), but because the amount of moisture is the same, about 22 grains, the relative humidity goes down (see Table 3-1).

Sensible Cooling

Opposite of sensible heating only would be sensible cooling only. Sensible cooling is heat removed from the air, which causes the air temperature to decrease as sensed by a thermometer. A horizontal line on the psych chart also represents a cooling process that removes sensible heat only. Figure 3-2: At condition point B the air is 100Fdb and 62.5Fwb and at condition point A it is 50Fdb and 40Fwb. As the air is cooled (heat removed), the dry bulb and wet bulb temperatures decrease but the dew point temperature (27Fdp) and the moisture content (22 grains) remain constant. The relative humidity also changes from approximately 8% to 40%. This change in relative humidity is because as the air gets cooler, the air's ability to hold moisture (per pound of air) lessens, but because the amount of moisture is the same, the relative humidity goes up (see Table 3-1).

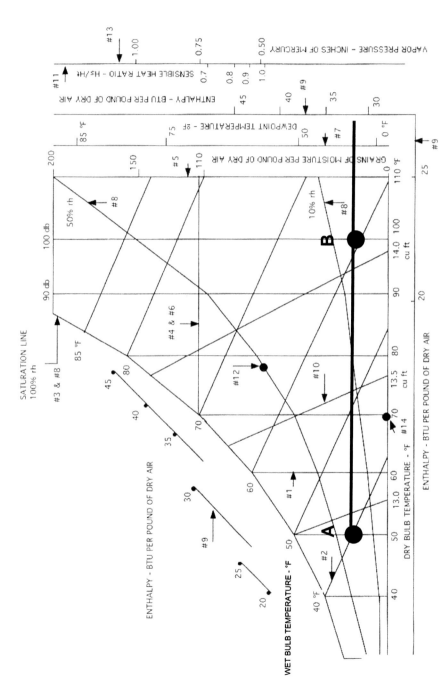

Figure 3-2. Representation of a Psychrometric Chart—Not to Scale

Humidification

Humidification occurs when air absorbs moisture as it passes through a water spray or through water on a wetted pad. To illustrate the concept of humidification see Figure 3-3 where air at 70Fdb and 10% rh enters a water spray at condition point A. The air leaves the spray at 70Fdb and 50% rh(B). In this example, the dry bulb temperature remains constant while the wet bulb temperature, dew point temperature, relative humidity and specific humidity increase. In typical psychrometric processes what actually occurs is either heating and humidification or cooling and humidification.

Dehumidification

The opposite of humidification is dehumidification. Dehumidification occurs when air passing through a chilled water coil, a refrigerant coil or a chemical absorbent/desiccant dehumidifier (used in ice rinks and supermarkets) releases moisture and is dehumidified. Figure 3-3 illustrates this concept: Air enters a dehumidifier at 70Fdb and 50% rh (condition point B). The air leaves the dehumidifier at condition point A (70Fdb and 10% rh). In this example, the dry bulb temperature remains constant while the wet bulb temperature, dew point temperature, relative humidity and specific humidity decrease. In a typical commercial and industrial cooling coil HVAC system, using either a chilled water coil or a refrigerant coil, the psychrometric process is both cooling and dehumidification simultaneously.

Heating and Humidification

Both heating and humidification occurs when air passes through a warm water spray or steam The air absorbs moisture and is humidified and heated simultaneously. Figure 3-3: Air at 70Fdb and 10% rh enters a warm water spray at condition point A. The air leaves the spray at 80Fdb and 50% rh (condition point C). The dry bulb temperature, wet bulb temperature, dew point temperature, relative humidity and specific humidity increase.

Cooling and Humidification

When air passes through cold water, either a spray or a wetted pad, it absorbs moisture and is humidified and cooled simultaneously. This is "evaporative cooling" and is effective in dry (arid) areas with low relative humidity. Figure 3-4: Air at 100Fdb, 70Fwb and 20% rh enters a cold water spray at point A. The air leaves the spray at 80Fdb, 70Fwb and 61%

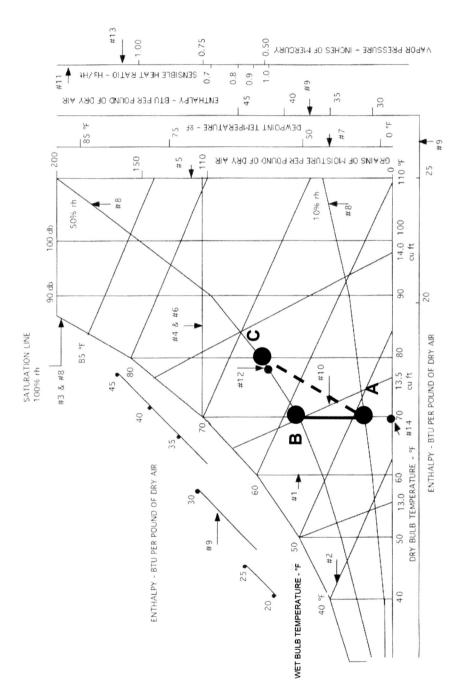

Figure 3-3. Representation of a Psychrometric Chart—Not to Scale

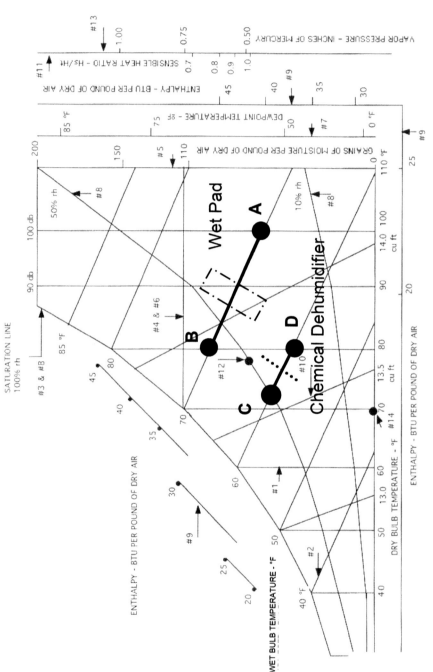

Figure 3-4. Representation of a Psychrometric Chart—Not to Scale

rh (B). The wet bulb temperature remains constant while the dry bulb temperature decreases. The dew point temperature, relative humidity and specific humidity increase.

Heating and Dehumidification

When air passes through a chemical dehumidifier the psychrometric process is heating and dehumidifying following along a constant wet bulb line. Figure 3-4: Air enters a chemical dehumidifier at 72Fdb, 60Fwb and 50% rh (condition point C). The air leaves the dehumidifier at condition point D (80Fdb, 60Fwb and 30% rh) The wet bulb temperature remains constant while the dry bulb temperature increases. The dew point temperature, relative humidity and specific humidity decrease.

Cooling and Dehumidification

Air is dehumidified and cooled simultaneously by passing it through sprays of cold water or over a cold surface such as an energized refrigerant coil or chilled water coil. The most common type of HVAC cooling is "mechanical cooling" using a cold surface. Figure 3-5: Air at 80Fdb and 50% rh enters a cooling coil (A). The air leaves the cooling coil at 55Fdb, 53Fwb and 90% rh (B). A line drawn between the two points illustrates the changes in the air as it goes through the cooling and dehumidifying process. This line is the "process line." The dry bulb temperature, wet bulb temperature, dew point temperature and specific humidity decrease. Relative humidity has increased.

HVAC CALCULATIONS AND THE PSYCHROMETRIC CHART

A condition point is the condition of the air at a given point in the HVAC system such as the supply air, room air/return air, outside air, mixed air, entering or leaving a heating or cooling coil, etc. To plot a condition point on the psych chart two properties of the air must be known. Typically, they are dry bulb temperature and wet bulb temperature or dry bulb temperature and relative humidity. Once the condition point is plotted most or all of the following properties of the air can be found on a psych chart: dry bulb temperature, wet bulb temperature, dew point temperature, relative humidity, moisture content, enthalpy, specific volume (density is calculated), and vapor pressure.

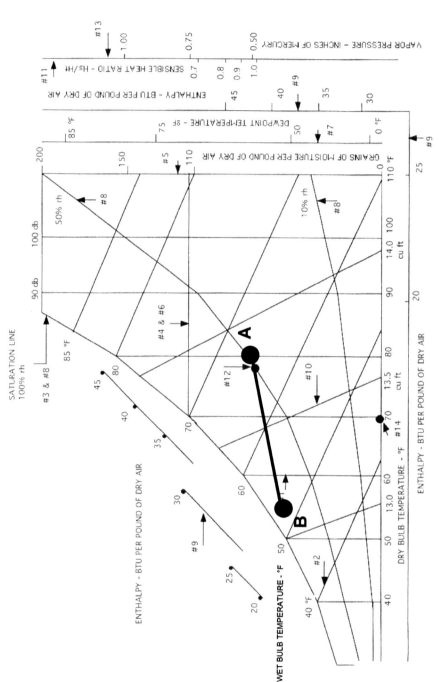

Figure 3-5. Representation of a Psychrometric Chart—Not to Scale

Determining Air Properties from the Psychrometric Chart
Exercise:

The return air from a conditioned space has these two properties: 75Fdb and 50% rh (Figure 3-6). Using a psychrometric chart, the other return air properties are:

Wet bulb: 62.5Fwb

Enthalpy: 28.1 Btu/lb of dry air

Dew point: 55Fdp

Specific humidity: 66 grains of moisture per pound of dry air

Vapor pressure: 0.44 inches of mercury

Specific volume: 13.7 cubic feet per pound of dry air

Density: 0.073 pounds per cubic feet (calculated)

Using the Psychrometric Chart
Exercise:

The HVAC engineer has determined that the optimum indoor temperature for a conditioned space in Wichita, Kansas is 75Fdb at 63Fwb. The design outside air is 98Fdb, 73Fwb. The conditioned space design sensible cooling load is 194,400 Btuhs and the total load is 228,700 Btuht. The design relative humidity leaving the cooling coil is 90%. Using a psychrometric chart plot and calculate the design conditions for air conditioning system AC-1 to provide the room requirements.

Plotting the Return Air Condition for Air Conditioning System AC-1
The return air condition is the room condition.

Figure 3-7: Point A, return air condition: 75Fdb, 63Fwb, 52% rh.

Calculating Conditioned Space Sensible Heat Ratio for AC-1
To calculate sensible heat ratio use:

SHR = Btuhs ÷ Btuht

Where:

SHR　 = sensible heat ratio

Btuhs　= sensible heat in Btu per hour in the conditioned space

Btuht　= total heat in Btu per hour in the conditioned space

Then:

SHR　 = 194,400 Btuhs ÷ 228,700 Btuht

SHR　 = 0.85

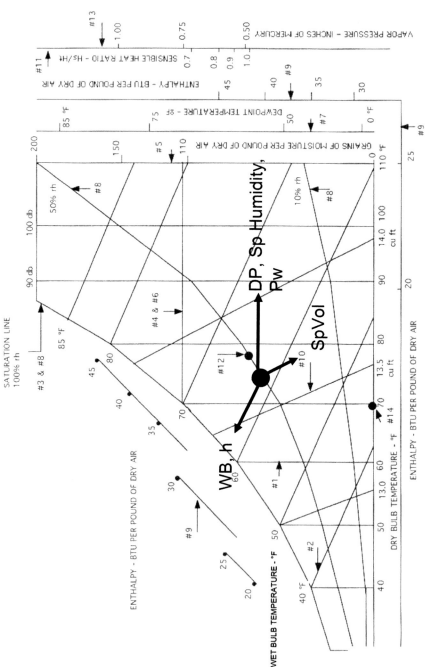

Figure 3-6. Representation of a Psychrometric Chart—Not to Scale

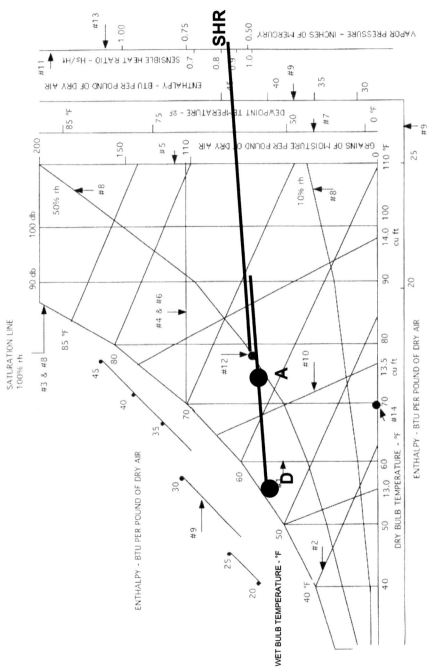

Figure 3-7. Representation of a Psychrometric Chart—Not to Scale

Plotting Sensible Heat Ratio Line for AC-1

Figure 3-7: Draw a line from the SHR reference dot (#12) to 0.85 on the SHR scale on the right side of the psych chart.

Plotting the Supply Air Condition for AC-1

Figure 3-7: Draw a line parallel to the SHR line through the return air condition point (A) to the 90% rh line. This is condition point D, the supply air condition: 57Fdb, 55Fwb, 90% rh.

Calculating Supply Air Flow for AC-1

To calculate supply air cubic feet per minute use:

cfm   = Btuhs ÷ (1.08 x TD)

Where:

cfm   =   cubic feet per minute (supply air flow)

Btuhs =   sensible heat in Btu per hour in the conditioned space

1.08   =   constant, 0.24 Btu/lb-F (specific heat of air) x 60 mm/hr x 0.075 lb/cf (weight of standard air)

TD   =   temperature difference between return air (75F) and supply air (57F)

Then:

cfm   = 194,400 Btuhs ÷ (1.08 x 18°F)

cfm   = 10,000

Plotting the Outside Air Condition for AC-1

The outside air conditions are:

Figure 3-8, condition point B, 98Fdb, 73Fwb, 31% rh.

Calculating Mixed Air Condition for AC-1

The total supply air for this system is 10,000 cubic feet per minute (cfm). The outside and return air requirements are 1500 cfm outside air (15%) and 8500 cfm return air (85%). The outside air and return air must add up to the total supply air (100%). To calculate mixed air temperature use:

MAT = (%OA x OAT) + (%RA x RAT)

Where:

MAT = mixed air temperature dry bulb

OAT = outside air temperature, dry bulb

RAT = return air temperature, dry bulb

Then:

MAT = (15% x 98 F) + (85% x 75 F) MAT = (14.7) + (63.8)

MAT = 78.5F

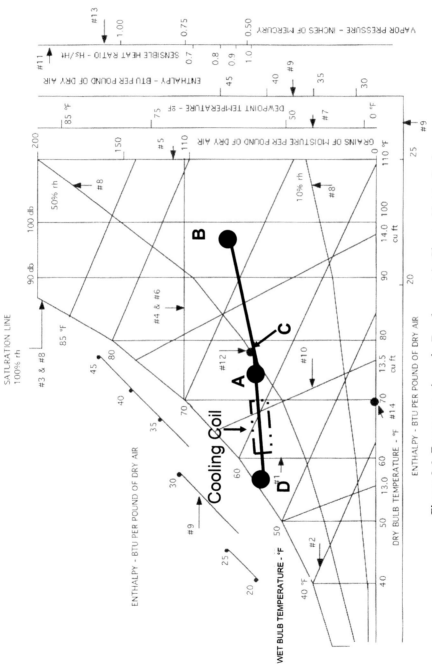

Figure 3-8. Representation of a Psychrometric Chart—Not to Scale

Plotting the Mixed Air Condition for AC-1

Figure 3-8, condition point C: Draw a line between the return air condition (point A) and the outside air condition (point B). Plot the calculated mixed air temperature (MAT) on line A-B. This is mixed air condition point C. Read the properties at point C: 78.5Fdb, 64.5Fwb, 48% rh.

Cooing Process Line for Air Conditioning System AC-1

The cooling process line illustrates the changes in the air as it goes through the cooling coil. Figure 3-8: Draw the process line between the mixed air condition, point C and the supply air condition, point D (Figure 3-7). The process line (C-D) shows a cooling and dehumidifying process. The air is cooled 21.5Fdb (78.5Fdb - 57Fdb) and dehumidified by removing 2 grains of moisture (70 grains - 68 grains) per pound of air.

Calculating Moisture Removed for AC-1

To calculate moisture removed in grains per hour use:

gr/hr   = cfm x 4.5 x Δgr/lb

Where:

gr/hr   = grains of moisture per hour
cfm   = cubic feet per minute (supply air flow)
4.5   = constant, 60 minutes per hour x 0.075 lb/cf (weight of stan-
           dard air)
Δgr/lb = moisture removed from the air as the air passes through the
           cooling coil

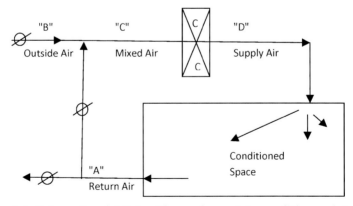

Figure 3-9. Schematic of AC-1 with psychrometric condition points corresponding to Figures 3-7 and 3-8.

Then:

gr/hr   = 10,000 cfm x 4.5 x 2 gr/lb

gr/hr   = 90,000

To calculate moisture removed in pounds per hour use:

lb/hr   = gr/hr ÷ 7,000 gr/lb (constant, 7000 grains in one pound)

lb/hr   = 90,000 gr/hr ÷ 7,000 gr/lb

lb/hr   = 12.86

To calculate moisture removed in gallons per hour use:

gal/hr = lb/hr ÷ 8.33 lb/gal (constant, weight of water)

gal/hr = 12.86 lb/hr ÷ 8.33 lb/gal

gal/hr = 1.54

Chapter 4

Heating and Ventilating Systems

HEATING SYSTEMS

For thousands of years man has used fire for warmth. In the beginning interior heating was just an open fire in a cave with a hole at the top. Later, fires were contained in hearths or sunken beneath the floor. Eventually, chimneys were added which made for better heating, comfort, health, and safety and also allowed individuals to have private rooms. Next, came stoves usually made of brick, earthenware, or tile. In the 1700s, Benjamin Franklin improved the stove, the first steam heating system was developed, and a furnace for warm-air heating used a system of pipes and flues and heated the spaces by gravity flow. In the 1800s, high-speed centrifugal fans and axial flow fans with small, alternating current electric motors became available and high-pressure steam heating systems were first used. The 1900s brought the Scotch marine boiler and positive-pressure hydronic circulating pumps that forced hot water through the heating system. The heating terminals were hot water radiators, which were long, low, and narrow, and allowed for inconspicuous heating when compared to tall, bulky, out-in-the-open steam radiators.

Centrifugal fans were added to furnaces in the 1900s to make forced-air heating systems similar to the ones used in today's residential and commercial systems. Larger commercial heating systems most often used today are low temperature (with boiler water temperatures generally in the range of 170F to 200F) or low pressure steam heating systems using boilers which operate around 15 pounds pressure [pounds per square inch, gauge (or psig) which is equivalent to approximately 30 pounds per square inch, absolute pressure, (psia)]; this is a 250F steam temperature.

COMBUSTION

Combustion is defined as a chemical reaction between a fossil fuel such as coal, natural gas, liquid petroleum gas, or fuel oil, and oxygen.

Fossil fuels consist mainly of hydrogen and carbon molecules. These fuels also contain minute quantities of other substances (such as sulfur) which are considered impurities. When combustion takes place, the hydrogen and the carbon in the fuel combine with the oxygen in the air to form water vapor and carbon dioxide.

If the conditions are ideal, the fuel-to-air ratio is controlled at an optimum level, and the heat energy released is captured and used to the greatest practical extent. Complete combustion (a condition in which all the carbon and hydrogen in the fuel would be combined with all the oxygen in the air) is a theoretical concept and cannot be attained in HVAC equipment. Therefore, what is attainable is called incomplete combustion. The products of incomplete combustion may include unburned carbon in the form of smoke and soot, carbon monoxide (a poisonous gas), as well as carbon dioxide and water.

BOILER HEATING SYSTEMS

Boiler heating systems provide heat to designated areas by transporting heat energy generated in the boiler. The two types of boiler heating systems are water heating and steam heating. The difference in heating systems is the medium used to transport the energy from the boiler to the area to be heated. Water is used to transport heat energy in the water heating system and steam is used to transport energy in the steam heating system.

Water Heating System
Water heating systems transport heat by circulating heated water to a designated area. Heat is released from the water as it flows through a coil (heat exchanger). After heat is released, the water returns to the boiler to be reheated and recirculated. Low water temperature water boilers are ≤ 250F and high temperature water boilers are greater than > 250F water temperature.

In the water heating system in figure 4-1 the return water (HWR) from the heating coil enters the boiler at 180F and is heated to 200F. The 200F supply water (HWS) is pumped from the boiler by the hot water pump (HWP) and enters the heating coil. The water flows through the coil giving off heat. Mixed air passes over the coil tubes and fins picking up heat from the water and is drawn through the supply air fan to maintain room air temperature at 70F. The air temperature leaving the coil is 105F.

The return air from the conditioned space mixes with the outside air at 30F to get a mixed air temperature of 62F. You can do this calculation by determining the amount of supply air to the conditioned space and the amount of return air. Calculate the percent return air and subtract from 100% to get percent of outside air. Use the MAT equation at the end of this chapter to find mixed air temperature. The 62F mixed air temperature enters the heating coil and is heated to 105F supply air temperature. Additional notes for Figure 4-1: The water pump should be installed on the leaving water (aka HWS) side of the boiler so it pumps the water out of the boiler. "Mixed air" may be called "conditioned air" after it leaves any energized (operating) heating or cooling coil. A "draw-through" system is when the fan is after the coil(s), if the fan is installed before the coil(s) it is a "blow-through" system.

When water is heated in the boiler some air is entrained in the water and is piped along with the water to the heat exchangers (coils). This air is removed from the water through automatic or manual air vents and air separators in the system (Figure 4-2).

The boiling point (aka boiling temperature) of water can be changed by changing the pressure on the water. If the pressure is to be changed the water must be in a closed vessel (boiler) and then the water can be boiled at a temperature of 212F or 250F or any other temperature desired. If the pressure on the water in the boiler is 14.7 psia the boiling temperature is 212F.

Sea level barometric pressure is 14.7 pounds per inch absolute (psia). Sea level barometric pressure is also 0 pounds per square inch gauge (psig). To change from pressure absolute to pressure gauge add or subtract 14.7. The equation is psia = psig +14.7. Sea level barometric pressure is also stated as 29.92 inches of mercury (inches Hg or in. Hg). The equations for changing inches of mercury to psi are: 1" Hg = 0.49 psi and 1 psi = 2.04" Hg. For estimation purposes only, round off 1" Hg to 0.5 psi or round off 1 psi to 2" Hg. Sometimes sea level barometric pressure, for estimation purposes only, is rounded off to 15 psia and 30" Hg. Also, it's common for the terms "pressure absolute" and "pressure gauge" to be changed around and said or written as "absolute pressure" and "gauge pressure."

Steam Heating System

In a steam heating system, water enters a heat conversion unit (the boiler, a closed vessel) and is changed into steam. When the water is boiled, some air in the water is released into the steam and is piped along with the steam to the heat exchanger (coil). At the coil, heat is released into the air flowing across the coil. As heat is removed the steam changes into condensate water. The condensate may be returned to the boiler by a gravity return system or by a mechanical return system using a vacuum pump (closed system) or condensate pump (open system). Some of the air in the piping system is absorbed back into the condensate water. However, much of the air collects in the heat exchanger. Steam traps (see below) are used to allow the air to escape, preventing the build-up of air which reduces the heat transfer efficiency of the system and may cause air binding in the heat exchanger.

Steam systems are classified as low, medium, intermediate or high pressure. It is important to note that low pressure steam contains more latent heat per pound than high pressure steam. Compare low pressure steam at 250F and 30 psia (946 Btu per pound) to high pressure steam at 700F and 3,094 psia (172 Btu per pound). This indicates that while high pressure steam may be required to provide very high temperatures and pressures for process functions, low pressure steam provides more economical heating operation.

Steam has some design and operating advantages over water for large heating systems. Heat capacity is one, prime mover is another. Steam at 212F when condensed releases or gives up approximately 1000 Btu per pound. On the other hand, a water heating system with supply water temperatures at 200F and return water temperatures at 180F only releases 20 Btu per pound (1 Btu/lbF). A water system uses a motor and pump to overcome system resistance (in pipes, valves, heat exchangers, etc.) to circulate the water through the system. Steam, based on its operating pressure, flows throughout the system on its own. Therefore, a motorized circulating pump is not needed although some systems do use small condensate return pumps.

Heat loss occurs through pipe radiation losses in both steam and water heating systems and at steam traps in steam heating systems. There are energy conservation opportunities in maintaining traps and proper insulation on steam and water pipes.

Boiling Temperatures and Pressures

In an open vessel, at standard atmospheric pressure (sea level), water vaporizes or boils into steam at a temperature of 212F. But the boiling temperature of water, or any liquid, is not constant. The boiling temperature can be changed by changing the pressure on the liquid. If the pressure is to be changed, the liquid must be in a closed vessel. In a water or steam heating system the vessel is the boiler. When the water is in the boiler it can be boiled at a temperature of 100F or 250F or 300F as easily as at 212F. The only requirement is that the pressure in the boiler be changed to the one corresponding to the desired boiling point. For instance, if the pressure in the boiler is 0.95 pounds per square inch absolute (psia), the boiling temperature of the water will be 100F. If the pressure is raised to 14.7 psia, the boiling temperature is raised to 212F. If the pressure is raised again to 67 psia, the temperature is correspondingly raised to 300F. A common low pressure HVAC steam heating system will operate at 15 pounds per square inch gauge pressure (psig), which is a pressure of 30 psia and a temperature of 250F.

The amount of heat required to bring the water to its boiling temperature is its sensible heat. Additional heat is then required for the change of state from water to steam. This addition of heat is steam's latent heat content or "latent heat of vaporization." To vaporize one pound of water at 212F to one pound of steam at 212F requires 970 Btu. The amount of heat required to bring water from any temperature to steam is called "total heat." It is the sum of the sensible heat and latent heat. The total heat required to convert one pound of water at 32F to one pound of steam at 212F is 1150 Btu. The calculation is as follows: the heat required to raise one pound of water at 32F to water at 212F is 180 Btu of sensible heat. 970 Btu of latent heat is added to one pound of water at 212F to convert it to one pound of 212F steam. Notice that the latent heat is over 5 times greater than sensible heat (180 Btu x 5.39 = 970 Btu). The total heat is 1150 Btu (180 + 970). See Figure 4-3.

Points in Figure 4-3 correspond to the following:

Point 1—One pound of ice (a solid) at 0F

Point 1 to Point 2—16 Btu of sensible heat added to raise the temperature of the ice from 0F to 32F. Specific heat of ice is 0.5 Btu/lb-F. Diagonal line from point to point is sensible heat and horizontal line from point to point is latent heat.

Point 2 to Point 3—Ice changing to water (a liquid) at 32F. It takes 144 Btu of latent heat to change one pound of ice to one pound of water. Total heat added is 160 Btu from point 1 to point 3.

Point 3 to Point 4—180 Btu of sensible heat added to raise the temperature of the water from 32F to 212F. Specific heat of water is 1.0 Btu/lb-F. Total heat added is 340 Btu from point 1 to point 4.

Point 4 to Point 5—Water changing to steam (a vapor) at 212F. It takes 970 Btu of latent heat to change one pound of water to one pound of steam. Total heat added is 1310 Btu from point 1 to point 5.

Point 5 to Point 6—X amount of Btu of sensible heat added to raise the temperature of the steam from 212F to X degrees Fahrenheit. This is called superheating the steam and the result is "superheated steam." For example, if the final temperature of the superheated steam is 250F then 19 Btu of sensible heat would have to be added (250F–212F = 38F. 38F x 0.5 Btu/lb-F specific heat for steam x 1 lb of steam = 19 Btu).

STEAM TRAPS

Steam traps are installed in locations where condensate is formed and collects, for example, all low points, below heat exchangers and coils, at risers and expansion loops, at intervals along horizontal pipe runs, ahead of valves, at ends of mains, before pumps, etc. The purpose of a steam trap is to separate the steam (vapor) side of the heating system from the condensate (water) side. A steam trap collects condensate and allows the trapped condensate to be drained from the system, while still limiting the escape of steam. The condensate may be returned to the boiler by a gravity return system, a mechanical return system using a vacuum pump (closed system), or condensate pump (open system).

Condensate must be trapped and then drained immediately from the system. If it isn't, the operating efficiency of the system is reduced because the heat transfer rate is slowed. In addition, the buildup of condensate can cause physical damage to the system from "water hammer." Water hammer can occur in a steam distribution system when the condensate is allowed to accumulate on the bottom of horizontal pipes and is pushed along by the velocity of the steam passing over it. As the velocity increases, the condensate can form into a non-compressible slug of water. If this slug of water is suddenly stopped by a pipe fitting, bend, or valve the result is a shock wave which can, and often does, cause damage to the system (such as blowing strainers and valves apart).

Steam traps also allow air to escape. This prevents the build-up of air in the system which reduces the heat transfer efficiency of the system and may cause air binding in the heat exchanger. In a steam heating system, water enters a heat conversion unit such as a heat exchanger or a boiler and is changed into steam. When the water is boiled, some air in the water is also released into the steam and is moved along with the steam to the heating coils or other heat exchanger. As the heat is released at the heat exchangers (and through pipe radiation losses) the steam is changed into condensate water. Some of the air in the piping system is absorbed back into the water. However, much of the air collects in the heat exchanger and must be vented.

Steam traps are classified as thermostatic, mechanical or thermodynamic. Thermostatic traps sense the temperature difference between the steam and the condensate using an expanding bellows, bimetal strip or other sensor to operate a valve mechanism. Mechanical traps use a float to determine the condensate level in the trap and then operate a discharge valve to release the accumulated condensate. Some thermodynamic traps use a disc which closes to the high velocity steam and opens to the low velocity condensate. Other types will use an orifice which flashes the hot condensate into steam as the condensate passes through the orifice.

BOILERS

Boilers are used in both hot water heating systems and steam heating systems. The hot water heating systems most often encountered in HVAC work will be low temperature systems with boiler water temperatures generally in the range of 170-200 degrees Fahrenheit. Most of the steam heating systems will use low pressure steam, operating at 15 psig (30 psia, and 250F). There are a great many types and classifications of boilers. Boilers can be classified by size, construction, appearance, original usage, and fuel used. Fossil-fuel boilers will be either natural gas-fired, liquid petroleum gas-fired or oil-fired. Some boilers are set up so that the operating fuel can be switched to natural gas (NG), liquid petroleum gas (LPG) or oil, depending on the fuel price and availability. The construction of boilers remains basically the same whether they're water boilers or steam boilers. However, water or steam boilers are divided by their internal construction into fire tube or water tube boilers.

FIRE TUBE BOILER

A fire tube boiler, as the name suggests, has the hot flue gases from the combustion chamber—the chamber in which combustion takes place—passing through tubes and out the boiler stack. These tubes are surrounded by water. The heat from the hot gases transfers through the walls of the tubes and heats the water. Fire tube boilers may be further classified as externally fired, meaning that the fire is entirely external to the boiler or they may be classified as internally fired, in which case, the fire is enclosed entirely within the steel shell of the boiler. Two other classifications of fire tube boilers are wet-back or dry-back. This refers to the compartment at the end of the combustion chamber. This compartment is used as an insulating plenum so that the heat from the combustion chamber, which can be several thousand degrees, does not reach the boiler's steel jacket. If the compartment is filled with water it is known as a wet-back boiler and conversely, if the compartment contains only air is called a dry-back boiler.

Still another grouping of fire tube boilers is by appearance or usage. The two common types used today in HVAC heating systems are the marine or Scotch marine boiler and the firebox boiler. The marine boiler was originally used on steam ships and is long and cylindrical in shape. The firebox boiler has a rectangular shape, almost to the point of being square. A Scotch marine fire tube boiler has the flame in the furnace and the combustion gases inside the tubes. The furnace and tubes are within a larger vessel, which contains the water and steam. Fire tube boilers are also identified by the number of passes that the flue gases take through the tubes. Boilers are classified as two-, three- or four-pass. The combustion chamber is considered the first pass. Therefore, a two-pass boiler would have one-pass down the combustion chamber looping around and the second pass coming back to the front of the boiler and out the stack. A three-pass boiler would have an additional row of tubes for the gas to pass through going to the back of the boiler and out the stack. A four-pass boiler would have yet another additional row of tubes for the gas to pass through going to the front of the boiler and out the stack. An easy way to recognize a two-, three- or four-pass boiler is by the location of the stack. A two- or four-pass boiler will have the stack at the front, while a three-pass boiler will have the stack at the back.

Fire tube boilers are available for low and high pressure steam, or hot water applications. The size range is from 15 to 1500 boiler horsepower (a boiler horsepower is 33,475 Btuh). HVAC fire tube boilers are typically used for low pressure applications.

WATER TUBE BOILER

In a water tube boiler, the water is in the tubes while the fire is under the tubes. The hot flue gases pass around and between the tubes, heating the water and then out the boiler stack. Most of the water tube boilers used in heating systems today are rectangular in shape with the stack coming off the top, in the middle of the shell. Water tube boilers produce steam or hot water for industrial processes, commercial applications, or other modest-size applications. They are used less frequently for comfort heating applications. Water tube boilers typically range from 25 boiler horsepower (836,875 Btuh or 836.88 MBh) to 250 boiler horsepower (8,368,750 Btuh, or 8368.75 MBh or 8.37 MMBh).

BOILER OPERATION

For a better understanding of boiler construction and operation, let's examine a four-pass, internally fired, fire tube, natural gas-fired, forced-draft, marine, wet-back boiler. The boiler consists of a cylindrical steel shell which is called the pressure vessel. It is covered with several inches of insulation to reduce heat loss. The insulation is then covered with an outer metal jacket to prevent damage to the insulation. Some of the other components are a burner, a forced-draft fan and various controls.

When the boiler is started it will go through a purge cycle in which the draft fan at the front of the boiler will force air through the combustion chamber and out the stack at the front of the boiler. This purges unwanted combustibles that might be in combustion chamber. An electrical signal from a control circuit will open the pilot valve allowing natural gas to flow to the burner pilot light. A flame detector will verify that the pilot is lit and gas will then be supplied to the main burner. The draft fan forces air into the combustion chamber and combustion takes place.

The hot combustion gases flow down the chamber and into the tubes for the second pass back to the front of the boiler. As the gases pass through the tubes they are giving up heat into the water. The gases enter into the front chamber of the boiler, called the header, and make another loop to the back of the boiler for the third pass. The fourth pass brings the hot gases back to the front of the boiler and out the stack. The temperature in the combustion chamber is several thousand degrees while the temperature of the gases exiting the stack should be about 320 degrees (or 150 degrees above the medium temperature).

BURNER

The function of the burner is to deliver, ignite and burn the proper mixture of air and fuel. The types of burners are varied and selection depends on the design of the boiler. Oil burners (Figure 4-8), except for small domestic types, deliver the fuel to the burner under pressure provided by the oil pump. The heavier oils, numbers 4, 5, and 6, generally require preheating to lower their viscosity so that they can be pumped to the burner. In addition, all oils must be converted to a vapor before they can be burned. Large commercial and industrial burners use two steps to prepare the oil for burning. The first step is called atomization which is the reduction of the oil into very small droplets. The second step is vaporization which is accomplished by heating the droplets. Oil burners are classified by how they prepare the oil for burning such as vaporizing, atomizing or rotary. Oil burners use the same methods of delivering air to the combustion chamber as do gas burners. They are either natural-, forced- or induced-draft. Regardless of what type of burner is used, proper combustion depends on the correct ratio of fuel-to-air.

Gas burners are classified as atmospheric or mechanical-draft burners. Atmospheric burners are sub-classified as natural-draft or Venturi burners. Mechanical-draft burners are either forced- or induced-draft burners. A typical gas burner used on large industrial and commercial boilers is a burner with a fan or blower at the inlet. This type of burner, which is a forced-draft burner, is called a power burner (Figure 4-10). It uses the blower to provide combustion air to the burner and the combustion chamber under pressure and in the proper mixture with the gas over the full range of firing from minimum to maximum. Another type of gas burner uses a blower at the outlet of the combustion chamber to create a slight partial vacuum within the chamber. This causes a suction which draws air into the chamber. This type of burner is an induced-draft burner.

FUEL-TO-AIR / AIR-TO-FUEL RATIO

A high fuel-to-air ratio causes soothing and lowers boiler efficiency. In certain conditions, it may also be dangerous if there's not enough air for "complete combustion" and dilution of the fuel. An improperly adjusted burner, a blocked exhaust stack, the blower or damper set incorrectly, or any condition which results in a negative pressure in the boiler room, can cause a high fuel-to-air condition. A negative pressure in the boiler room can be the result of one or a combination of conditions such as an exhaust fan creating a negative pressure in the boiler room, a restricted combustion air louver into the boiler room, or even adverse wind conditions.

High air to fuel ratios also reduce boiler efficiency. If too much air is brought in (excess air), the hot gases are diluted too much and move too fast through the tubes before proper heat transfer can occur. High air volumes are typically caused by improper blower or damper settings.

ELECTRIC BOILER

Electric boilers produce heat by electricity and operate at up to 16,000 volts. Electric boilers are typically compact, clean and quiet. They have replaceable heating elements, either electrode or resistance-coil. With the electrode type boiler, the heat is generated by electric current flowing from one electrode to another electrode through the boiler water. Resistance-coil electric boilers have the electricity flowing through a coiled conductor similar to an electric space heater. Resistance created by the coiled conductor generates heat. Resistance-coil electric boilers are not as common as electrode electric boilers.

Electric boilers are an alternative to oil or gas boilers where these boilers are restricted by emission regulation and in areas where the cost of electric power is minimal. Electric boilers can be fire tube or water tube and supply low or high pressure steam or hot water. Sizes range from 9 kW to 3,375 kW output, which is 30,708 Btuh to 11,515,500 Btuh (1 kW = 3412 Btuh).

HEAT AND FLUID FLOW CALCULATIONS FOR HEATING SYSTEM

Looking at the heating system, Figure 4-1, calculate gpm of water flow if the heating coil load is 243,810 Btuh and TD is 20F (200F EWT–180F LWT).

Btuh = gpm x 500 x TD

Where:

Btuh = Btu per hour

gpm = volume of water flow, gallons per minute

500 = constant (60 min/hr x 8.33 lb/gal x 1 Btu/lbF)

TD = temperature difference of the water entering (EWT) and leaving (LWT) the coil.

Then:

gpm =Btuh ÷ (500 x TD)

gpm =243,810 ÷ (500 x 20)

Answer:

24.4 gpm of water flows through the heating coil.

Now calculate the air TD across the heating coil if:

LAT - leaving air temperature coil is 105F

EAT - entering air temperature coil is 62F. EAT is MAT.

RAT - return or room air temperature is 70F

OAT - outside air temperature is 30F

198,450 Btuh is the Sensible Room Heating Load.

The math is:

198,450 = 5250 cfm SA x 1.08 x 35 TD (105-70)

243,810 Btuh is the Sensible Coil Heating Load.

The difference of 45,360 Btuh (243,810-198,450) is the additional heat required for the outside air.

The math is:

45,360 = 1050 cfm OA x 1.08 x 40 TD (70-30)

Then:

 TD = Btuh ÷ (1.08 x cfm)
 TD = 243,810 ÷ (1.08 x 5250) Answer:
 TD = 43F

Then:

 62F EAT + 43F TD = 105F LAT)
 243,810 Btuh is the Coil Sensible Heating Load.

The math is:

 243,810 = 5250 cfm x 1.08 x 43 TD (105-62)

The mixed air temperature (MAT also called EAT) was calculated using this equation:

MAT = (%OA x OAT) + (%RA x RAT)

Where:

MAT = mixed air temperature OAT = outside air temperature
RAT = return air temperature, also called room air temperature

Then:

MAT = (20% x 30F) + (80% x 70F)
MAT = (6) + (56)

Answer:

MAT = 62 F

VENTILATING SYSTEMS

In occupied buildings carbon dioxide, human odors and other contaminants such as volatile organic compounds (VOC) or odors and particles from machinery and other process functions need to be continuously removed or unhealthy conditions will result. Ventilation is the process of supplying "fresh" outside air to occupied buildings in the proper amount to offset the contaminants produced by people and equipment.

Ventilation systems have been around for a long time. In 1490,

Leonardo Da Vinci designed a water driven fan to ventilate a suite of rooms. In 1660, a gravity exhaust ventilating system was used in the British House of Parliament. Then, almost two hundred years later, in 1836, the supply air and exhaust air ventilation system in the British House of Parliament used fans driven by steam engines. Today, local building codes, association and testing organization guidelines, and government or company protocols stipulate the amount of ventilation required for buildings and work environments. Ventilation guidelines in the USA are approximately 15 to 25 cfm (cubic feet per minute) of air volume per person of outside air (OA).

Ventilation air may also be required as additional or "make-up" air (MUA) for kitchen exhausts, fume hood exhaust systems, and restroom and other exhaust systems. Maintaining room or conditioned space pressurization (typically +0.03 to +0.05 inches of water gauge) in commercial and institutional buildings is part of proper ventilation.

Figure 4-1 shows a system that requires, at a minimum, 20% of the total supply air to be ventilation outside air (OA) and 80% return air (RA). The outside air is brought into the mixed air plenum by the suction action of the supply air fan and proper operation of the OA dampers. In some cases a dedicated OA fan may be needed and/or additional controls or monitoring devices. The outside air coming through the outside damper is mixed with the return air from the conditioned space. The OA dampers are set for a 20% minimum and the return air dampers control the amount of return air. In this example if the amount of OA goes higher than 20% (because the automatic temperature controls require it) then the exhaust air (EA) dampers open to let some of the return air (in the return air fan plenum) escape to the outside, which maintains the proper pressure in the conditioned space. Exhaust air (EA) dampers are also called relief air (RA) dampers or spill dampers. There's more on heating and ventilating systems in future chapters.

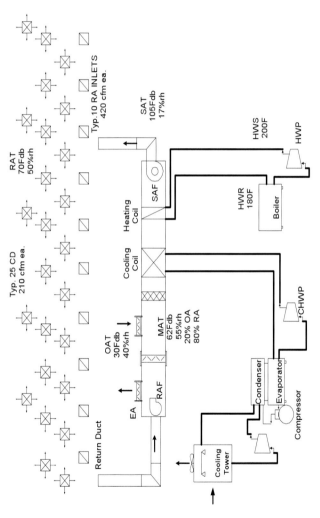

Figure 4-1. Water Heating System and Air Distribution.

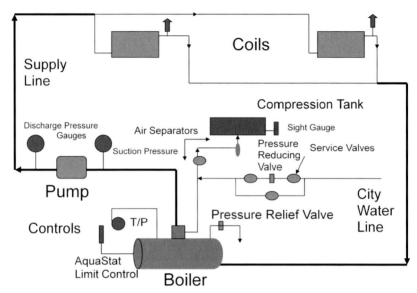

T/P—Temperature/Pressure Gauge(s) Air Vent

Figure 4-2. Water Heating System.

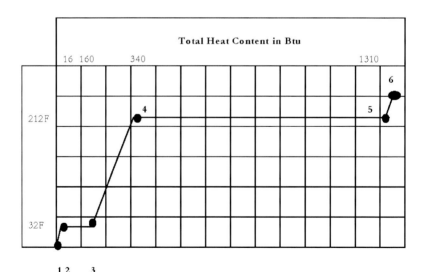

Figure 4-3. Btu Change from One Pound Ice to Water to Steam to Superheated Steam.

Figure 4-4. Large, 4-Pass, Fire Tube Boiler.

Figure 4-5. Small Oil-Fired, 2-Pass, Fire Tube Boiler

Figure 4-6. Fire Tube Boiler. Turbulators inside tubes (a device to spin the flue gas inside the tube to increase boiler efficiency).

Figure 4-7. Fire Tube Boiler. Combustion chamber, sight glass at opposite end of boiler to view flame, fire tubes to right.

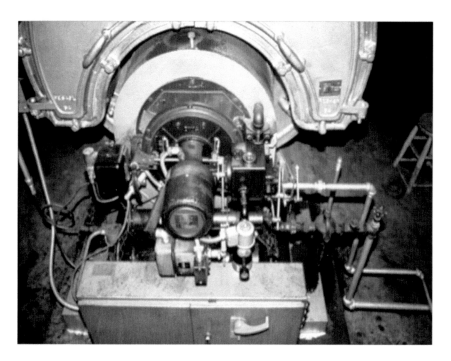

Figure 4-8. Fire Tube Boiler. Oil Burner. Atomizer Nozzle

Figure 4-9 Water Tube Boiler. Gas-Fired.

Figure 4-10. Water Tube Boiler. Power Gas Burner.

Chapter 5

Air Conditioning Systems

Brooklyn, New York, was the place, and 1902 was the year the first truly successful air conditioning system for room temperature and humidity control was placed into operation. But first it took the engineering innovations of Willis Carrier to advance the basic principles of cooling and humidity control and design the system. Cooling air had already been done successfully but it was only part of the air conditioning problem. The other part was how to regulate space humidity. Carrier recognized that drying the air could be accomplished by saturating it with chilled water to induce condensation. In 1902, Carrier built the first air conditioner to combat both temperature and humidity. The air conditioning unit was installed in a printing company and chilled coils were used in the machine to cool the air and lower the relative humidity to 55%. Four years later, in 1906, Carrier was granted a patent for his air conditioner the "Apparatus for Treating Air." However, Willis Carrier did not invent the very first system to cool an interior structure nor interestingly, did he come up with the term "air conditioning." It was Stuart Cramer, a textile engineer, who coined the term "air conditioning." Mr. Cramer used "air conditioning" in a 1906 patent for a device that added water vapor to the air.

In 1911, Mr. Carrier, who is called the "father of air conditioning," presented his "Rational Psychrometric Formulae" to the American Society of Mechanical Engineers. Today, the formula is the basis in all fundamental psychometric calculations for the air conditioning industry. Though Willis Carrier did not invent the first air conditioning system, his cooling and humidity control system and psychrometric calculations started the science of modern psychrometrics and air conditioning. As already mentioned, air "cooling" was only part of the answer. The big problem was how to regulate indoor humidity. Carrier's air conditioning invention addressed both issues and has made many of today's products and technologies possible. In the 1900s, many industries began to flourish with the new ability to control the

indoor environmental temperature and humidity levels in both occupied and manufacturing areas. Today, air conditioning is required in
most industries and especially in ones that need highly controllable
environments, such as clean environment rooms (CER) for medical or
scientific research, product testing, and sophisticated computer and
electronic component manufacturing.

HEAT AND FLUID FLOW
CALCULATIONS FOR AIR CONDITIONING SYSTEMS

Looking at the air conditioning system in Figure 5-1, calculate
mixed air temperature. MAT = (%OA × OAT) + (%RA × RAT)

Where:
MAT = mixed air temperature
OAT = outside air temperature
RAT = return air temperature

Then:
MAT = (20% × 90F) + (80% × 75F)
MAT = (18) + (60)

Answer:
MAT = 78F

Calculate cfm if the Room Cooling Load is 113,400 Btuh (Sensible
Heat). There is a 20 TD difference between the air leaving the coil at
55F and room design temperature set at 75F.

Using:
Btuh = cfm × 1.08 × TD

Then:
cfm = Btuh ÷ (1.08 × TD)
cfm = 113,400 ÷ (1.08 × 20)

Answer:
cfm = 5250

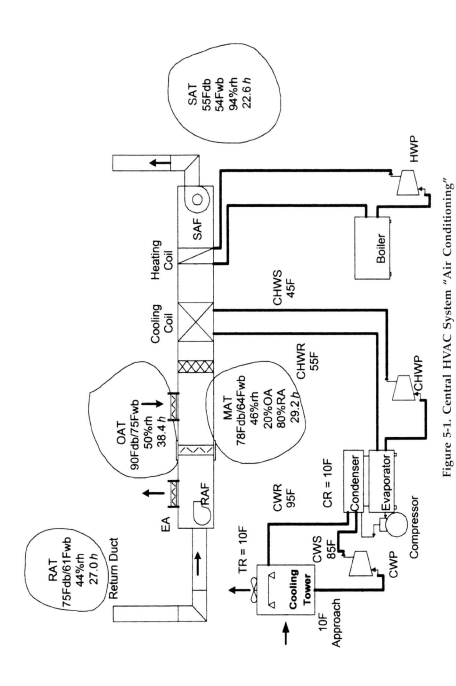

Figure 5-1. Central HVAC System "Air Conditioning"

Now calculate total heat removed by the cooling coil using this equation:

Btuht = cfm × 4.5 × Δh

Where:

Btuht = Btu per hour total heat (wet cooling coil)
cfm = volume of airflow
4.5 = constant, 60 min/hr × 0.075 lb/cf
Δh = Btu/lb change in total heat content of the supply air (from wet bulb and dry bulb temperatures and a psychrometric chart)

The enthalpies (in Btu/lb from the psychrometric chart) are:

78Fdb/64Fwb = 29.156 Btu/lb
55Fdb/54Fwb = 22.627 Btu/lb
Δh = 6.529 Btu/lb

Then:

Btuht = cfm × 4.5 × Δh
Btuht = 5250 × 4.5 × 6.529

Answer:

Btuht = 154,248 (Coil Load)

And since:

12,000 Btuh equals 1 ton of refrigeration (TR)
154,248 Btuh ÷ 12,000 Btuh per ton is 12.854 TR ($\approx$13 TR)

Now calculate the water flow through the coil using:

Btuh = gpm × 500 × TD

Then:

gpm = Btuh ÷ (500 × TD)
gpm = 154,248 ÷ (500 × 10) (TD = 55F - 45F)

Answer:

gpm = 30.8 (water flow through the cooling coil)

THE AC REFRIGERATION CYCLE

Let's go through the air conditioning system when it is in the cooling mode. Let's say that it is a summer day and the outside air is 90 degrees. The outside air damper is open to allow 200 cfm of outside air to mix with 1000 cfm of return air. The return air temperature is 75F. The temperature of the mixed air is 78F. The mixed air comes through the filter section where it is cleaned and enters the coil to be cooled. The coil in this example is a refrigerant evaporator coil (Figure 5-2). The other type of coil used in HVAC systems uses cooled water to bring the temperature of the mixed air down. This coil is called a chilled water coil. Both types of coils are also termed "heat exchangers" and they can be located on either side of the fan. Let's take a look at a simple mechanical refrigeration cycle and see what happens to enable the mixed air to be cooled down to 55F. This 55F air leaving the evaporator coil is now called the supply air. The volume of supply air is 1200 cfm.

The purpose of the refrigeration cycle is to remove unwanted heat from one place and discharge it into another place. In our HVAC system the unwanted heat is in the conditioned space. This heat in the conditioned space is picked up by the supply air and brought back through the duct system to the evaporator coil. Now let's start our refrigeration cycle. To begin, a mechanical refrigeration cycle is a completely closed system consisting of four different stages: expansion, evaporation, compression, and condensation. Contained in this closed system is a chemical compound called a refrigerant. The system is closed so that the refrigerant can be used over and over again, for each time it passes through the cycle it removes some heat from the supply air and discharges this heat into the outside air. The closed cycle also keeps the refrigerant from becoming contaminated, as well as, controlling its flow.

The expansion stage is a good place to start our trip through the refrigeration cycle. This stage consists of a pressure reducing device (also called a metering device, MD) such as an expansion valve, capillary tube or other device to control the flow of refrigerant into the evaporator coil. Our system has a thermal expansion valve abbreviated TXV. The refrigerant enters the expansion stage as a high-pressure, high-temperature liquid at 90F. It goes through the metering device and leaves the expansion stage as a low-pressure, low-temperature liquid. This low-temperature liquid refrigerant, let's say that it is 40F (its boiling temperature at this pressure), enters the evaporator coil. This begins the evaporation stage of the

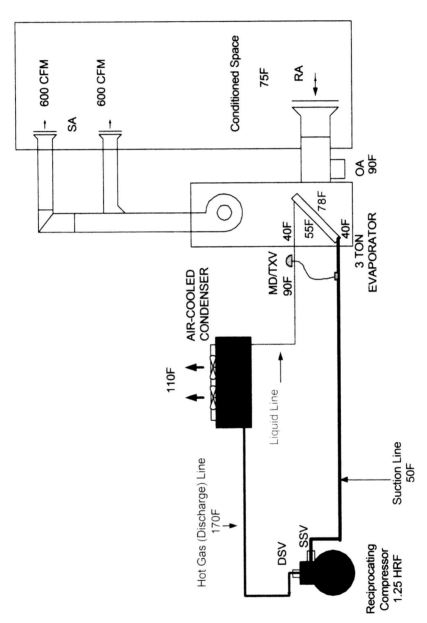

Figure 5-.2 Air Conditioning System Example

cycle. At same time that the 40F liquid refrigerant is passing through the tubing of the evaporator coil the 78F mixed air is passing over the same tubes. In order for heat to flow there must be a difference in temperature. Heat always flows from a higher level or temperature to a lower level or temperature. The air passing over the evaporator coil is warmer than the liquid refrigerant in the tubes. Therefore, heat will be picked up by (or transferred to) the refrigerant. In other words, the air is cooled and the refrigerant is heated. This heating of the refrigerant causes it to boil off and change state from a liquid to a vapor just as adding heat to water will cause it to boil off and change state to steam.

The difference between the refrigerant in our system and water, which incidentally is also a refrigerant (refrigerant-718), is that the boiling point of our refrigerant is minus forty degrees below zero (-40), while the boiling point of water is 212 degrees above zero. Both these boiling points occur at sea level. It is important to understand that the boiling point of a liquid will change in the same direction as the pressure to which the liquid is subjected. For example, water at sea level, 14.7 pounds per square inch, boils at 212F, while water subjected to 25 pounds per square inch of pressure boils at approximately 240F. Since our closed refrigeration system is under pressure, in other words greater than atmospheric, we have elevated the boiling point of the refrigerant to approximately 40F above zero. As the refrigerant passes through the evaporator tubes the boiling process continues. As long as the refrigerant is changing state from a liquid to a vapor the temperature remains at 40F. However, once all the liquid has been changed to a vapor, and this occurs near the end of the evaporator, the vapor can now absorb additional heat. This process is called superheating the vapor, or simply, superheat.

Our system will pick up about 10 degrees of superheat and the refrigerant, which is now a low-pressure, low-temperature vapor, will flow through the suction line and enter the compression stage at 50F. The compression stage consists of an electrically driven mechanical compressor. The compressor has two main functions within the refrigeration cycle. One function is to pump the refrigerant vapor from the evaporator so that the desired temperature and pressure can be maintained in the evaporator. The second function is to increase the pressure of the refrigerant vapor through the process of compression, and simultaneously increase the temperature of the vapor. This change in pressures also causes the refrigerant to flow through the system.

Let's say that our compressor increases the pressure of the vapor so that the corresponding temperature of the vapor will be 120F. This is the condensing temperature, that is, the temperature in the condenser. This high-pressure, high-temperature vapor leaves the compressor and enters the condensation stage. In our example, the actual temperature of the refrigerant in the hot gas or discharge line is 170F. The temperature of the refrigerant will cool down from 170F to 120F as it goes through the hot gas line and in the condenser. This loss of heat, in this case 50F of sensible heat, is called "desuperheating."

The condensation stage in our refrigeration system consists of an air-cooled condenser coil and a fan. Some systems however, use a pump and a water-cooled condenser. Our air-cooled condenser has a fan or blower, sometimes called the outdoor fan, which draws outside air across the condenser coil. The temperature of the refrigerant vapor flowing through the condenser tubes is 120F. At the same time, the 90F outside air is passing over the condenser tubes. As before, heat travels from a higher temperature to a lower temperature. Since the air passing over the condenser coil is cooler than the refrigerant in the tubes, heat will be picked up by the outside air. In other words, the refrigerant is cooled and the air is heated. The condenser is said to be discharging or rejecting its heat into the atmosphere.

Let's back up for a minute. Where did we get this heat that is in the condenser? Well, about 75% of it is the unwanted heat from the conditioned space. The other 25% is heat from the compression stage. So now we have taken the unwanted heat from one place, the conditioned space, and discharged it to another place, the outside.

In order for the refrigerant to be able to pick up more heat from the supply air it must once again become a low-temperature liquid. The cooling of the vapor in the condenser causes the refrigerant to change state from a vapor to a liquid. This process is called condensation. As the refrigerant vapor passes through the tubes the condensation process continues. As long as the refrigerant is changing state from a vapor to a liquid the temperature remains at 120F.

However, once the entire vapor has been changed to liquid, the liquid can reject additional heat. As the refrigerant, which is now a high-pressure, high temperature liquid (120F @ 260 psig) flows through the liquid line to the pressure reducing device it continues to give up heat. This is called "subcooling." The liquid refrigerant will enter the expansion stage's pressure reducing device (metering device) at approximately 90F. The liquid was subcooled 30F. Only liquids can be subcooled and only vapors can be superheated or desuperheated. When the liquid refrigerant goes through the metering device the pressure on the refrigerant is reduced to 70 psig. This reduction in pressure (from 260 psig to 70 psig) reduces the boiling point of the liquid refrigerant to 40F. However, the temperature of the liquid refrigerant at 90F is above the new boiling point (40F). Because the liquid refrigerant is hotter than its boiling point a part of the liquid refrigerant begins to boil off. This boiling off of the liquid refrigerant is called flashing. The liquid refrigerant which is boiled off or flashed, changes state to a vapor or gas. This vapor is called "flash gas." When a part of the liquid refrigerant is flashed, it removes heat from the remaining liquid. This flashing continues until the remaining liquid refrigerant is cooled down to the boiling point which corresponds to the pressure on the liquid (40F @ 70 psig). About 18% of the liquid is flashed off to a vapor and is not available to pick up heat (i.e., latent heat of vaporization) but can pick up sensible heat in the evaporator stage. The vapor and the remaining liquid (82%) enters the evaporation stage and the cycle starts over. The AHU has taken 1200 cfm of mixed air at 78F and cooled it down to 55F supply air.

AIRFLOW

Supply air moves through ductwork because of a difference in pressures. Just as heat moves from a higher level to a lower level, so do fluids. Fluids move from a higher pressure to a lower pressure. Air is a vapor and as such is a compressible fluid.

Remember, we said that the refrigerant vapor moved through the system because the pressure on one side of the compressor was higher than on the other side. The same is true for the air. The fan produces a pressure at the discharge of the fan that is higher than the pressure in the conditioned space. For example, the pressure in the conditioned space is atmospheric pressure while the pressure at the fan discharge is greater than atmospheric pressure.

To continue, the air moves through the ductwork until it reaches the supply air outlets in the conditioned space. As the 55F supply air is discharged it mixes with the warmer room air. Also as the supply air comes in contact with the greater mass of room air the velocity of the air slows down. After circulating through the room the air exits the room by way of the return air inlet. This amount of air, the 1200 cfm of supply air continuously flowing through the room, will result in about 7.5 complete air changes per hour. Once again, the air flows through the ductwork because of a difference in pressures. In this case, the room air is at atmospheric pressure, or slightly above, because of room pressurization and the inlet to the fan is less than atmospheric pressure so the air flows towards the fan. The return air carrying the heat removed from the conditioned space mixes with the outside air, which also contains some heat. This mixture goes through the filter section and into the cooling coil and our cooling cycle starts over again.

FOUR TYPES OF
AIR CONDITIONING COOLING SYSTEMS

Water	to	Water
Air	to	Air
Air	to	Water
Water	to	Air

AC COOLING SYSTEM #1

Heat Rejection Side		Heat Pickup Side
Water	**to**	**Water**

A water-to-water cooling system (Figure 5-3) has a water-cooled condenser and cooling tower on the heat rejection side. Chilled water coil(s) in the air handling unit(s) (AHU) or fan-coil unit(s) (FCU) are on the heat pickup side.

AC COOLING SYSTEM #2

Heat Rejection Side		Heat Pickup Side
Air	**to**	**Water**

An air-to-water cooling system (Figure 5-4) has an air-cooled condenser on the heat rejection side. Notice that while the condenser is rated for 3.75 tons or 45,000 Btuh (3.75 ton × 12,000 Btuh/ton) the evaporator is rated for 3 tons or 36,000 Btuh. Why is the condenser rated for more Btuh than the evaporator? Because of the additional heat from the compressor, 25% more Btuh from "heat of compression." If a condenser rated at 36,000 Btuh was installed the system would not work. Chilled water coil(s) in the air handling units (AHU) or fan-coil units (FCU) are on the heat pickup side.

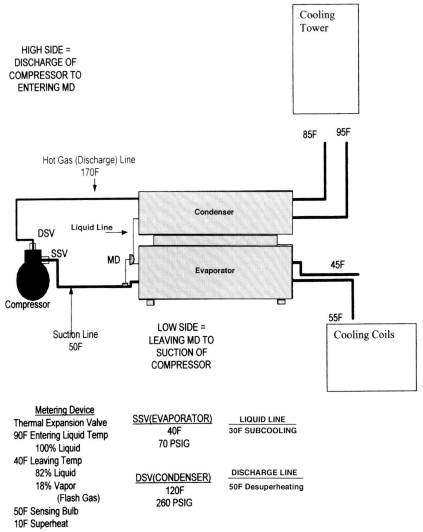

HIGH SIDE =
DISCHARGE OF
COMPRESSOR TO
ENTERING MD

Cooling Tower

85F   95F

Hot Gas (Discharge) Line
170F

Condenser

DSV   Liquid Line

SSV   MD

Evaporator   45F

Compressor

55F

Suction Line
50F

LOW SIDE =
LEAVING MD TO
SUCTION OF
COMPRESSOR

Cooling Coils

Metering Device
Thermal Expansion Valve
90F Entering Liquid Temp
100% Liquid
40F Leaving Temp
82% Liquid
18% Vapor
(Flash Gas)
50F Sensing Bulb
10F Superheat

SSV(EVAPORATOR)
40F
70 PSIG

DSV(CONDENSER)
120F
260 PSIG

LIQUID LINE
30F SUBCOOLING

DISCHARGE LINE
50F Desuperheating

Figure 5-3. Water-to-Water AC System

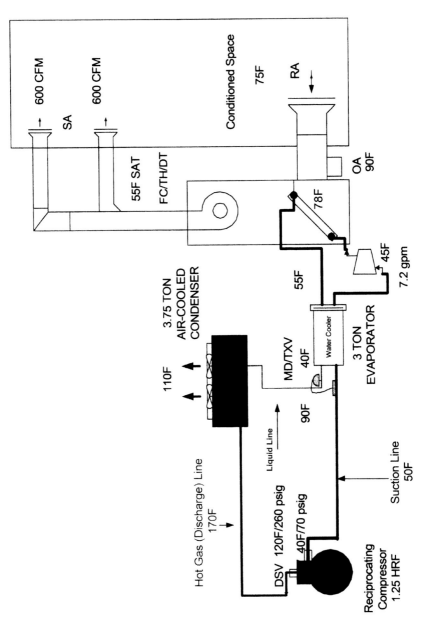

Figure 5-4. Air-to-Water AC System

AC COOLING SYSTEM #3

Heat Rejection Side		Heat Pickup Side
Air	to	**Air**

An air-to-air cooling system (Figure 5-5) has an air-cooled condenser on the heat rejection side. On the heat pickup side there is an evaporator (refrigerant DX) coil in the air handling units (AHU) or fan-coil units (FCU). In this system the condenser is rated for 3 tons for a 3 ton evaporator. This is because the condenser is rated at 15,000 Btuh per ton (12,000 × 1.25HRF). When we say a 3 ton or X ton system, we are talking about the evaporator or heat pickup side of the system. The heat rejection side will always be more tonnage.

AC COOLING SYSTEM #4

Heat Rejection Side		Heat Pickup Side
Water	to	**Air**

A water-to-air cooling system (Figure 5-6) has a water-cooled condenser and cooling tower on the heat rejection side. On the heat pickup side there is an evaporator (refrigerant DX) coil in the air handling units (AHU) or fan-coil units (FCU).

PROBLEM

Calculate heat removed by the cooling tower in Figure 5-7.

Btuh = gpm × 500 × TD

gpm = Btuh ÷ (500 × TD)

gpm = 192,810* ÷ (500 × 10)

(TD = 95F − 85F)(This TD is the tower range, TR)

gpm = 38.6

*Includes heat of compression from the compressor

HRF (Heat Rejection Factor) = 1.25

(1.25 × 154,248 Btuh coil load = 192,810 Btuh tower load)

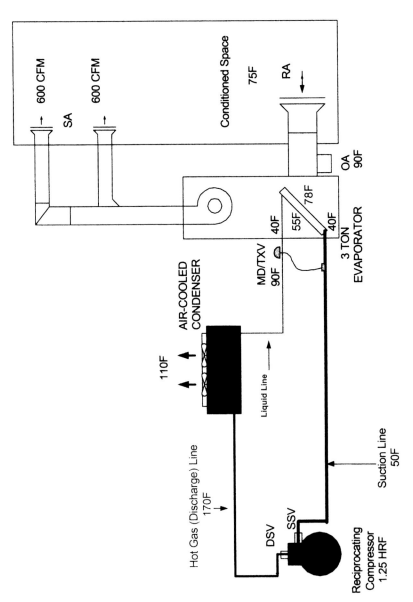

Figure 5-5. Air-to-Air AC System

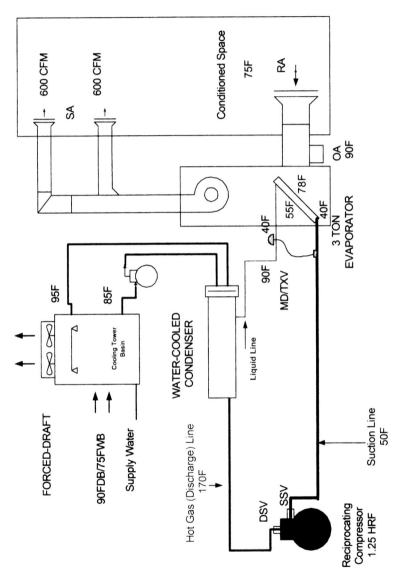

Figure 5-6. Water-to Air AC System

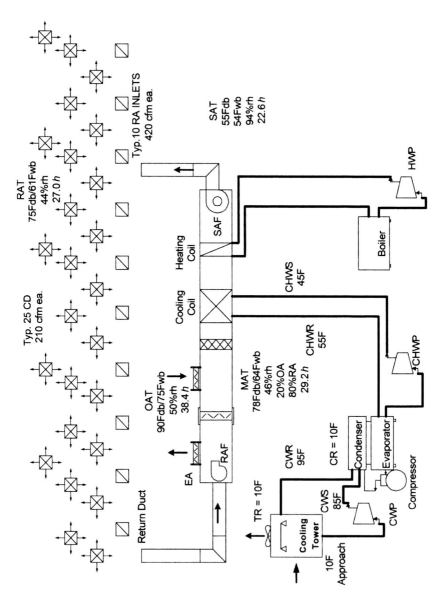

Figure 5-7. Cooling tower. TR is tower range (95F - 85F), CR is condenser rise (85F to 95F).

AIR CONDITIONING COMPONENTS
VAPOR-COMPRESSION SYSTEM

Evaporators (heat picked up from the conditioned space)
  Direct or Dry Expansion (DX)
  Flooded

Condensers (heat rejected to the outside air)
  Air-cooled
  Water-cooled
  Evaporative (combination of air and water cooled)

Compressors (pump)
  Reciprocating (up to 200 tons of refrigeration, TR)
    Constant volume
  Centrifugal (80 to 10,000 TR)
    Variable volume

Metering Devices (flow control)
  Thermal Expansion Valve (TXV) (TEV)
  Automatic Expansion Valve (AXV) (AEV)
  Float Valve
    (High Side or Low Side, Flooded Systems)
  Capillary Tube
  Hand Valve

EVAPORATORS

     HVAC evaporator temperatures are usually between 34 and 45
degrees Fahrenheit. This is true whether the air conditioning unit's
cooling coil is a direct expansion refrigerant coil supplying cold air to
the conditioned space or a water cooler supplying chilled water to the
cooling coils. Operating at less than 34F increases the likelihood of frost-
ing up the direct expansion refrigerant coil or freezing the water in the
water cooler and water coil. However, operating at higher evaporator
temperatures reduces the horsepower-per-ton ratio of the compressor
(Table 5-1).

Figure 5-8. Forced Air Cooling Tower

CONDENSERS

Condensers may be either air-cooled or water-cooled. The compressor's discharge pressure depends on how rapidly the condenser cooling medium, that is, the air or the water, will carry away the heat of the refrigerant vapor. This heat transfer rate depends on both the temperature of the condenser cooling medium and the volume of flow of the medium across or around the heat transfer surfaces of the condenser. The importance of lower condenser temperatures is that the lower the refrigerant temperature that can be maintained in the condenser, the lower the condenser pressure will be and the smaller the horsepower-per-ton ratio of the compressor (Table 5-1).

When the coils become dirty the dirt acts as an insulator reducing heat transfer. If this occurs on the evaporator the evaporator temperature is lowered. If the condenser coil is dirty the temperature inside the condenser is increased. Service technicians can help maintain good system performance by being aware of the evaporator and condenser temperatures. Efficiency is increased by increasing evaporator temperature and decreasing condenser temperature. One way this can be done is by improving heat transfer by keeping the evaporator coil and condenser coil clean (Table 5-1).

Table 5-1. Reduction in system efficiency when condensers and evaporators are not maintained.

Operating Condition	Evaporating Temperature	Condensing Temperature	Tons Ref.	Brake HP	BHP per ton	Increased BHP per ton
Normal Operation	45F	105F	17.0	15.9	0.93	
Dirty Condenser	45F	115F	15.6	17.5	1.12	20%
Dirty Evaporator	35F	105F	13.8	15.3	1.10	18%
Dirty Cond. & Evap.	35F	115F	12.7	16.4	1.29	39%

COMPRESSORS

The function of the compressor in the refrigeration system is to pump refrigerant vapor around the system and compress the low pressure refrigerant vapor to a higher pressure. The compressor must be capable of pumping the refrigerant vapor from the evaporator as fast as it vaporizes. If it doesn't, the accumulated refrigerant vapor will increase the pressure inside the evaporator. If this happens, the boiling point of the liquid refrigerant will be raised and the cooling process will stop. The second function of the compressor is to compress the refrigerant vapor changing it from a low pressure vapor to a higher pressure vapor. This process of compression adds heat to the vapor changing it from a low temperature vapor to a higher temperature vapor… a temperature higher than the condensing medium, water or air. This is important so the heat from the refrigerant can be rejected into the lower temperature condensing medium. The higher pressure supports the higher temperature. For example, in Figure 5-3, 70 psig supports a temperature of 40F, while 260 psig supports a temperature of 120F.

METERING DEVICES

The main types of metering devices are thermal expansion valve (abbreviated TXV or TEV), automatic expansion valve (AXV or AEV), float valve (on the high side or low side of flooded systems), and capillary tube (cap tube). A metering device is a pressure-reducing device; it reduces the pressure in the system from high to low. It is also a flow-control device. It controls the flow of refrigerant into the evaporator coil so that if the refrigerant is boiling off too soon in the evaporator, the metering device will open up to allow more refrigerant in. If the refrigerant is boiling off too late in the evaporator, the metering device will close to allow less refrigerant into the evaporator. In the systems illustrated in this chapter, the metering devices are TXVs. A sensing bulb from the TXV is attached to the suction line as it leaves the evaporator. The suction line is insulated and the sensing bulb is underneath the insulation and is in direct contact with the suction line pipe. The superheat, 10 degrees in the examples, is set on the TXV. If the entire refrigerant boils off too soon there is more evaporator for the refrigeration gas to flow though and pick up additional sensible heat adding to the superheat. When the sensing bulb senses the superheat is greater than 10 degrees the metering device will open up to allow more refrigerant into the evaporator. On the other hand, if the entire refrigerant boils off too late there is less evaporator for the refrigeration gas to flow though and pick up sensible heat, reducing the required superheat. When the sensing bulb senses the superheat is less than 10 degrees, the metering device will close to allow less refrigerant into the evaporator. The entire refrigerant will then boil off sooner and pick up the required sensible heat to have 10 degrees superheat at the sensing bulb.

Table 5-2. Refrigeration Troubleshooting Chart

| Pressures | | Temperatures | | Amperage | System |
Discharge	Suction	Superheat	Subcooling	Reading	Problem
Low	Low	High	High	Low	TXV Sensing Bulb Lost Charge or
Low	Low	High	High	Low	Low Outside Ambient Air Temperature or
Low	High	High	High	Low	Inefficient Compressor
High	High	Low	High	High	Refrigerant Overcharge
Low	Low	High	Low	Low	Refrigerant Undercharge
Low	Low	Low	High	Low	Insulated Evaporator
Low	Low	High	High	Low	Restricted Liquid Line
High	High	Low	Low	High	Insulated Condenser or
High	High	Low	Low	High	TXV Sensing Bulb Mounted Loose or
High	High	Low	Low	High	TXV Sensing Bulb Poorly Insulated

Chapter 6

Fans and Pumps, Drives and Motors

PRIME MOVERS

The prime movers of air and water in HVAC systems are motor driven air fans and water pumps. This chapter is an introduction to fan and pump operation, types, construction, pressures, characteristics and applications. Air and water are fluids, air is compressible and water is not. The physical laws of air and water are the same and the prime movers and other components are essentially the same, the names and terms change (fan wheel, pump impeller, air duct, water pipe, air damper, water valve, cubic feet per minute for air volume, gallons per minute for water volume, etc.). As you read through this book you will see and understand the similarities and differences between air and water movement.

FAN OPERATION

As a fan rotates, centrifugal force throws the air outward reducing the pressure at the inlet of the fan wheel. Air is forced through the fan by the greater atmospheric pressure at the inlet. The air leaves the fan wheel at a relatively high velocity. Then, in the fan housing, the velocity is reduced and converted into pressure. The size of the fan wheel and its rotational speed determine the pressure developed by the fan.

Fans are constant volume machines. However, fans in HVAC systems are either constant volume or through controls made variable volume. A constant volume fan is driven by a constant speed motor. A variable volume fan is coupled with a variable frequency motor to change the speed of the fan, or automatic dampers, vanes or blades are used to restrict the amount of air into or out of the fan.

Fans can be operated in series and/or in parallel. When they operate in series, the flow capacity remains the same but the pressure and horsepower are increased. With parallel fan operation the pressure remains constant, while the capacity and horsepower are increased.

Fan pressures are measured in inches of water gauge (in. wg) or inches of water column (in. wc), same thing just different name. Fans for HVAC applications are low pressure machines, typically 6 in. wg or less.

FAN PRESSUYRES

The pressures in fans (and air ducts) are static pressure (SP or Ps), velocity pressure (VP or Pv), and total pressure (TP or Pt). Static pressure is due to the fan's outward push of air on the supply side and the inward pull on its return side. Velocity pressure is due to the air movement within the fan (or duct). Total pressure is the sum of static pressure and velocity pressure (TP = SP + VP). For example a typical fan (or duct) pressure might be: 2.0 in. wg SP, 0.15 in. wg VP and therefore, 2.15 in. wg TP. Total pressure is always the greatest pressure. Generally speaking for relative values of static pressure to velocity pressure, static pressures are typically 2 to 4 inches water gauge and velocity pressures 0.04 to 0.25 in. wg. In field work we use manometers to measure static pressure and occasionally total pressure in fans. In ducts we measure static pressure and velocity pressure. My book, Testing and Balancing HVAC Air and Water Systems, 5th edition has more detailed information on testing air systems.

FAN TYPES

Heating, ventilating and air conditioning fans are divided into three general categories: axial, centrifugal, and special design.

Axial Fans

Axial fans are fans in which the airflow within the fan wheel is parallel to the fan shaft. Axial fans are classified as propeller, tube axial, and vane axial. The following table provides an overview of axial fans.

Key: P Propeller fan   TA Tube Axial fan   VA Vane Axial fan

Fan and Wheel Construction	Fan Type
Simple ring enclosure	P
Cylindrical tube	TA
Tube with straightening vanes	VA
Single thickness blades	P, TA
Airfoil blades	TA, VA
Long blades and small hub	TA
Short blades and large hub	VA
2 or more blades lightweight construction	P
4 to 8 heavy constructed blades	TA, VA
Adjustable pitch blades	VA

Static Pressure Range	Fan Type
Low pressure. Generally, 0.75 in. wg or less	P
Medium pressure. Typically to 3 in. wg	TA
Medium to high pressure	VA

Airflow and Discharge	Fan Type
Airflow is parallel to the shaft	ALL
Circular or spiral pattern	P, TA
Vanes straighten out spiral pattern	VA

Efficiency	Fan Type
Low	P
Medium	TA
High	VA
Maximum efficiency near full airflow	ALL

Horsepower Characteristics	Fan Type
Lowest at maximum airflow	ALL
Highest at minimum airflow	ALL
Horsepower increases as static pressure increases	ALL

Performance Curve Characteristics
Tube Axial and Vane Axial: The performance curve has a dip to the left

of peak pressure caused by "aerodynamic" stall. Pressures in this area should be avoided.

Typical Applications

Propeller Fan: Delivers large volumes of air at low pressures for general air circulation applications or exhaust without any attached ductwork. Propeller fans produce large volumes of airflow at low pressures. A typical commercial application of propeller fans would be general room air circulation or exhaust ventilation. Very large propeller fans are sometimes used in cooling towers. The housing for a typical propeller fan is normally a simple ring enclosure and the fan will usually have two or more single thickness blades. Propeller fans are generally not very efficient. A characteristic of propeller fans is that the operating horsepower, which is called brake horsepower, is lowest at maximum airflow and highest at minimum airflow. An example of this characteristic is the typical box-type home fan. If you look at this type of fan you'll notice that the first position on the air volume switch is "off." The next position is "high," then "medium," and then "low." This means that when the fan is turned on the electrical current draw and the horsepower will be at its lowest. In the last position, or the "low" position, the current and horsepower are the highest.

Tube Axial Fan: Medium pressure in medium to high air volume applications using ducted systems such as fume hood exhausts, paint spray booths and drying ovens. Tube axial fans are heavy duty propeller fans. The wheel of the tube axial fan is enclosed in a cylindrical tube and is similar to the propeller type wheel. The main exception is that the wheel has more blades, usually 4 to 8, and the blades are of much heavier construction. The tube axial fan is more efficient than the propeller fan and is most efficient when it's operating at its highest air volume. Like the propeller fan, the tube axial fan's operating horsepower is lowest at maximum airflow and highest at minimum airflow.

Vane Axial Fan (Figure 6-1): Medium to high pressure in medium to high air volume applications where good downstream air distribution is needed. Vane axial fans are basically tube axial fans with straightening vanes. They're used in HVAC ducted systems in office buildings or other commercial applications to provide airflow to the conditioned space. Also used for ventilation and exhaust in higher volume applications. The housing is a cylindrical tube similar to the tube axial fan with the addition of air straightening vanes. The straightening vanes straighten out the spiral motion of the air and improve the efficiency of the fan. The vane axial fan has the highest efficiency of all the axial type fans. The wheel of the vane axial fans has shorter blades and a larger hub than the tube axial and, like the propeller fan and the tube axial fan, the operating horsepower is lowest at maximum airflow and highest at minimum airflow. The static efficiencies of the VA fan are typically 70% to 72% and the operating range of the VA fan is about 60% to 90% of full flow CFM. The fan blades are often airfoil shapes and may be available with adjustable pitch. The variable pitch vane axial fan (VPVA) is similar in construction to the fixed pitch vane axial fan (FPVA). The main difference in the two fans is that the VPVA is fixed speed with automatic variable pitch blades adjusted by controls whereas the FPVA has fixed blades and the speed is adjustable either by a sheave change or variable frequency drive controls.

Figure 6-1. Axial Fan. Inline exhaust air fan. Possible system effect caused by elbow and square-to-round duct transition on both sides of the fan.

Centrifugal Fans

Centrifugal fans are fans in which the airflow within the fan wheel is radial or circular to the fan shaft. Five classifications of centrifugal fans are forward curved (FC, Figure 6-2), backward curved (BC), backward inclined (BI), airfoil (AF, Figure 6-3) and radial (R). The following table provides an overview of centrifugal fans.

Fan and Wheel Construction	Fan Type
24-64 shallow curved blades	FC
Blades curve in the direction of rotation	FC
Wheel is usually 24" in diameter or smaller	FC
Multiple wheels on a common shaft	FC
Lightweight construction	FC
10 to 16 curved blades	BC
Blades curve away from rotation	BC, AF
Medium to heavyweight construction	BI, BC, AF
10 to 16 flat blades	BI
Blades incline away from rotation	BI
10-16 aerodynamically shaped blades	AF
Blades curve away from rotation	AF
6 to 10 flat blades	R
Blade tips are radial to the center of the wheel	R
Heavyweight construction	R

Static Pressure Range	Fan Type
Low to medium	FC
Medium to high	BC, BI, AF
High	R

Airflow	Fan Type
Airflow is radial to the shaft	ALL
Discharge	Fan Type
Top or bottom horizontal	ALL
Up or down blast	ALL
Top or bottom angular down	ALL
Top or bottom angular up	ALL

Efficiency	Fan Type
Low to medium	FC
Medium	BC, BI
Most efficient of the centrifugal fans	AF
Least efficient of all centrifugal fans	R

Horsepower Characteristics

Forward curved and radial fans are "overloading," meaning that the horsepower increases continuously as air quantity increases and static pressure decreases. Backward curved, backward inclined and air-

Figure 6-2. Forward Curved Fan. Building corridor heating system. Hot water coil above fan. Air filter below fan. Fan motor seen on left side.

Figure 6-3. Airfoil Fan. Fixed pitch motor and fan sheaves. Five belts. Rotation is clockwise as seen from the drive side. Inlet is opposite the drive side for this single inlet, single wide fan.

foil fans are "non-overloading," meaning that the horsepower increases with an increase in air quantity but only to a point to the right of maximum efficiency and then gradually decreases.

Performance Curve Characteristics

Forward Curved: Highest efficiency occurs to the right of peak pressure when the fan is delivering 40% to 50% of full volume. There is a dip in the pressure curve left of the peak pressure point, which under certain conditions may cause the fan to operate at one point and then another resulting in fan pulsations. The forward curved fan pressure curve is less steep and the efficiency is less than backward curved, backward inclined or airfoil fans

Backward Curved, Backward Inclined, Airfoil Fan: There is no dip in the pressure curve left of the peak pressure point and therefore has a more stable and predictable operation. Highest fan efficiencies occur 50% to 65% of full volume.

Radial: May have a dip to the left of peak pressure but not as great as with the forward curved fan. Usually not enough of a dip to cause difficulty.

Typical Applications

Forward Curved Fan: The forward curved fan (aka "squirrel cage" fan) is generally used in residences and in small commercial and industrial low to medium pressure and volume applications. The fan housing is of light weight construction. The fan wheel has 24 to 64 shallow blades that curve toward the direction of rotation. The wheel is usually 24 inches in diameter or smaller. A characteristic of this type of fan is that the operating horsepower is low when the fan's airflow is also low but continues to increase as the airflow increases. There may also be multiple wheels on a common shaft to increase total volume output (fans in parallel add their volumes, e.g., if one fan wheel delivers 100 cubic feet per minute, two fan wheels of the same size will produce 200 cubic feet per minute, etc.).

Backward Curved Fan, Backward Inclined Fan: Medium to high pressure and volume industrial applications where dust and other particles might cause erosion to airfoil blades (see airfoil fan below). The wheel in the backward curved or backward inclined fan has 10 to 16 blades that curve or incline (lean) away from the direction of rotation. The operating horsepower of this type of fan increases with an increase in airflow (but only to a point), and then gradually decreases. Because of this characteristic, the backward curved and backward inclined fans are called "non-overloading" fans. In other words, the fan motor, if selected properly, will not draw more electrical current than its nameplate rated current regardless of the airflow. Backward fans are more efficient than forward curved fans but less efficient than airfoil fans.

Airfoil Fan: Medium to high pressure and volume industrial and commercial applications. An airfoil fan has the best efficiency of all the centrifugal fans. The fan wheel has 10 to 16 aerodynamically shaped blades similar to an airplane wing which curve away from the direction of rotation. As with the backward curved and BI fans, the operating horsepower increases with an increase in airflow (but only to a point), and then gradually decreases. Therefore, airfoil fans are also "non-overloading" fans.

Radial Fan: A radial fan is used in heavy-duty industrial applications that require high pressures and velocities such as waste collection and material-handling applications (sawdust collection is an example). The fan housing is made of heavy construction. The fan wheel tends to be heavily built with the blades being thick and narrow. The fan wheel has 6 to 10 "paddle wheel" blades that may be coated with special materials for protection. The radial fan is the least efficient of all the centrifugal fans. The operating horsepower increases continuously as the airflow increases. Therefore, the radial fan, like the forward curved fan, is an "overloading" type of fan.

SPECIAL DESIGN FANS

   Centrifugal Power Roof Ventilator (Figure 6-4)
   Axial Power Roof Ventilator
   Tubular Centrifugal Fan (TCF)

Centrifugal power roof ventilator and axial power roof ventilator are small fans used to exhaust air from restrooms, attic spaces, and other small areas. They operate at low static pressure, horsepower, and efficiency.

Tubular Centrifugal

Wheel Construction: The wheel is housed in a cylindrical tube with backward inclined or airfoil blades. Basically it is a hybrid in that it is a

Figure 6-4. Centrifugal Power Roof Ventilator. Kitchen Exhaust.

centrifugal fan wheel in a vane axial fan housing.

Static Pressure Range: Low.

Airflow and Discharge: The air is discharged radially from the wheel then changes directions by 90 degrees to flow through a guide vane section and then flows parallel to the fan shaft.

Efficiency: Lower than backward curved or backward inclined fan.

Horsepower Characteristics: The horsepower curve increases with an increase in air quantity (but only to a point) to the right of maximum efficiency and then gradually decreases. "Non-overloading."

Performance Curve Characteristics: Generally, no dip in the pressure curve left of the peak pressure point, although some may have a dip similar to the axial fan. Stable and predictable operation similar to the backward bladed fan except for lower capacities, pressures and efficiencies.

Typical Applications: Used in return systems where saving space is a consideration.

OTHER FAN CATEGORIES

Fans are further categorized by class of construction, direction of fan wheel rotation, width of fan wheel, arrangement of drive components, and direction of air discharge.

Class of Construction: Class of construction or pressure classes are based on pressure developed by the fan. The pressure classes are I, II, III and IV with each higher class representing heavier fan construction, and higher pressure, speed and performance capabilities of the fan.

Direction of Fan Wheel Rotation: For centrifugal fans direction of fan wheel rotation is described as either clockwise (CW) or counter-clockwise (CCW). To determine fan rotation, view a centrifugal fan from the drive side, not the inlet side. The drive side is intended to mean the side that is driven by the motor, but it really varies with the fan configuration. On a single inlet, single wide (SISW) fan, the drive side is the side opposite the inlet. On a double inlet, double wide (DIDW) fan, the drive side is the side that has the drive. On fans with dual drives the side with the higher horsepower rating is considered the drive side. Axial fans will normally have an arrow on the housing which indicates direction of rotation. Rotation should be observed and if necessary, corrected. Correct fan rotation is important. On a centrifugal fan that's rotating backwards, the airflow can be reduced by 50% or more. An axial fan rotating backwards will not produce airflow in the proper direction.

Width of Fan Wheel: Axial fans do not use fan width designations, however centrifugal fans are designated either single inlet, single wide, or double inlet, double wide. Single inlet, single wide (SISW) fans have one fan wheel and a single entry. Since in SISW fans the motor is opposite the inlet, the fan bearings are not in the air stream. This makes SISW fans more common in smaller sizes where inlet duct needs to be attached, and in applications where the air temperature is relatively high or the air is dirty. Double inlet, double wide (DIDW) fans have two single wide fan wheels mounted back-to-back on a common shaft in a single housing. Air enters both sides of the fan. DIDW fans are generally used for moving large volumes of air in open inlet systems.

Arrangement of Drive Components: The arrangement of the drive components describes the drive placement in the fan. Arrangements are numbered from 1 to 10 and are found in fan manufacturer catalogs.

Direction of Air Discharge: For a centrifugal fan, direction of air discharge is, as with fan rotation, viewed from the drive side (as if the fan was sitting on the ground). Angular discharges, either up or down positions without being specified are assumed to be either 45, 135, 225 or 315 degrees from the vertical and in the direction of rotation. An up blast discharge is 360 degrees and is designated simply "UB," while a top angle up blast 45 degrees from vertical with a clockwise rotation is designated TAU CW 45. As you can see, when describing fans it is best to make a complete call out of the fan, such as clockwise top angle upblast 30 degrees from vertical (TAU CW 30). Axial fans do not have discharge direction designations. Centrifugal fan air discharge designations are:

Clockwise (CW)

Counterclockwise (CCW)

Top Horizontal Discharge (TH or THD)

Top Angular Down Discharge (TAD or TADD)

Down Blast Discharge (DB or DBD)

Bottom Angular Down Discharge (BAD or BADD)

Top Angular Up Discharge (TAU or TAUD)

Up Blast Discharge (UB or UBD)

Bottom Angular Up Discharge (BAU or BAUD)

Bottom Horizontal Discharge (BH or BHD)

FAN CURVE

A fan curve shows performance of a fan at various static pressures and volumes of air flow. It is developed from actual tests. The tests are performed with constant fan speeds. Measurements and calculations are made for torque, horsepower, dry bulb and wet bulb temperatures, barometric pressures, inlet and outlet fan pressures and air flow. The values are then plotted to develop the fan performance curve.

The curve may show static pressure (SP), static efficiency (SE), total pressure (TP), total efficiency (TE, aka ME for mechanical efficiency), brake horsepower (BHP), rotational speed (RPM) and air volume (CFM). Airflow static pressure is plotted on a chart with the static pressure on the left vertical axis and the airflow on the bottom horizontal axis. At the top left of the chart is the block tight static pressure (BTSP) condition at maximum static pressure and zero airflow. The fan performance curve will then fall to the lower right of the chart to a point of wide open cubic feet per minute (WOCFM). Fan associations and manufacturers have established procedures and standards for the testing and rating of fans. The procedure requires the testing of the entire range of a fan's performance from free delivery (maximum airflow in cubic feet per minute, cfm) to no delivery (zero cfm). Measurements at each of the fan operating points include: discharge and inlet pressures, wet and dry bulb temperatures and barometric pressures (to calculate air density), and torque and speed (to calculate horsepower input). These measured and calculated values are then plotted to develop the fan performance curve. Testing continues and performance curves are established for other fan speeds. This defines the fan airflow delivery capability at various speeds and static pressures. If the curve is developed for one fan speed only, the air density, wheel size, and fan speed are usually stated on the curve and are constant for the entire curve.

To find air volume at a measured fan static pressure draw a line from the static pressure line (static pressure units are inches of water column, in. wc) to the rpm (aka fan performance) curve. At this point, draw a line down to intersect the air volume line (in cubic feet per minute, cfm). This is the air volume produced by the fan. Using Figure 6-5, start from the 3 inch fan static pressure point on the vertical axis and draw a line horizontally to insect with the 1000 revolutions per minute (rpm) fan speed curve. At this point draw a line vertically down to intersect the horizontal fan flow rate line and read 6000 cubic feet per minute. The operating point

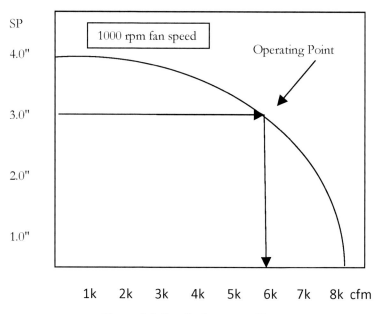

Figure 6-5. Fan Performance Curve

for the fan is the intersection of the fan static pressure and the fan performance curve. Any change to the fan speed or air distribution system (closing or opening of dampers, changes in the condition of the filters, etc.) can be calculated using "fan laws" and then graphically depicted on the fan curve. For example, a decrease in system resistance will mean an increase in cfm, while an increase in system resistance results in a decrease in cfm. The system resistance increases or decreases as the square of the air volume increases or decreases (Fan Law 2). To increase or decrease the air volume, a physical change must be made to either the duct system (e.g., opening or closing a damper) or fan speed, or both.

Although fan performance curves can be useful in commissioning and troubleshooting fans you should be aware that because of installation conditions, which almost never duplicate the ideal conditions under which fans are tested by the manufacturer, fan performance as determined by field test may be less than shown by manufacturer's tests. This is known as "system effect" aka system effect factor.

Fan Surge

Fan surge is an unstable operating condition. It occurs as the fan's operation moves towards block tight static pressure. The airflow out of the fan falls below that needed to maintain the required static pressure difference between the inlet and outlet of the fan. When the static pressure difference falls below requirement air will surge back into the fan reducing duct static pressure. This reduction in duct static pressure allows the fan to regain its differential static pressure and the airflow and the system momentarily comes back to normal until the duct static pressure increases which causes the airflow drop and the process starts again. The resulting fluctuations in fan static pressure, duct static pressure, and airflow causes turbulence, vibration and noise in the fan and ductwork and reduces the efficiency of the fan.

A surge line is plotted on the left side of the fan curve to indicate the performance area where the surge can occur and the fan becomes unstable. Surge is greatest for airfoil fans and least for forward curved fans. Do not select or operate fans to the left of the surge line as indicated on individual fan performance curves. Fans selected to the right of the surge line operate in a stable condition. That being said, if you try to operate the fan too far to the right down the fan curve it may also be unstable there as well. Try to select and operate fans (and pumps) in the middle one third of the curve.

FAN PERFORMANCE TABLE

To make fan selection comparisons as simple as possible most fan manufacturers publish fan performance tables (Figure 6-6) aka fan rating tables. The fan manufacturers provide the fan curve information in tabular form. These tables generally show the cubic feet per minute (cfm),

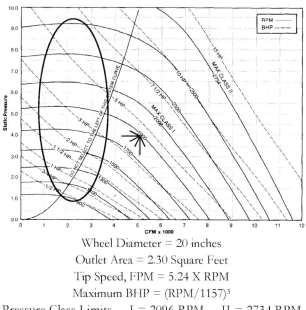

Wheel Diameter = 20 inches
Outlet Area = 2.30 Square Feet
Tip Speed, FPM = 5.24 X RPM
Maximum BHP = (RPM/1157)^3
Pressure Class Limits   I = 2096 RPM   II = 2734 RPM

CFM	OV	STATIC PRESSURE									
		3"		3$^1/_2$"		4"		5"		5"	
		RPM	BHP	RPM	BHP	RPM	BHP	RPM	BHP	RPM	BHP
3250	1413	1507	2.09	1611	2.49	1707	2.90				
3500	1522	1528	2.23	1627	2.63	1724	3.05	1814	3.49		
3750	1630	1555	2.37	1648	2.78	1740	3.22	1831	3,67	1916	4.14
4000	1739	1582	2.53	1674	2.96	1761	3,40	1847	3.86	1933	4.34
4250	1848	1619	2.72	1701	3.14	1787	3.59	1868	4.06	1949	4.55
4500	1957	1657	2.92	1735	3,34	1814	3,79	1895	4.25	1971	4.77
4750	2065	1696	3.13	1773	3,57	1846	4.02	1922	4.51	1998	5.02
5000	2174	1735	3.36	1811	3.82	1884	4.28	1952	4.76	2025	5.27
5250	2283	1774	3.60	1550	4.07	1922	4.56	1990	5.05	2055	5.55
5500	2391	1817	3.86	1890	4.35	1960	4.84	2028	6.35	2092	5.86

Figure 6-6. Fan Performance Table for a 20" Airflow Single Wide Fan. From the table if the fan volume is 5000 cfm the outlet velocity is 2174 fpm (5000 ÷ 2.3 sf outlet area). If the fan operates at 4" SP the RPM is 1884 and the BHP is 4.28 for this Class I fan. Note surge line (Do Not Select To The Left Of This System Curve) on left side of fan curve above. Arrows indicate operating point.

static pressure (SP), revolutions per minute (rpm), outlet velocity (ov), brake horsepower (bhp), blade configuration, wheel configuration, fan wheel diameter, outlet area, tip speed equation, maximum brake horsepower equation, and pressure class limits.

You can use the tables to help determine how the fan is operating under field conditions by measuring the fan speed and the fan static pressure and entering this information on the table. If the measured conditions are within the scope of the table the approximate CFM and brake horsepower can be determined.

Fans that have high rotating speeds and operate at high pressures are built to withstand the stresses of centrifugal force. However, if the fan is rotating too fast, the wheel could fly apart or the fan shaft could whip. Therefore, for safety reasons, the performance tables list the maximum RPM for each class of fan. Maximum allowable fan speed should be checked before increasing the fan RPM to ensure that the new operating condition doesn't require a different class fan.

PUMP OPERATION

HVAC pumps are centrifugal, electrically power-driven machines (Figures 6-7 and 6-8) used to overcome system resistance and produce required water flow. As a pump impeller is rotated, centrifugal force

Figure 6-7. Centrifugal Direct Drive HVAC Pump. Left to right: Strainer, suction pipe, pump, motor. Off top of pump is discharge pipe, pressure gauge and discharge valve.

throws the water outward from the impeller. The centrifugal force and other design characteristics reduce the pressure (a partial vacuum is created) at the inlet of the impeller and allow more water to be forced in through the pump suction opening by atmospheric or external pressure. This makes the pump's discharge pressure higher than the pump's suction pressure. After the water enters the pump's suction opening, there's a further reduction of pressure between this opening and the inlet of the impeller. The lowest pressure in the system is at the pump inlet. The water leaves the impeller at a relatively high velocity. Then, in the pump casing, the velocity is reduced and converted into static pressure (aka "static head pressure" or "static head"). The size of the pump impeller and its rotational speed determines the static head pressure developed

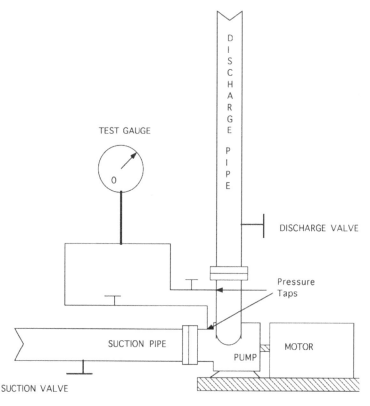

Figure 6-8. Centrifugal Pump. Direct drive with pump directly coupled to motor. Discharge valve can be modulated between open and closed. The suction valve is always open when the pump is operational.

by the pump.

A typical HVAC centrifugal pump will have a volute or spiral casing with one or more closed backward curved radial flow impellers. If the pump has one impeller it is a single stage pump. If it has two or more impellers in series on a common shaft it is a multistage pump. The inlet to the pump may be on just one side. This is a single inlet pump. If the inlet to the pump is on both sides of the pump it is a double suction inlet pump. The suction inlet pipe may be the same size or larger than discharge pipe. A larger inlet reduces the resistance on the water entering the pump (see cavitation).

Most HVAC water pumps are constant volume machines and are coupled directly (direct drive) to a constant speed motor. Some direct drive pumps are driven by a variable frequency-variable speed motor. Varying the speed of the motor changes the speed of the pump. Varying the pump speed makes the pump variable volume. Some pumps are belt driven. These pumps may be either constant speed and volume or variable speed and volume.

HVAC pumps are used to move heated water from a boiler to heating coils and back, chilled water to and from cooling coils, heated or cooled water to and from storage tanks and various heat exchangers and water to and from condensers and cooling towers. In some design applications multiple centrifugal pumps may be selected to operate in series or parallel, or series and parallel, to increase flow rate or pumping pressure or both. When pumps operate in series, the flow capacity remains the same but the pressure and horsepower are increased. With parallel pumping the pressure remains constant, while the capacity and horsepower are increased (Table 6-1). Note: Centrifugal fans operating in parallel or series have the same flow-pressure-horsepower characteristics as centrifugal pumps.

Table 6-1. Pumps Operating in Series and Parallel

Pump #	Flow	Pressure	Power
1	400 gpm	100 feet of head	30 horsepower
2	400 gpm	100 feet of head	30 horsepower
1 & 2 in series	400 gpm	200 feet of head	60 horsepower
1 & 2 in parallel	800 gpm	100 feet of head	60 horsepower

When two pumps are operating in parallel and one pump is removed from operation, the flow rate (gallons per minute) from the single pump increases from the flow rate that it delivers when both pumps are operating. On the other hand, when two pumps are operating in series and one pump is removed the flow rate of the single pump decreases from the flow rate that it delivers when both pumps are operating.

PUMP CURVE

A pump curve is developed by the pump manufacturer and is a graphic representation of the performance of a specific pump. The curve is for a constant speed. Most curves will show head pressure, flow volume, pump efficiency, horsepower/brake horsepower, pump inlet and discharge size, rotational speed, maximum and minimum impeller diameter, and net positive suction head required. Head pressure is indicated on the vertical scale (y-axis) and volume of flow (gallons per minute, gpm) is on the horizontal scale (x-axis). Most HVAC pump curves are for direct drive pumps and are referenced to impeller size since the pump speed stays constant with motor speed. The pressure is normally plotted against flow rate. This type of pressure-capacity curve is used because it gives a general description of pump operation without being affected by water temperature or density. Some pump curves, however, are for a specific fluid, water temperature, or density. Pressure versus flow is shown on these curves as psi versus flow rate (pounds per square inch verses flow in gallons per minute or pounds per square inch verses flow in pounds per hour).

The pump flow rate (volume, gallons per minute, gpm) can be plotted on the pump curve by measuring pump head pressure. Draw a line from the head pressure line (head pressure units are feet of water, ft of H2O) to the impeller curve. This intersection is the pump operating point. At this point, draw a line down to intersect the water volume line (in gallons per minute, gpm). This is the water volume produced by the pump. Example: Using Figure 6-9 draw a line horizontally from the 150 ft of head point on the vertical axis to insect with the 10 inch pump impeller curve. At this point draw a line vertically down to intersect the horizontal pump flow rate line and read 600 gallons per minute.

The operating point for the pump is the intersection of the head pressure and the pump performance curve. Any change to the pump or

Figure 6-9. Pump Performance Curve

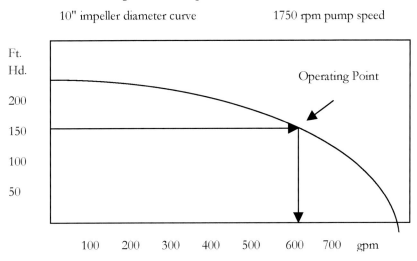

the water distribution system (closing or opening of valves, changes in the condition of the coils, etc.) can be calculated and graphically depicted using the pump curve. A decrease in system resistance will mean an increase in gpm, while an increase in system resistance results in a decrease in gpm. The system resistance increases or decreases as the square of the water volume increases or decreases.

If the operating head exceeds the design head, the water volume, horsepower and net positive suction head required will be less than design. If, on other hand, the operating head is lower than design head, the water volume, horsepower and net positive suction head required will be greater than design. To increase or decrease the water volume, a physical change must be made to either the piping system (e.g., opening or closing a valve) or the pump impeller size (or speed for variable speed drive system). Any change to the pump impeller, pump speed or the distribution system can be calculated using the pump laws, and graphically depicted on the pump curve.

DRIVES

If you're going to take fan drive component information first stop the motor and put your own personal padlock on the motor disconnect switch so that only you have control over starting the fan. Next, remove the belt guard (some fans do not have belt guards or a housing around the fan wheel) and read and record the information from the motor and fan sheaves and the belts. Also, measure the shaft sizes and the distance between the center of the fan and motor shafts. This is a good time to also measure and record the slide adjustment on the motor frame. The motor slide is for adjusting belt tension. For instance, if a sheave needs changing and there is space available on the motor frame, you may be able to move the motor forwards or backwards, so that the old belts are easier to remove and new belts are easier to put on and adjust to proper tension. If the adjustment space is not adequate, a change in sheave size will mean that you'll have to install a different size belt.

The fan sheave is the driven pulley on the fan shaft. The motor sheave is the driver pulley on the motor shaft. The motor share of may be either a fixed or adjustable groove sheave. Adjustable groove sheaves, or simply adjustable sheaves are also known as variable speed or variable pitch sheaves.   An adjustable sheave means that the belt grooves on the sheave are movable. A fixed sheave means that the belt grooves are not movable. Fixed sheaves are typically used for fans. And generally, after fans have been air balanced for the proper airflow, adjustable motor sheaves are replaced with fixed sheaves. The reason for replacing is, size for size, fixed sheaves are less expensive than adjustable sheaves and there's less wear on the belts.

Some of the terms that we need to define refer to belts. Let's start with V-belts. There are two types of V-belts generally used on HVAC equipment. Light duty, fractional horsepower (FHP) belts, sizes 2L through 5L, and heavier duty industrial belts, sizes "A" through "E." Fractional horsepower belts are generally used on smaller diameter sheaves because they're more flexible than industrial belts for the same equivalent cross-sectional size. For example, a 5L belt and a "B" belt have the same cross-sectional dimension, but because of its greater flexibility the 5L belt would generally be used on light duty fans that have smaller sheaves. Note: the 4 L belt and the "A" belt have the same cross-sectional dimension. The general practice in HVAC design is to use belts of smaller cross-sectional size with smaller sheaves instead of large belts and large sheaves for the drive components. Multiple belts are used to avoid excessive belt stress. The term pitch diameter is a measurement that refers to where the middle of the V-belt rides in the sheave groove.

DRIVE INFORMATION

Now we have an understanding of some important drive component terms, let's go back and continue to get information from the sheaves and belts (Figure 6-10). After you have the belt guard off check the outside of the sheave for a stamped part number. The part number indicates the sheave size. For example, on the motor sheave, you might find 3MVP60B74P. Looking in the manufacturers catalog you'd find that the numbers and letters indicate that the motor sheave has 3 fixed grooves (to accommodate three belts). "M" is the designation for companion sheaves so when purchasing a set of motor and fan sheaves each one would be in the "M" category. VP indicates that it is a variable pitch (aka variable speed) sheave which can have either a "B" belt with a pitch diameter range of 6.0 to 7.4 inches or an "A" belt with a pitch diameter range from 5.9 to 7.0 inches and the bushing size is a P. Looking at the bushing bore table we will see that the P is for a P2 bushing and we can purchase the correct bushing to fit shaft sizes from three-quarter inches to one and three-quarter inches. Table 6-2 shows catalog information for a fixed pitch "M" companion sheave and bushing. A 3MVB154R is selected.

Figure 6-10. Fixed pitch fan sheave left and variable pitch motor sheave right. There is belt dust on top of the motor and under the fan sheave. My guess is the fan was going too fast and this motor sheave was opened too far and the belts are riding too low and are out of alignment. Therefore, the belts are deteriorating. The VP sheave needs to be replaced with the proper size fixed companion sheave and new belts installed correctly.

Table 6-2. Manufacturer part information catalog for "M" companion fixed pitch sheave.

SHEAVE SPECIFICATION			
Pitch Diameter		Part Number	
"A" Belts	"B" Belts	Sheave	Bushing
10.6"	11.0"	3MVB11OQ	Q1
12.0"	12.4"	3MVB124Q	Q1
13.2"	13.6"	3MVB136Q	Q1
15.0"	15.4"	3MVB154Q	Q1
15.0"	15.4"	3MVB154R	R1

BUSHING BORE SPECIFICATION	
Bushing	Bore Range
P1	$1/2"$-1-$3/4"$
Q1	$3/4"$ -2-$11/16"$
Q2	$1"$-2-$5/8"$
R1	1-$1/8"$ -3-$3/4"$

MOTORS AND MOTOR NAMEPLATE

Motors used on HVAC equipment will be either single-phase or three-phase alternating current induction motors. The number of motor phases will generally be determined by size. Smaller motors, 1/2 horsepower and below, will be single-phase, while larger motors will use three-phase current. Most motors will have a nameplate with information that includes, manufacturer, model number, serial number, horsepower, phase, voltage, amperage, rpm, service factor and motor enclosure.

Horsepower is a unit of power. The horsepower listed on the motor nameplate is known as nameplate horsepower (nhp). Brake horsepower (bhp) is not listed on the nameplate but it is an important calculated number which gives the total horsepower applied to the drive shaft of any piece of rotating equipment. Brake horsepower is always less that nameplate horsepower. Phase (PH) is the number of separate voltages supplying the motor (single-phase or three-phase).

The nameplate voltage (V, voltage, volts) and nameplate amperage (A, amperage, amps), or full load amperage, (FLA), are the rated operating voltage and full load operating current at rated horsepower. Many of the motors on HVAC units will be dual voltage motors. This means that depending on how the motor is wired, it is capable of operating from either of the voltages listed on the motor nameplate. For example, single-phase motors may be listed at 110/220 volts or 115/230 volts. A three-phase motor may list dual voltages of 220/440 volts, 230/460 volts or 240/480 volts. A dual voltage motor will also list dual amperage on the nameplate. For example, a three-phase, 30 horsepower, dual voltage motor has the following ratings: 230/460 volts, 80/40 amps. This means that if the motor is wired for 230 volts, the full load amps will be 80. Notice that volts and amps are inversely proportional. In other words, when the voltage doubles the amperage reduces by one half. Locked rotor amperage (LRA) is the starting current drawn from the line with the rotor locked and with rated voltage supplied to the motor. This amperage will generally be 5 to 6 times the full load amps.

The rated revolutions per minute (RPM) on the nameplate is the speed that the motor will turn when it is operating at nameplate conditions. Some motors will have two to four different speeds available. The motor speed can be typically be changed by moving an external selector switch. Motor speeds can also be changed automatically by adding a variable frequency drive (VFD) aka variable speed drive (VSD). As the frequency of the motor is changed the motor speed changes in direct proportion. VFDs are used on motors controlling variable flow air and water systems.

The service factor (SF), is the number by which the horsepower or amperage rating is multiplied to determine the maximum safe load that a motor may be expected to carry continuously at its rated voltage and frequency. For example, a 50 horsepower motor might have a service factor of 1.10. The service factor would allow this motor to operate safely at 55 horsepower (50 x 1.10). Typical service factors are: 1.0, 1.10, 1.15 and 1.25.

The type of motor enclosure (ENCL) is another important factor to understand about HVAC motors. The "open protected" (or "open") motor is open for air passage through the windings for cooling purposes. The "totally enclosed" motor has a special housing design to prevent water damage to a motor installed in a wet atmosphere (such as a cooling tower). An "explosion proof' motor is designed to prevent a spark from the motor entering an area where it could cause an explosion. Fully loaded, open motors are generally designed for a maximum temperature rise of 104F above ambient air temperature. Totally enclosed and explosion proof motors are generally designed to operate at a maximum temperature rise of approximately 130F. There is also an internal maximum allowable temperature rise that is indicated on the motor nameplate. Maximum internal temperatures can be over 200F. However, even though the motor is not overheated and the surface temperature is well below 200F, the motor could still be too hot to touch. Therefore, not being able to hold one's hand on the motor does not necessarily mean that the motor is overheated.

Some motor nameplates will also list power factor (PF) and efficiency (EFF). Power factor is the ratio of real power to apparent power (watts divided by volt-amperes), expressed as a decimal e.g., 0.90. Efficiency (Eff) is a non-dimensional measure of the performance obtained from the ratio of the output to its input, expressed as a decimal, e.g., 0.92. It is always less than 100%.

MOTOR CONTROLLERS and MOTOR PROTECTION

Motor controllers start and stop the motor. HVAC motor controllers are grouped into three categories: manual starters, contactors and magnetic starters. Manual starters are motor-rated switches that have provisions for overload protection. They are generally limited to motors of 10 horsepower or less and are normally located close to the motor. Contactors are electro-mechanical devices that "open" or "close" contacts to control motors and can be remotely and automatically operated. Magnetic starters (aka "mags" or "mag starters") are contactors with an overload protection relay.

Overload protection (OLP) devices, aka thermal overload protection, heaters or thermals, prevent motors from overheating. If a motor becomes overloaded or one phase of a three-phase circuit fails (single-phasing*), there will be an increase in current through the motor. If this increased current drawn through the motor lasts for any appreciable time and it is greatly above the full load current rating, the windings will overheat and damage may occur to the insulation, resulting in a burned-out motor. Because most motors experience various load conditions from no load to partial load to full load to short periods of being overloaded, their overload protection devices must be flexible enough to handle the various conditions under which the motor and its driven equipment operate. Single-phase motors often have internal thermal overload protection. This device senses the increased heat load and breaks the circuit, stopping the motor. After the thermal overload relays have cooled down a manual or automatic reset is used to restart the motor. Other single-phase and three-phase motors require external overload protection.

---

*Single-phasing is a condition which results when one phase of a three-phase motor circuit is broken or opened. Motors will not start under this condition, but if already running when it goes into the single-phase condition the motor will continue to run with a lower power output and possible overheating.

Chapter 7

Air Distribution

After the fan produces airflow the other air distribution components convey the air to and from occupied and non-occupied work and process spaces to heat, cool, ventilate and remove contaminates. These components include ductwork, dampers, diverters, air valves, terminal boxes, supply outlets, and return and exhaust inlets. Air moves through the ductwork because of a difference in pressures. Just as heat moves from a higher level to a lower level, so do fluids.

Fluids move from a higher pressure to a lower pressure. Air is a vapor and as such is a compressible fluid. Air moves through the duct system because the pressure on one side (discharge) of the fan is higher than on the inlet or suction side of the fan. The fan produces a pressure at the discharge of the fan that is higher than the pressure in the conditioned space, i.e., the pressure in the conditioned space is atmospheric pressure while the pressure at the fan discharge is greater than atmospheric pressure.

DUCTWORK

Ductwork or duct is a passageway used for conveying air. Duct is fabricated from various materials including:

Aluminum sheet metal

Aluminum flex

Black iron

Fiberglass board or fiberboard to reduce heat transfer and noise transfer

Galvanized sheet metal

Galvanized sheet metal wrapped with insulation to reduce heat transfer

Galvanized sheet metal lined with insulation to reduce heat and noise transfer

Plastic-wrapped insulated wire flex

Resin

Stainless steel

Most HVAC duct is made from various thicknesses of galvanized sheet metal. The thickness of metal is called its gauge. The higher the gauge the thinner the metal. The gauges used for typical sheet metal duct are 26, 24 and 22.

Duct is fabricated in various shapes and sizes to fit the architecture and construction of the building. Ducts may be insulated, either lined or wrapped, to provide a thermal barrier to keep the supply air from losing heat when in the heating mode or gaining heat when in the cooling mode. Vapor-barrier insulation is used to wrap cooling duct to keep it from "sweating." Condensation will form on the duct if it is not insulated when there is 10 degree or more temperature difference between the cool air inside the duct and the warm, ambient air. Another reason for insulating duct is sound attenuation.

Duct may be round (Figure 7-1), rectangular (Figure 7-2), or flat-oval, a round duct that is spread or flattened to form an oval or a rectangle with rounded corners (Figure 7-3). Duct sizes are normally given in inches. Generally, HVAC round duct is manufactured in one-inch-diameter increments from 3 inches to 10 inches. Above ten inches, standard round

Figure 7-1. Round cooling duct wrapped with vapor-barrier insulation. Flex (flexible) duct attached to takeoff from round duct to top of ceiling diffusers.

Figure 7-2. Rectangular exhaust air duct, no insulation.

Figure 7-3. Bottom and side view of flex duct attached to flat oval connection on VAV box.

duct is made in two inch increments (12, 14, 16, etc.).

Rectangular (and square) ducts are generally made in even sizes such as 14 inches x 12 inches, 12" x 12," 24" x 20," 46" x 30," etc. The first number given (spoken or written on a blueprint or other drawing) is the side of the duct that is being viewed. For example, looking at a duct on a drawing as being viewed from the top of the duct (top view is assumed unless otherwise stated) has a call out of 24 inches x 18 inches (24 inches wide and 18 inches high). A side or elevation view of the same duct would be designated as 18 inches x 24 inches. Duct lengths may be given in inches or feet. For example, a call may be made for either a 48 inch length (or joint) of duct or a 4 foot joint of duct.

HVAC duct is designed, fabricated, installed and classified as either low, medium, or high pressure duct based on pressure generated by the fan to move the air through the restrictions in the air distribution system. Low pressure duct is up to 2 inches of water pressure and 2,500 feet per minute air velocity. Medium pressure duct is between 2 and 4 inches of water pressure and between 2,000 and 4,000 feet per minute air velocity. High pressure duct is above 6 inches of water pressure and 2,000 feet per minute air velocity.

Air moves through the ductwork until it reaches the supply air outlets in the conditioned space. As the supply air is discharged it mixes with the mass of room air and the velocity of the supply air slows down. After circulating through the room, the air exits the room by way of the return air inlet and goes back to the air handling unit. As long as the fan is on, the supply air continuously flows through the room, resulting in about 5 to 10 complete air changes per hour for office areas.

When the system is in the cooling mode the return air carries the heat removed from the conditioned space. The return air mixes with the outside air, which also contains some heat. This mixture then goes through the filter section and into the cooling coil and the cooling cycle starts over again. When the system is in the heating mode (wintertime condition) the return air is cooler than the supply air. Heat from the supply air has been released into the conditioned space. The "cool" return air mixes with the outside air, which is cool to cold. This mixture then goes through the filter section and into the heating coil and the heating cycle starts over again.

Duct systems are either single path (single duct) or dual path (single or dual duct). A single path system is one in which the air flows through various configurations of heating and cooling coils which are located in series to each other. Single zone heating and cooling units and multizone or variable air volume terminal reheat units are single path systems. A dual path system is one in which the air flows through heating and cooling coils which are located essentially parallel to each other. The heating and cooling coils may be side-by-side or stacked one on top the other. The heating coil is located in a plenum or chamber called the hot deck and the cooling coil is located in the cold deck. Some systems may not have a heating coil, but instead bypass return air or mixed air into the hot deck. Examples of dual path systems are constant volume dual duct mixing box systems and variable air volume dual duct terminal box systems.

AIRFLOW CONTROL

Airflow in HVAC systems is controlled by using dampers, diverters, air valves, terminal boxes, supply air outlets and return and exhaust air inlets.

Damper

A damper is a device to control (regulate) the volume of airflow and is constructed with the appropriate strength and rigidity for the operating pressures of its respective duct system. There are three main categories of dampers: manual volume dampers, automatic temperature control dampers, and gravity-controlled backdraft dampers. Flow characteristics of dampers can be inconsistent and may vary by manufacturer, and from one system to another. The actual effect of closing a particular damper in a particular system can only be determined in the field by measurement. Dampers are further categorized as single or multiple blade. Two common types of single blade damper are butterfly and gate (Figure 7-4). The butterfly damper has a flat blade attached to a rod. The damper is placed in the center of the duct with the rod protruding through the duct. A locking handle on one end of the rod controls the position of the damper. The gate damper (aka blast gate or "guillotine" damper) protrudes through the duct and is moved into the duct to control air volume.

The classifications of multiple blade (aka multi blade) dampers are parallel blade and opposed blade. The blades of parallel blade dampers (Figure 7-5) rotate parallel to each other, producing a "diverting" air pattern when partially closed (Figure 7-6). This diverting pattern moves the air to the side or to the top or bottom of the duct. If the damper is placed too close upstream to a coil, the diverting flow pattern may adversely affect the heat transfer of the coil. If the damper is placed too close upstream to a branch duct takeoff, the diverting flow pattern may reduce airflow into the branch duct. Parallel blade dampers (PBD) are best used in mixing applications but may be used for volume control. Parallel blade dampers and opposed blade dampers may be manual or automatic. An opposed blade damper (OBD) is a multi blade damper with a linkage which rotates the adjacent blades in opposite directions to each other (Figure 7-7). This means that the blade openings become increasingly narrow as the damper closes, resulting in a uniform, "non-diverting" airflow pattern. Opposed blade dampers have a better flow characteristic than parallel blade dampers. They are generally recommended for large duct systems for volume control and mixing applications.

Manual Volume Damper (MVD)

A manual volume damper (aka air balancing damper) is used to control the quantity of airflow in the distribution duct by adding resistance to restrict airflow. MVDs are installed in constant and variable air volume systems; single path, dual path, or multizone. The single blade butterfly damper is the most common type of manual damper used for volume control, but multiple blade and gate dampers are also used. Single and multiple blade dampers are used in supply, return and exhaust systems. The blast gate damper is typically used in exhaust applications such as in the exhaust duct from a laboratory fume hood. The proper selection and placement of volume dampers equalizes the pressure drops in the different air paths allowing the system to be balanced in the least amount of time with the least amount of resistance and air noise. If volume dampers are not properly selected, placed, installed and adjusted, they may not control the air as intended. They may also add unnecessary resistance to the system and create noise problems.

The resistance a volume damper creates in a duct or duct system is determined by how complicated the system is. For instance, if the system is very simple and the damper is a large part of that resistance, then any movement of the damper will change the resistance of the entire system and good control of the airflow will result. If, however, the damper resistance is very small in relation to the entire system, poor control will be the case. For instance, partial closing of a damper will increase its resistance to airflow, but depending on the resistance of the damper to the overall system resistance, the reduction in airflow may or may not be in proportion to closure. In other words, closing a damper 50% doesn't necessarily mean that the airflow will be reduced to 50%. For example, a damper when open might be 10% of the total system resistance. When this damper is half closed the airflow will be reduced to 80% of maximum flow. However, a similarly built damper in another duct system is 30% of the total system resistance when open. When this damper is half closed the airflow is reduced to 55% of maximum.

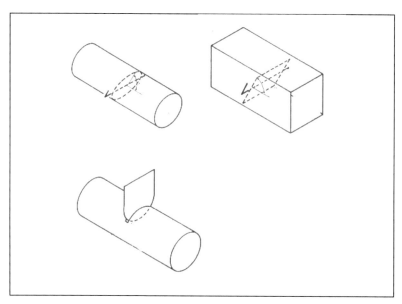

Figure 7-4. Single Blade Dampers (Manual Volume Dampers, MVD). Top: Butterfly damper in round and rectangular duct. Bottom: Gate damper in round duct.

Figure 7-5. Parallel blade dampers in mixed air plenum. This arrangement of PBD does not allow for good mixing of OA (left) and RA (bottom). Possible air stratification. Both sets of dampers are ATCD, note pneumatic damper motor in center of OA dampers and far end of RA dampers.

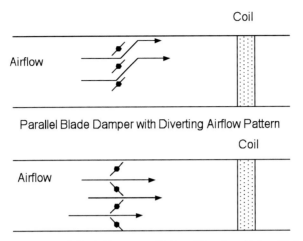

Parallel Blade Damper with Diverting Airflow Pattern

Opposed Blade Damper with Non-Diverting Airflow Pattern

Figure 7-6. PBD and OBD Airflow Pattern.

Figure 7-7. Opposed blade dampers in mixed air plenum. Good mixing of OA (top) and RA (right). Both sets of dampers are ATCD.

The relationship between the position of a damper and its percent of airflow is termed its "flow characteristic." Opposed blade dampers are generally recommended for large duct systems because they introduce more resistance to airflow for most closed positions, and therefore, have a better flow characteristic than parallel blade dampers. However, flow characteristics of dampers aren't consistent and may vary from one system to another. The actual effect of closing a damper can only be determined in the field by measurement.

Proper location of balancing dampers not only permits maximum air distribution but also equalizes the pressure drops in the different airflow paths within the system. Manual dampers should be provided in each duct takeoff to control the air to grilles and diffusers. They should also be in main and branch duct. Manually operated opposed blade or single blade quadrant type volume dampers should also be installed in every zone duct of a multizone system.

Do not install single blade or multiple blade manual volume dampers immediately behind diffusers and grilles because when throttled the dampers create noise at the outlet. Proper selection and installation of manual volume dampers in the takeoffs eliminates the need for volume controls at grilles, registers and diffusers. For small duct, a single blade damper is satisfactory. Generally, for large duct, manual volume dampers should be multiple blade. Every damper, single or multi blade, should have a locking handle (Figure 7-8).

Figure 7-8. Manual Volume Damper Locking Quadrant. Mark damper shaft for damper position.

Other Flow Control Dampers

Automatic temperature control dampers (ATCD) are controlled by the temperature requirements of the system and are usually multiple blade parallel or opposed blade dampers. They are controlled for either two-position or modulating control action. The control may be direct action or reverse action, normally open or normally closed, analog or digital. The control source may be electrical or compressed air (pneumatic).

Gravity-controlled backdraft dampers open when the air pressure on the upstream (entering air) side of the damper is greater than the downstream (leaving air) pressure. They close from gravitational force when there's no airflow. Backdraft dampers are used in exhaust systems, ventilation systems and the inlet and discharge of fans.

Diverter

A diverter (Figure 7-9) is a manually movable device used in low pressure systems to divert air. Typical applications are diverting air from a larger duct, such as a main duct, into branch ducts or from branch ducts into takeoffs. A diverter used to divert air from a branch duct into a takeoff is an extractor.

A diverter used in a main duct to divert air more or less air into branch ducts is known as a splitter. For example, a main duct "tees" into two branch ducts. The air volume in the main duct is 1000 cubic feet per minute and the air balancing documents call for 700 cfm to be directed down one branch duct and 300 cfm down the other duct. The splitter in the main is moved in the appropriate direction to accomplish this.

Air Valve

An air valve is a device to control volume of airflow. Air valves are automatically controlled and are used in higher pressure systems and terminal boxes. Typical application is in variable air volume terminal box systems (Figure 7-10). Air valves have better flow characteristics than dampers and are used to give a more positive control in the system or terminal. The control source for the air valve may be electricity or compressed air. The control may be direct action or reverse action, normally open or normally closed, analog or digital.

Terminal Box

A terminal box is a unit which may by design control return air, supply air, temperature and humidity to and from the conditioned

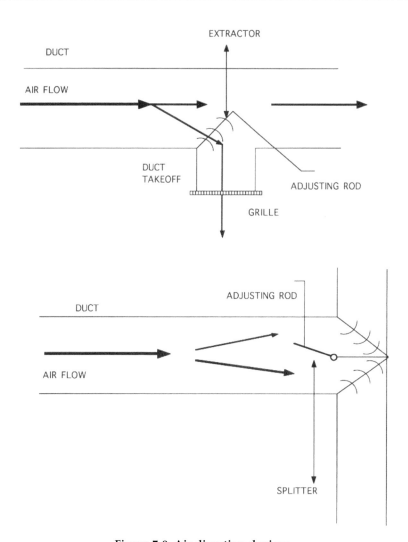

Figure 7-9. Air diverting devices.

space. Depending on design and use the internal components of a terminal box may include: air volume device (damper or air valve), sound attenuator, thermal insulation and heat exchanger. The air volume through the box may be factory set but can also be adjusted in the field. A terminal box operates off static pressure in the duct system and static pressure or velocity pressure in the box. The box reduces the inlet pressures to a level consistent with the low pressure, low velocity duct connected to

Figure 7-10. Air Valve. Installed in a VAV terminal.

its discharge. Each box has a minimum inlet static pressure requirement (typically 0.75" to 1") to overcome the pressure losses through the box plus any losses through the discharge duct, manual volume dampers and outlets.

Classifications for terminal boxes include constant air volume, variable air volume, single inlet, dual inlet or double duct, medium pressure, high pressure, pressure dependent, pressure independent, cooling only, heating and cooling, cooling with reheat, system powered, fan powered, induction, and bypass. Other designators include: pneumatic, electric, direct digital control, direct acting, reverse acting, normally open, and normally closed.

Constant Volume Terminal Box

A constant volume terminal box (Figure 7-11) delivers a constant quantity of air. A single duct terminal box is supplied with cool, conditioned air through a single inlet duct. Air flowing through the box is controlled to maintain a constant volume to the conditioned space. A reheat coil (water, steam or electric) may be installed in the box or immediately downstream from it. A room thermostat controls the reheat coil. A dual duct terminal box is supplied by separate hot and cold inlet ducts. Warm air is supplied to the box through the "hot" duct. The warm air may be either heated air or return air from the conditioned space.

Cool air is supplied through the "cold" duct. The cool air may be cooled and dehumidified, when the refrigeration unit is operating, or simply cool outside air brought in by the economizer cycle. Dampers in the inlet ducts mix warm and cool air as needed to properly condition the space and maintain a constant volume of discharge air. The "mixing" dampers are controlled by the room thermostat.

Variable Air Volume Terminal Box

Some variable air volume systems are designed so that the total volume of all the VAV terminal boxes is greater than the maximum out-

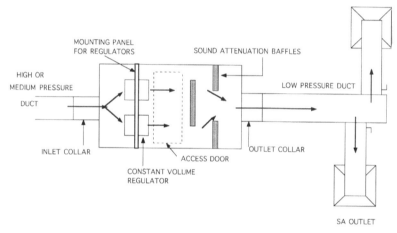

SINGLE DUCT CONSTANT VOLUME PRESSURE INDEPENDENT TERMINAL BOX

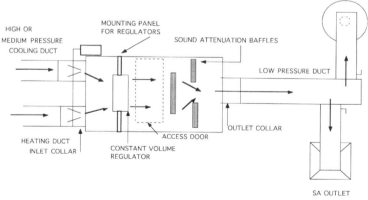

DUAL DUCT CONSTANT VOLUME PRESSURE INDEPENDENT TERMINAL BOX

Figure 7-11. Constant Volume Single and Dual Duct Terminal Boxes.

put of the fan. This is known as "diversity." For example, the boxes in a VAV system total 10,000 cfm but the maximum output of the fan is 8000 cfm. The fan volume to box/outlet volume diversity is 20%.

A variable air volume terminal box varies the amount of air delivered to the conditioned space as the heat load varies. The operating costs of a variable air volume system are reduced when the total volume of air is reduced throughout the system. For example, a variable frequency drive (VFD) controls fan motor speed. When the supply fan reduces its volume output to accommodate the needs of the VAV boxes the brake horsepower to operate the fan goes down by approximately the square of the reduction in air volume. So using VAV systems means energy and cost savings. The exception to this is the variable air volume bypass box system which is actually a constant volume system at the central fan.

A single duct variable air volume terminal box is supplied with cool air through a single inlet duct. The volume of air through the box is varied by the throttling action of an automatically controlled damper or air valve. A dual duct variable air volume terminal box is supplied with conditioned air (cooled and heated) through separate hot and cold inlet ducts. There are a variety of control schemes to vary the air volume and discharge air temperature out of dual duct boxes.

A variable air volume terminal box may be controlled as direct acting and normally closed or reverse acting and normally open. A terminal box is direct acting and normally closed when a rise in space temperature causes an increase in control pressure to the box damper controller to open the damper to supply more cool air to the space. A terminal box is reverse acting and normally open when a rise in space temperature causes a decrease in control pressure to the box damper controller to open the damper to supply more cool air to the space.

Pressure Dependent Variable Air Volume Terminal Box

A pressure dependent VAV terminal box means that the controls for the box only position the motorized volume damper in response to a signal from the conditioned space thermostat. Additionally, the quantity of air passing through the box is dependent on the inlet static pressure. To give a comparison of the controlling ability of a pressure dependent box to a pressure independent box let's use this example: The inlet static pressure is 0.03 inches wg at a pressure dependent box. At this inlet pressure the air volume is 70 cfm. At 0.3 inches of inlet static pressure the air volume is 200 cfm. Compare to a pressure independent box that has an air volume of 725 cubic feet per minute at 0.3 in. wg inlet static pressure and 775 cfm at 4.0 inches wg inlet static pressure. Pressure dependent boxes do not control airflow as well as pressure independent boxes.

Pressure Independent Variable Air Volume Box

A pressure independent variable air volume box, in addition to the motorized volume damper or air valve, has a flow sensing device inside the box (Figures 7-12, 7-13 and 7-14). The sensing device and its controller regulate the air volume through the boxes in response to the room thermostat. The controller is mounted on the outside of the box with connections to the sensing device, volume damper or air valve and the room thermostat. The controller may be electric, electronic DDC or pneumatic. The sensing devices in these boxes maintain air volume at any point between maximum and minimum settings, regardless of the box's inlet static pressure, as long as the pressure is within the design operating range. The air volume will vary from a design maximum airflow down to a minimum airflow. Minimum airflows are generally around

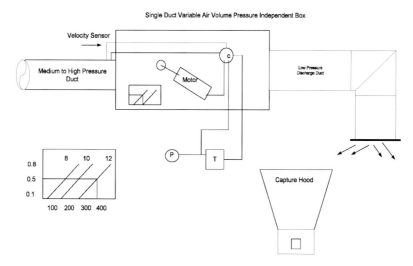

Figure 7-12. VAV Box. From left: Inlet duct with Velocity Pressure Sensor. Box with chart indicating Pressure (vertical scale) and CFM flow (horizontal scale). Diagonal lines on chart are box inlet size. C: controller, P: control power (pneumatic or electric), T: conditioned space thermostat. Low pressure duct attached to box into conditioned space. Airflow typically measured with a capture hood.

Figure 7-13. Single Duct VAV Box. From left: Insulated galvanized duct, flex duct, control box on VAV box, reheat coil (hot water), sealed low pressure duct.

Figure 7-14. Dual Duct VAV Box. From bottom left: Pneumatic controller for cold duct damper, cold duct inlet, hot duct inlet and controller for hot duct damper. Top center: Velocity pressure sensor in box discharge.

50% to 25% of maximum. Minimum airflow is based on local code and custom.

To further illustrate the pressure independent control let's say that the temperature rises in the space and the room thermostat (responding to the load conditions in the space) sends a signal to the controller. The controller responds by actuating the volume damper or air valve (volume control device) to open for more cooling. The sensing device determines the differential pressure in the box and transmits a signal to the box controller which regulates the airflow within the preset maximum and minimum range. The controller opens the volume control device more. As the temperature in the space drops, the volume control device closes. If the box also has a reheat coil, the volume control device, on a call for heating, will close to its set minimum position for that box and the reheat coil will be activated. Because of its pressure independence, assuming the box is working properly, the airflow through the box is not affected by pressure changes in the system as other VAV boxes modulate open or closed.

Induction and Bypass Variable Air Volume Box

A variable air volume ceiling induction terminal box has a volume device. It is located in the inlet duct to control the flow of primary supply air into the box. An induction damper allows return air from the ceiling plenum into the box to mix with the primary supply air. When the thermostat in the conditioned space calls for cooling, the volume damper opens and the induction damper closes. As the space cools down, the volume damper throttles back and the induction damper opens to maintain a relatively constant mixed airflow into the conditioned space. When the induction damper is wide open, the volume damper is throttled to a maximum of about 75% to allow for the proper induction-primary air ratio.

A variable air volume bypass terminal box is supplied with a constant air volume but supplies a variable air volume to the conditioned space. The supply air enters the box and can exit either into the conditioned space through the discharge ductwork or back to the return system through a bypass damper. A room thermostat regulates the automatic temperature control damper so that the conditioned space receives either all the cool supply air entering the box or only a part of it. There is no reduction in the main supply air volume feeding the box. This type of system has no central fan energy savings.

Variable Air Volume Fan Powered Terminal Box

A variable air volume fan powered terminal box (Figure 7-15) has a small centrifugal secondary fan and a return air opening from the ceiling plenum. The secondary fan may be constant volume and operate continuously (series system), it may shut off (intermittent, parallel system) or it may operate in a variable volume mode. When the room thermostat signals for cooling, the box operates as would the standard VAV box. However, when the room thermostat signals that the room is overcooled, the primary air volume is reduced and the secondary fan draws warm air from the ceiling plenum. The cool primary air from the main system mixes with the ceiling return air. A system of dampers regulates the air volume, direction of airflow, and mixing of the air streams. If the room thermostat continues to signal for heat, the primary air volume damper continues to close and more ceiling air is drawn into the box. This maintains the air volume at a relatively constant rate

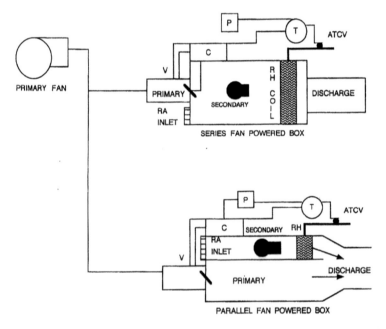

Figure 7-15. Fan Powered VAV Boxes. V: velocity sensor, RA: return air, C: controller, P: power source, T: thermostat, ATCV: automatic temperature control valve, RH: reheat coil (hot water), Secondary: forward curved fan, Ceiling return air damper:▤ Primary air damper:❱

but varies the temperature of the air into the conditioned space. In areas where there is a greater heating load, a reheat coil is installed in the box or discharge duct.

A variable air volume fan powered terminal box has several advantages over the standard VAV box such as heating capabilities and a relatively constant airflow to the conditioned space. Fan powered VAV boxes are used around the space perimeter or other areas where: air stagnation is a problem when the main supply air is reduced, there are seasonal heating and cooling requirements, heat is required during the unoccupied hours when the central fan is off, the heating load can be satisfied with re-circulated ceiling return air.

The following information is a comparison of Series and Parallel Fan Powered Variable Air Volume Terminal Boxes.

Series Fan Powered Box

Secondary fan selected for cooling load (typically 100% of cooling)

Secondary fan control is interlocked with central fan

Secondary fan operates continuously under greater loads

Secondary fan static pressure overcome losses in low pressure distribution system

Central fan static pressure overcomes losses in air volume control device (damper or air valve). Therefore, box inlet static pressure is adequate to overcome damper or air valve only.

Air valve or damper selected for cooling load

Parallel Fan Powered Box

Secondary fan selected for heating load (typically 60% of cooling)

Secondary fan control from T-stat signal. Not interlocked with central fan

Secondary fan operates fewer hours under lighter loads

Secondary fan static pressure overcomes losses in low pressure distribution system

Central fan static pressure overcomes losses in damper/air valve and low pressure distribution system and therefore box inlet static pressure overcomes damper/air valve plus low pressure distribution system

Air valve/damper selected for cooling load

Variable Air Volume Fan Powered Bypass Terminal Box

A variable air volume fan powered bypass terminal box acts the same as the conventional bypass box, with the addition of a secondary fan in the box. The bypass box uses a constant volume supply primary fan but provides variable air volume to the conditioned space. The supply air comes into the box and can exit either into the conditioned space through the secondary fan and the discharge ductwork or back to the return system through a bypass damper. The fan in the box circulates the primary air or return air into the room. The conditioned space receives either all primary air, all return air, or a mixture of primary and return air, depending on the signal from the room thermostat. Since there's no reduction in the main supply air volume feeding the box this type of system has no savings of primary fan energy.

Variable Air Volume System Powered Terminal Box

A variable air volume system powered terminal box uses static pressure from the primary supply air duct to power the variable air volume controls. The minimum inlet static pressure required with this type of box is usually higher than other variable air volume systems in order to operate the VAV controls and still provide the proper airflow volume throughout the distribution system.

Supply Air Outlet

A supply air outlet is an opening placed in the supply duct that allows the conditioned air into the space. There are various types of supply air outlets including ceiling diffusers, spot or directional outlets, grilles, registers, and supply openings.

Ceiling Diffuser

A ceiling diffuser is a supply air outlet located in the ceiling. Ceiling diffusers will typically have pattern deflectors arranged to promote the mixing of the supply air with the room air to produce a horizontal air pattern. This horizontal flow pattern, called surface effect, is caused by the inducement or entrainment of room air when the outlet discharges air directly parallel with and against a ceiling. The air then tends to flow along the ceiling. A high degree of surface effect is required for cooling applications, especially variable air volume systems, because it helps to reduce the dumping of cold air. Dumping is defined as the rapidly falling action of cold air caused by a variable air volume box or other device reducing its air velocity. Surface effect, however, contributes to "smudging." Smudging is the term used for the black markings on ceilings and supply air outlets made by suspended dirt particles in the room air. The dirt particles are entrained in the mixed air stream and then deposited on the ceilings and outlets. Ceiling diffusers are generally categorized as rectangular, square, perforated face, round, linear slot, and light troffer.

A rectangular or square ceiling diffuser (Figure 7-16) is used in open and lay-in ceilings and may have adjustable vanes to change the airflow pattern to a one-, two-, three-, or four-way throw. A perforated face diffuser is similar in construction to the standard square ceiling diffuser with a perforated face plate attached to diffuse the air.

A round ceiling diffuser (Figure 7-17) is used in open and lay-in ceilings and supplies air in all directions and may have adjustments (cones) to change the airflow pattern to more horizontal or more vertical. A directional spot diffuser is a spherical diffuser that swivels in all directions to supply air to a specific area.

A linear slot ceiling diffuser (Figure 7-18) is manufactured in various lengths and numbers of slots. 1, 2, 4 and 5 foot lengths with 1, 2, 3 or

Figure 7-16. Square Ceiling Diffuser.

Figure 7-17. Round Ceiling Diffuser.

Figure 7-18. Linear Slot Ceiling Diffuser.

4 slots are common. Slots may be adjustable for different throw patterns (horizontal right or left or both, or vertical).

A light troffer is a type of ceiling diffuser which fits over a fluorescent lamp fixture and supplies air through a slot along the edge of the fixture. Some supply air to only one side of the fixture, while other types supply air to both sides of the fixture. Figure 7-19.

Directional Spot Outlet

This type of air outlet discharges a high velocity air jet that can be directed to condition a specific space or area. Figure 7-20.

Grille and Register

A grille is a duct-, wall-, ceiling- or floor-mounted louvered covering for an air opening. A register (Figure 7-21) is a grille with a built-in or attached damper assembly. The damper is used to control air volume. To control airflow pattern, some grilles have a removable louver. Reversing or rotating the louver changes the air direction. Grilles are also available with adjustable horizontal and/or vertical bars so the direction, throw and spread of the supply air stream can be controlled.

Return Air Inlet

A return air inlet is an opening placed in the ceiling, wall, floor or return duct that allows air from the conditioned space to return back to the HVAC unit. Various types of return air inlets are used including

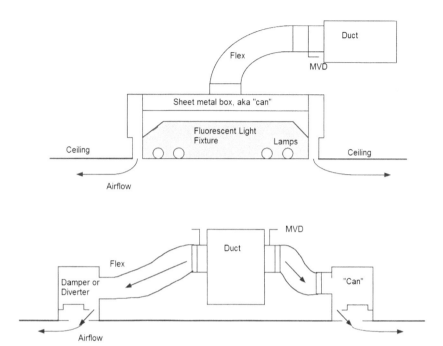

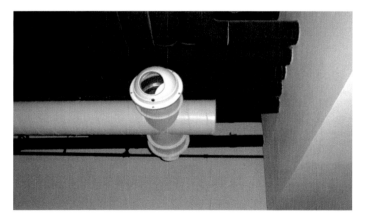

Figure 7-19. Top: Light troffer sits atop fluorescent light fixture. May be single- or double-sided (double shown). Bottom: Linear Slot Ceiling Diffuser (aka linear diffuser, LD or linear air diffuser, LAD) on each side of duct. Diffuser may be single- (shown) or multiple-slotted. Diverter in diffuser is moveable to cause air to go horizontally along ceiling or drop vertically.

Figure 7-20. Spot or Directional Outlet.

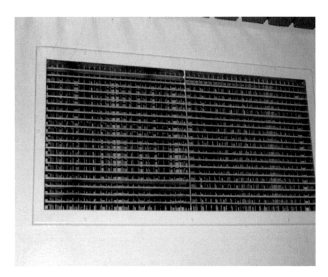

Figure 7-21. Register.

grilles, registers, plastic lay-in ceiling grids (aka "egg-crates") and return openings. Dampers may or may not be installed in the inlets nor the return air duct.

Miscellaneous

Figure 7-22 has my drawings of supply outlets and return inlets and VAV CDs. A VAV CD may be used with a constant air volume system when one or a few conditioned spaces need more cooling during certain periods of the day. For example, a small meeting room, break room or office with intermittent use. Some systems are entirely VAV CDs with a bypass or relief duct to a ceiling return plenum and back to a constant air volume fan. The CD on the left shows airflow from the room circulating between the lower and upper plates. If the sensed room temperature is above the set point of the temperature controller for the device the top plate will drop down and more cool air will be allowed into the room. When the sensor in the CD senses that the room is overcooled the controller will draw the top plate up reducing airflow to the conditioned space. The variable air volume CD on the right has a perforated face with a sliding volume damper above it. When the temperature in the room indicates its overcooled the temperature sensor in the CD will move the plate over the perforated face reducing airflow as appropriate to the conditioned space and vice versa.

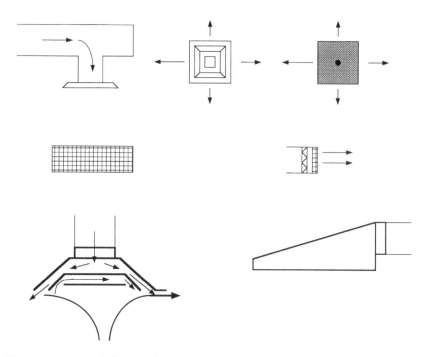

Figure 7-22. Top left to right.: Duct to ceiling diffuser. Four-way throw CD changeable to 1-, 2-, or 3-way throw, perforated face CD. Center left: sidewall grille supply or return, or lay-in egg-crate return air inlet. Center right: Register with grille face and OBD. Bottom: VAV CDs. Coned (left) and perforated face (right).

Filter

Air filter are typically made from pleated paper (Figure 7- 23) or fiberglass to remove solid particulates such as dust, debris, pollen, mold, and bacteria from the mixed air. Some filters have elements with a static electric charge to attract dust. A charcoal filter is used where odors are a concern. A chemical air filter consists of an absorbent or catalyst for the removal of airborne molecular contaminants such as volatile organic compounds or ozone. Filters made for cleanrooms or other applications which are more than 99% efficient are known as High Efficiency Particulate Air (HEPA) and Ultra Low Penetration Air (ULPA) filters. H o w often filters need to be change will depend on the type of HVAC and the cleanliness of the environment in which the system is located. In order for filters to function properly they must be installed correctly. To help with this, most filters will have an arrow showing direction of airflow. To help determine when filters need changing, a differential pressure gauge can be installed with measuring points on either side of the filters. This gives a pressure drop across the filters. When the pressure drop reaches a given limit the filters need to be changed. Since airflow capacity decreases and static pressure increases over time filters actually become more efficient as the filter loads. The filter should be changed once it has reached its capacity. The filter frame should be tight against all sides of the filter housing so that no air bypasses the filters.

Figure 7-23. Air Filters. Yellow, pleated-paper filters on the entering air side of a cooling coil in a large air handling unit.

Chapter 8

Water Distribution

After the pump produces water flow the water distribution components convey the water to heat and cool the occupied and non-occupied work and process spaces. The components include: temperature, flow, volume, and pressure measuring stations, pipe systems, filtration, flow control, water flow balancing stations, pressure control components, air control components, and heat conversion equipment.

HYDRONIC PIPE SYSTEMS

HVAC hydronic pipe systems heat and cool with water. Pipe systems can be classified in many ways including: open, closed, series loop direct return, reverse return, one-pipe, two-pipe, three-pipe, four-pipe, and primary or primary-secondary and combination as a pipe system may combine several of the piping arrangements mentioned. The pipe circuits provide heated or chilled water to coils in central air handling units, fan-coil units, ducts, terminal boxes, unit heaters (Figure 8-1), valence units, and fin-tube radiation. Hydronic systems for central air handling units are typically four-pipe heating and cooling circuits.

Figure 8-1. Unit heater with propeller fan. Unit heaters can be piped with heating water or steam.

Closed System and Open System

A closed pipe system is one in which there's no break in the piping circuit and the water is "closed" to the atmosphere. An open pipe system is one in which there's a break in the piping circuit and the water is "open" to the atmosphere. Figure 8-2.

An example of an open system is a water-cooled condenser and cooling tower. Examples of closed pipe systems are a chilled water system transporting water to and from the water chiller and to and from cooling coils, or a heated water system transporting water from the water boiler to heating coils and then back to the boiler.

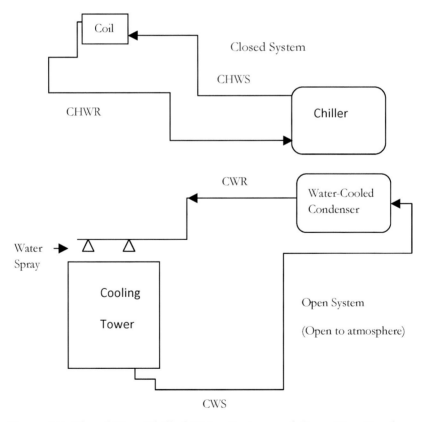

Figure 8-2. Closed Pipe Chilled Water System and Open Pipe Condenser Water System. CHWR: Chilled Water Return. CWS: Condenser Water Supply.

One-Pipe System

One pipe systems have been used for heating in residences, small commercial buildings and industrial buildings. A one pipe system uses a single loop main distribution pipe. Each terminal coil is connected by a supply and return branch pipe to the main. System Efficiency: A diverting tee should be installed in either the supply branch, return branch or sometimes both branches. Select diverting tees to create the proper amount of resistance in the main to direct water to the coil. If the diverting device is not installed, the water circulating in the main will tend to flow through the straight run of a normal tee and not be directed into the coil. This is because the coil has a higher pressure drop than the main. If this happens the coil is "starved" for water. Additionally, installing separate control valves and service valves in the branches will provide a better system. If the system has control valves and there are too many terminals, the coils farthest from the boiler may not receive water at a temperature that is high enough to maintain desired space temperatures.

Two-Pipe System

Two pipe systems have a supply pipe and a return pipe to each coil. There are separate automatic control valves and manual service valves for each water coil. For larger water systems, use two-pipe arrangements to maintain the water temperature to each coil equal to the boiler temperature.

Three-Pipe System

Three-pipe systems have two supply mains and one return main to each HVAC unit. There is only one water coil in each unit. One supply main provides chilled water from the chiller to the coil. The other main supplies heated water from the boiler to the coil in the unit. The chilled water and heated water supplies are not mixed. Each water coil has a three-way valve which switches to deliver either chilled water or heating water (but not both) to the coil. The return main receives water from every coil. This means that in some instances the return pipe may transport a mixture of both chilled and heated water. This results in a waste of energy because the chiller receives warmer water and the boiler receives cooler water than design. Therefore, both the boiler and the chiller must work harder to supply their proper discharge temperature. System Efficiency: Retrofit or replace three-pipe systems.

Four-Pipe System

Four-pipe systems consist of two, two-pipe systems (Figure 8-3). One two-pipe system is for chilled water and the other two-pipe system is for heating water. The HVAC unit typically has two separate water coils, one for heating water and one for chilled water. There is no mixing of the chilled and heated water. Each coil has its own automatic control valve, either two-way or three-way, and manual service valves. Some HVAC units have only one coil. The coil is supplied with both heated and chilled water. A three-way valve is installed in the supply line to the coil. The chilled water and heated water supplies are not mixed. The three-way valve switches to deliver either chilled water or heating water (but not both) to the coil. In the return line from the coil a three-way, two-position valve diverts the leaving heated or chilled water to the correct return main.

Direct Return System and Reverse Return System

Both direct return and reverse return systems (Figure 8-4) have a supply pipe and a return pipe to each water coil and typically there will

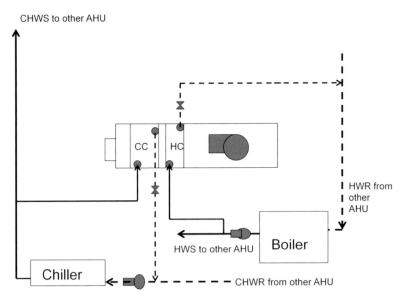

Figure 8-3. Four-pipe System with 2-way ATCV. AHU is a blow-thru system. CC: Cooling Coil, HC: Heating Coil, HWS: Heating Water Supply, HWR: Heating Water Return, CHWS: Chilled Water Supply CHWR: Chilled Water Return.

separate automatic temperature control valves and manual service and balancing valves for each water coil. For direct return systems the water coils are piped so that the first coil supplied by the boiler (or chiller) is also the first coil returned back to the boiler (or chiller). The reverse return systems have the water coils piped so that the first coil supplied is the last coil returned. The direct return system requires less main pipe than a reverse return systems and therefore has a lower initial piping cost. Reverse return systems require more piping than direct return systems and therefore have a higher initial piping cost but are considered to be self-balancing or require less balancing. System Efficiency: Flow meters and balancing valves are needed throughout to water balance either system, including bypasses.

Primary-Secondary System

In a primary-secondary system the primary pump circulates the water through the primary (main) loop and the secondary pump moves water through the secondary loop. A primary-secondary system can have one or many secondary circuits. In order to overcome the pressure

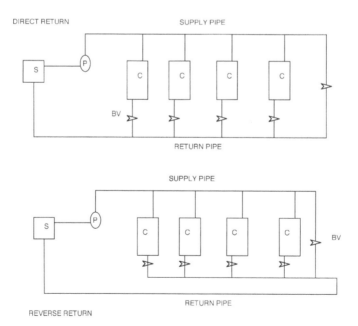

Figure 8-4. Direct Return System and Reverse Return System. S- Chiller or Boiler, P-Pump, C-Coil, BV-Manual Balancing Valve

loss in the secondary circuit and provide water flow to the coil a secondary pump is installed. When the two circuits are interconnected the primary pump and the secondary pump have no effect on each other i.e., flow in one will not cause flow in the other. A crossover pipe, which may vary in length to a maximum of about two feet, is installed between the primary and secondary loop. The cross-over pipe has a negligible pressure drop which ensures the isolation of the two loops. The water flow in the secondary loop may be less than, equal to, or greater than the flow in the primary loop. System Efficiency: Primary-secondary systems reduce pumping horsepower requirements and balance valve settings while increasing system control. As an example, a primary only system has a pump operating at 70% efficiency (a common efficiency). It is moving 1,825 gallons per minute against a system resistance of 153 feet of head. This requires 101 brake horsepower (bhp). If the system is retrofitted to a primary-secondary system the primary system resistance reduces to 83 feet of head. The power required from the primary pump is now only 55 brake horsepower. The secondary pump, also operating at 70% efficiency, requires 25 bhp for a total brake horsepower of 80, a 21 bhp savings. If this retrofitted system operates 5000 hours per year at an electric cost of 10 cents per kiloWatt-hour the savings is over $11,000 per year.

FILTRATION

A water filter is called a strainer. Inside the strainer body is a mesh screen. The screen is in the shape of a sleeve or basket and is designed to catch sediment or other foreign material in the water. If the system is an open system the screen must be periodically removed and cleaned. If the system is closed the screen must be removed and cleaned during startup. If the screen is not cleaned there will be a higher than normal pressure drop across the strainer and a lower water flow.

System efficiency: There will also be a higher than normal pressure drop when the screen has too fine a mesh. If the system is closed and the strainer has a fine mesh screen (construction screen) replace it with a larger mesh screen during startup. In addition to pump strainers, individual fine mesh strainers may also be installed before automatic control valves or spray nozzles (which operate with small clearances and require protection from materials that might pass through the pump strainer.) To avoid cavitation, strainers placed in the suction side of the pump must be properly sized and kept clean. For instance, you may be able to remedy cavitating condenser pumps by removing the strainer altogether or moving the strainer to the pump discharge. Only use strainers where necessary to protect the components in the system. For example, in the case of cooling towers, the strainer in the tower basin may provide adequate protection and you may not need a condenser pump strainer.

FLOW CONTROL

Water flow in heating and cooling applications is controlled through the use of various types of manual and automatic valves. A manual valve is classified as a gate valve, check valve, balancing valve, flow control valve, or combination valve. An automatic control valve is classified according to type of construction, control, and flow. Construction classifications are two-way valve, either single-seated or double-seated and three-way single-seated mixing valve or double-seated diverting valve. Control classifications are modulating or two-position. And flow classifications are constant flow (a three-way valve) and variable flow (a two-way valve).

Gate Valve

A gate valve is a service valve used for tight shutoff to isolate part or all of the system in order to service or remove equipment. A gate valve regulates flow only to the extent that they are either fully open or fully closed. Even though a gate valve has a low pressure drop it cannot be used for throttling. The internal construction of a gate valve is such that when the plug is only partly opened, the resulting high velocity water stream will cause erosion of the valve plug and seat. This is known as "wire drawing." The erosion of the plug and seat will allow water leakage when the valve is used for tight shutoff.

Check Valve

A check valve is used to limit the direction of flow, that is, allow water to flow in one direction only. It is in installed on the discharge of the pump to prevent backflow. The operation of a check valve is such that when there is water pressure in the correct direction, the water forces the gate in the valve to open. The gate will close due to gravity (swing check valve) or spring action (spring-loaded check valve) when the system is off or when water pressure is in the wrong direction.

Balancing Valve and Flow Control Valve

A balancing valve is a manual valve. A flow control valve may be either manual or automatic. Balancing and flow control valves are used to regulate flow rate (gallons per minute).

Plug Valve

A plug valve is used primarily to balance water flow, but it may also be used for shutoff. A plug valve has a low pressure drop and good throttling characteristics. Some plug valves have adjustable memory stops. The memory stop is set during the final water balance so if the valve is closed for any reason it can later easily be reopened to the final balance setting.

Calibrated Balancing Valve

A calibrated balancing valve (CBV) is a plug valves with pressure taps in the valve casing at the inlet and outlet. It has been calibrated by the manufacturer for flow versus pressure drop. A graduated scale or dial on the valve shows the degree the valve is open. Calibration data which shows flow rate in gallons per minute (gpm) versus measured pressure drop is provided by the manufacturer.

Combination Valve

A combination valve is also called a multipurpose or triple-duty valve. This valve regulates flow and limits direction. It comes in a straight or angle pattern and combines a check valve, calibrated balancing valve and shutoff valve into one casing. The valve acts as a check valve preventing backflow when the pump is off and can be closed for tight shutoff for servicing. A combination valve also has pressure taps for connecting flow gauges and reading pressure drop. A calibration chart is supplied with the valve for conversion of pressure drop to gpm for water balancing. A combination valve typically has a memory stop.

Ball Valve and Butterfly Valve

A ball valve has a low pressure drop with good flow characteristics and is often used for water balancing. A butterfly valve also has a low pressure drop and is sometimes used as a water balancing valve. However, a butterfly valve does not have the good throttling characteristics of a ball or plug valve.

Globe Valve

A globe valve is normally used in water make-up lines. Although a globe valve is used for throttling flow it has a high pressure drop and therefore should not be used for water balancing.

Two-way Valve

A two-way valve is an automatic temperature control valve used to regulate water flow that controls the heat transfer (heating or cooling) in the water terminal (coil). They close off when heat transfer is not required and open up when heat transfer is needed. Single-seated, two-way control valves (Figure 8-5) are the type most used in HVAC systems.

Double-seated, two-way valves may be used when there is a high differential pressure and tight shutoff is not a requirement. The flowthrough double-seated valves tends to close one port while opening the other port. This design creates a balanced thrust condition which enables the valve to close off smoothly without water hammer, despite the high differential pressure. A two-way valve is a variable flow valve.

Three-way Valve

A three-way control valve is an automatic temperature control valve which may be either single-seated (mixing valve) or double-seated (diverting valve). The single-seated, mixing valve (Figures 8-6 and 8-7) is the most common. A three-way valve is a constant flow valve. A mixing valve has two inlets and one outlet. A double-seated, diverting valve has one inlet and two outlets. The terms "mixing" and "diverting" refer to the internal construction of the valve, not the valve application. The determination of which valve to use is based on where the valve will be installed so that the plug will seat against flow. Substituting one type of valve for the other in a system (or installing either design incorrectly) will tend to cause chatter see System Efficiency below). Either valve may be installed for a flow control action (bypassing application, Figure 8-8) or a temperature control action (mixing application, Figure 8-9) depending on its location in the system.

In a mixing application a circuit pump is installed between the valve common port and the coil. To illustrate the basic operation let's say that the circuit pump is constant volume and set up to move 10 gpm

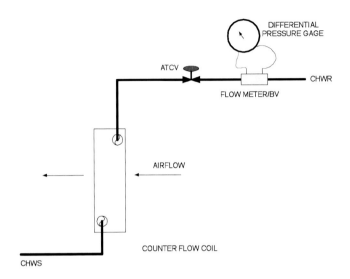

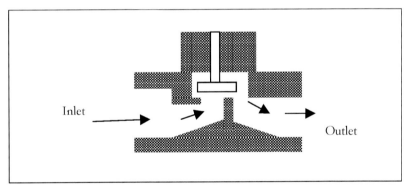

Figure 8-5. Two-way Single-Seated Automatic Temperature Control Valve. Must be piped so water flow is against seating of plug. Valve is controlling water flow through a chilled water coil. The coil is piped counter flow. There is a flow meter and manual balancing valve in the return pipe.

of water through a heating coil. Scenario 1: The thermostat calls for full heating. The boiler primary pump supplies 180F water to the valve. The valve is direct acting and the NO port is full open and the NC port is full closed. The circuit pump is on and moves 10 gpm of 180F water though the coil. Air across the coil removes heat and the water leaves the coil at160F. This return water goes back to the boiler to be reheated. Scenario 2: The thermostat calls for no new heating. The boiler primary pump supplies 180F water to the valve. The NO port is full closed so no new

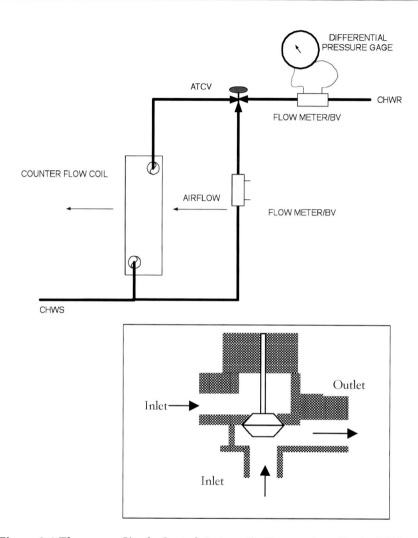

Figure 8-6 Three-way Single-Seated Automatic Temperature Control Valve. Must be piped so water flow is against seating of plug. Valve is controlling water flow through chilled water coil and the bypass. The coil is piped counter flow. There is a flow meter and manual balancing valve in the return pipe and the bypass pipe.

Figure 8-7. Pneumatically controlled 3-way mixing valve in a bypass application. Manual air vent at top of return pipe. Water is supplied at the bottom of the coil and returned from the top. Bypass pipe goes from hot water supply pipe up to the bottom port on the 3-way valve.

180F is supplied to the coil. This water goes to the next coil in the system or a bypass and back to the boiler. The NC port is full open so the secondary pump moves 10 gpm of recirculated water though the coil. Air across the coil continues to remove heat from the water. In other words, the coil's leaving water does not go back to the boiler to be reheated but is recirculated through the secondary pump and coil and the temperature drops with each passage. Scenario 3: The thermostat calls for some new heating. The boiler supplies 180F water to the valve. In this scenario both the NO port and the NC port are half open. The secondary pump still moves 10 gpm, 5 gpm of 160F return/recirculated water and 5 gpm of new supply 180F water though the coil. The mixed water temperature into the coil is 170F (5 gpm @180F and 5 gpm @160 gpm). Ten gpm enters the coil and 10 gpm leaves the coil with 5 gpm going back to the boiler and 5 gpm being recirculated.   Another type of modulating three-way valve is used in the supply line to coils in a three pipe system. This

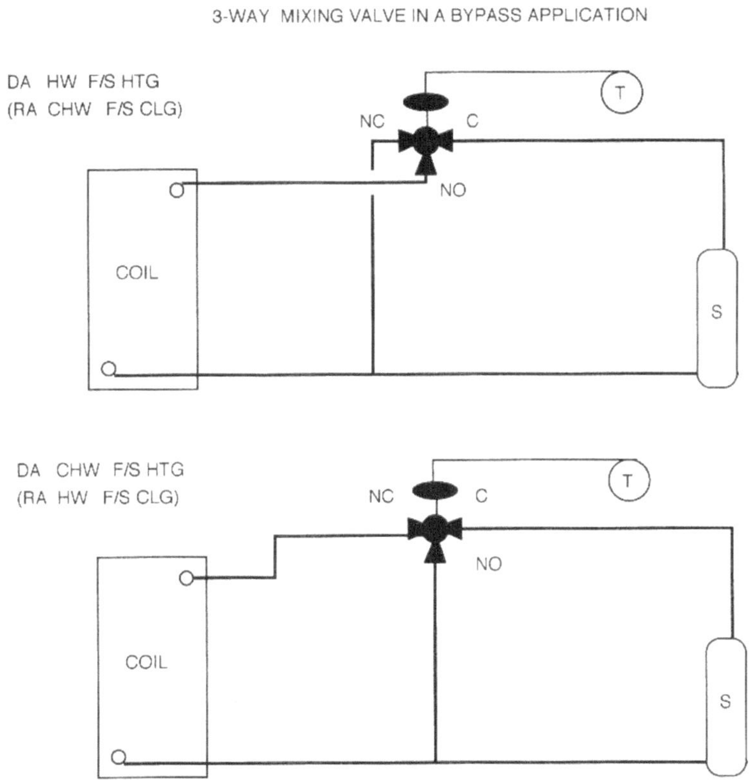

3-WAY MIXING VALVE IN A BYPASS APPLICATION

DA HW F/S HTG
(RA CHW F/S CLG)

NC   C

NO

T

COIL

S

DA CHW F/S HTG
(RA HW F/S CLG)

NC   C

NO

T

COIL

S

Figure 8-8. Mixing valve in bypass application. DA or RA: Direct Acting or Reverse Acting thermostat. Hot Water or Chilled Water. F/S: Fail Safe to Heating or Cooling condition. S: Water Source (boiler or chiller). NC: Normally Closed, C: Common, or NO: Normally Open valve port.

valve has two inlets and one outlet. One inlet is supplied with heating water and the other is supplied with chilled water. The valve varies the quantity of heating and chilled water, but does not mix the two streams. Depending on the thermostatic controls in the occupied space, the valve opens to allow either heating water only or chilled water only into the coil. This same type of three-way modulating valve is used in the supply line to a four pipe, one coil system. The return line also has a three-way valve, but it is two-position only. The return valve has one inlet and two outlets. Depending on temperature of the water entering the coil, the water leaving the coil enters the valve and is diverted to either the heating water return main or the chilled water return main.

3-WAY MIXING VALVE IN A MIXING APPLICATION

DA HW F/S HTG
(RA CHW F/S CLG)

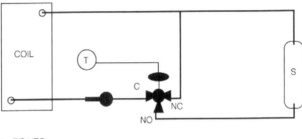

DA CHW   F/S HTG
(RA HW F/S CLG)

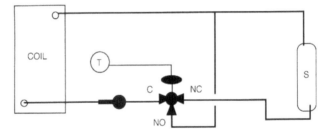

Figure 8-9. Mixing valve in mixing application. DA or RA: Direct Acting or Reverse Acting thermostat. Hot Water or Chilled Water. F/S: Fail Safe to Heating or Cooling condition. S: Water Source (boiler or chiller). NC: Normally Closed, C: Common, or NO: Normally Open valve port. Circuit pump.

Two-way and three-valve automatic valves must be installed with the direction of flow opposing the closing action of the valve plug so the water pressure tends to push the valve plug open. If the valve is installed the opposite way the valve may chatter. Chattering occurs when the valve plug (in an incorrectly installed valve) modulates to the almost full closed position. The velocity of the water around the plug becomes very high because the area through which the water flows has been reduced. This high velocity (and resulting high velocity pressure) overcomes the spring resistance and forces the plug closed. When the plug seats flow is stopped and the velocity and velocity pressure go to zero. At this point, the spring force takes over and opens the plug. When the plug is opened (to the almost closed position) the cycle is repeated and chattering is the result. System Efficiency: Change-out or retrofit 3-way constant flow valves to 2-way variable flow valves.

MEASURING STATIONS

Measuring stations or ports are found in various parts of the pipe system to measure temperature, pressure and flow at pumps, large heat exchangers, pipes, coils, etc.

Temperature

Temperature measuring stations are installed at various points in the pipe system. It is typically on the entering and leaving sides of chillers, condensers, boilers and coils. In many cases this measuring station can be used for either temperatures or pressures. An analog temperature measuring station or thermometer well is sometimes installed in the pipe so that a test thermometer can be inserted into the pipe to indirectly measure the temperature of the water. The well holds a heat conducting oil. Heat from the water is conducted through the well wall into the oil and onto the thermometer. In order to make good contact with the water, a temperature well must extend far down into the pipe. A thermometer well should be installed vertically. If this is not possible, it should be installed not more than 45° from vertical. If the well is installed below 45°, the well will be dry, air will act as an insulator and the temperature reading will be incorrect.

Pressure

Analog pressure measuring test wells can also be installed in the piping system on entering and leaving sides of chillers, condensers, boilers and coils. A hand held pressure probe, which is connected to a differential gauge or test gauge, is inserted into the well. In many cases this measuring station can be used for either temperatures or pressures.

Flow

A flow measuring station aka flow meter is a permanently installed device used to measure flow through pumps, heat exchangers, pipes and coils. Types include annular, orifice plate, Venturi, and calibrated balancing valve. A data chart which shows flow rate in gallons per minute (gpm) versus measured pressure drop is furnished with each flow meter. A flow measuring station must be installed away from any source of flow disturbance such as elbows, reducers, valves, etc., to allow the turbulence to subside and the water flow to regain uniformity. The manufacturer will specify the lengths of straight pipe upstream and downstream of the meter needed to get accurate, reliable readings. Straight pipe lengths vary with the type and size of flow meters. Typical specifications are between 5 to 25 pipe diameters upstream and 2 to 5 pipe diameters downstream of the flow meter.

An annular flow meter has a multi-ported flow sensor installed in the pipe. The holes in the sensor are spaced to represent equal annular areas of the pipe. The flow meter is designed to sense the velocity of the water as it passes the sensor. The upstream ports sense high pressure and the downstream port senses low pressure. The resulting differential pressure is measured with an appropriate differential pressure gauge.

An orifice plate is has a fixed circular opening installed in a flow pipe. The orifice is smaller than the pipe's inside diameter which produces a measurable "permanent" pressure loss. The pressure drop across the orifice occurs as the water passes through the orifice and an abrupt change in velocity causes turbulence and friction. A differential pressure gauge is connected to the pressure taps and flow is read.

A Venturi flow meter operates on the same principle as an orifice plate flow meter, but the shape of the Venturi allows gradual changes in velocity. The "permanent" pressure loss is less than the loss created by an orifice plate. The pressure drop is measured with a differential gauge. A balancing valve is needed with an annular flow meter, orifice plate, Venturi and other types of flow meter.

A calibrated balancing valve is designed to do both duties of a flow meter and a balancing valve. The manufacturer calibrated the valve by measuring pressure drop through the valve at various positions against known flow quantities. Calibration data which shows flow rate in gallons per minute (gpm) versus measured pressure drop is provided with the valve. Pressure drop is measured with a differential gauge.

BALANCING STATION

A balancing station is an assembly to both measure and control water flow. It has a measuring device and a volume control device. A balancing station must be installed with the recommended lengths of straight pipe entering and leaving the station. A calibrated balancing valve is one type of balancing station.

PRESSURE CONTROL

A pressure valve and a pressure tank are used to maintain the proper water pressure throughout the hydronic system. Two types of pressure valves are used and both are abbreviated PRV, pressure reducing valve and pressure relief valve. The two pressure control tanks used are the expansion tank and the compression tank. Though different, both types of tanks are commonly referred to as an expansion tank.

Pressure Valve

A pressure reducing valve is installed in the make-up water pipe to the system, for example, to reduce the pressure of city water down to the pressure needed to completely fill the system. A pressure reducing valve typically comes set at 12 psi (about 28 feet, 12 psi x 2.31feet/psi). This is adequate pressure for one or two story buildings. For a three story or taller building, a pressure reducing valve is adjusted for a minimum of an additional 5 psi pressure at the highest terminal. As an example, an HVAC system in San Diego, California, has its pump, boiler and compression tank located on the ground floor of a three-story building. Fan-coil units are located on all three floors. The elevation of the fan-coils on the third floor is 30 feet. The pressure reducing valve is set for 18 psi (30 feet is approximately 13 psi) + 5 psi.

A pressure relief valve (Figure 8-10) is a safety component installed on a boiler or other similar equipment (water heater, etc.) to protect the system and human life. It is preset by the manufacturer to open at a predetermined pressure which is less than the maximum pressure rating so that the operating pressure does not exceed the design limits of the system. If the set pressure is exceeded the relief valve is forced open and a portion of the water is diverted out of the system.

Pressure Tank

After a water system is constructed it is filled with water from the city supply main, or other source (well, river, etc.). The system pressure reducing valve is adjusted and the system is tested for integrity and proper pressure and flow. After the system has passed the testing phase and has been balanced for the correct water flow through the various components it is ready for normal operation.

Water in the system will expand when heated and then contract when it cools. Heated expanding water will increase the pressure inside

Figure 8-10. Boiler Pressure Relief Valve.

a closed system and if there is no pressure relief the pressure could break a pipe or damage other components. Water pressure tanks are used to maintain the proper pressure in a system and accommodate this fluctuation in water expansion and contraction while controlling pressure changes in the system. Expansion tanks are used in open systems and compression tanks are used in closed systems.

An expansion tank is simply an open vessel used in an open water system to compensate for the normal expansion and contraction of water. As the water temperature increases, the water volume in the system increases, the water expands in the expansion tank, and the tank water level rises. Corrosion problems are associated with expansion tanks as a result of the exposure to the air and evaporation or boiling of the water. Because of this, expansion tanks may be limited to installations having operating water temperatures of 180F or less.

A compression tank (Figure 8-11) is a closed vessel containing water and air or an air bladder. The tank is generally filled with water to about two-thirds full. The air in the compression tank or the bladder acts as a cushion to keep the proper pressure on the system. It accommodates the fluctuations in water volume and controls pressure changes in the system. Pressures in the water system will vary from the minimum

Figure 8-11. Compression tank, aka expansion tank, with air separator (white) piped in bottom. Water level sight glass at front center of tank.

pressure required to fill the system to the maximum allowable working pressure created by the boiler.

If the air in the compression tank leaks out, water will begin to fill the tank. This condition is called a "water-logged" tank. Water-logging can happen when the air leaks out of the compression tank and the pressure on the system is reduced below the setpoint on the pressure reducing valve. The pressure reducing valve will then open to allow in more water to fill the tank until the setpoint on the pressure reducing valve is reached. When the tank becomes water-logged the fluctuations in water volume and the proper system pressures cannot be maintained. A water-logged tank must be drained and the air leaks found and sealed. If the tank remains water-logged when the water in the system is heated the water will expand to completely fill the tank. Since there is no longer an air cushion, and nowhere else for the water to go, every time the boiler fires the pressure relief valve on the boiler will open to spill water in order to relieve the pressure in the system. When the pressure relief valve opens and reduces the pressure in the system, the pressure reducing valve opens to bring fresh water into the system. This cycle continues. Each time fresh water comes into the system, it also brings in unwanted air.

Compression tanks must be installed on the suction side of the pump. If the compression tank is installed on the discharge of the pump, the pump must operate at the pressure set by the pressure reducing valve. This may create a pressure condition at the pump inlet which will be lower than the vapor pressure of the water. If the inlet pressure is lower than the water vapor pressure, there is the possibility of pump cavitation. In addition, the water pressure at the air vent may be negative instead of positive. This will bring air into the system and possibly create air and corrosion problems. The location where the compression tank connects to the system is called "the point of no pressure change."

AIR CONTROL

Most HVAC water systems are closed and are typically filled with water from a city main or similar source and this source also supplies the system with new water to replace water lost through leakage, shut down for repairs, evaporation, etc. To prevent air problems, such as air locks in the bends of the piping and coils, or corrosion in the system the source water should be introduced into the system at some point either in the air line to the compression tank or at the bottom of the compression tank.

In addition to the air that is already in the system when water is heated air entrained in the water is released. In the paragraphs above I said that a compression tank can become water-logged if air leaks from the tank. A compression tank can also become water-logged if there is inadequate air control in the system. For example, when the boiler is off the water in the compression tank cools. The cool water absorbs air from the compression tank. The water flows back into the boiler by gravity. When the water is heated the air is again released and vented out. After several cycles, all the air is removed from the tank and the tank fills with water. Now the water has nowhere to go except the pressure relief valve on the boiler, which will begin opening on every boiler firing cycle to relieve the pressure caused by the water expansion.

However air gets into the system it must be removed. In a closed water system that has been correctly designed, installed, and operated, air in the system travels through the pipes and is vented out at the high points of the system or collected in the compression tank. Air separators, automatic air vents and manual air vents are types of air control devices designed to free entrained air and remove it from a water system.

Air Separator

The types of air separators include: centrifugal, dip tube and in-line air separating tank. The centrifugal air separator uses centrifugal force and low water velocities for air separation. As water circulates through the air separator, centrifugal motion creates a vortex or whirl-pool in the center of the tank and sends heavier, air-free water to the outer part of the tank. The lighter air-water mixture moves to a low velocity air separation and collecting screen located in the vortex. The entrained air collects and rises into the compression tank.

The boiler dip tube air separator is a tube in the top or top side of the boiler. When the water is heated, air is released and collects at a high point in the boiler. The dip tube allows this collected air to rise into the compression tank.

The inline, low velocity, air separating tank with dip tube type of air separator is used when a boiler isn't available or useable as the point of air separation.

Air Vent

An air vent may be automatic or manual. One type of automatic air vent is the hydroscopic air vent which contains a material that expands when wet and holds the air vent valve closed. When there is air in the system the hydroscopic material dries out causing it to shrink and open the air vent valve.

The float type of automatic air vent has a float valve that keeps the air vent closed when there's water in the system and vent. When there is air in the system it rises into the air vent replacing the water. The float drops and opens the air vent valve.

Manual valves are installed at the high points and bends in the system. The vent is manually opened periodically on a schedule to allow entrained air to escape.

HEAT CONVERSION EQUIPMENT

Heat conversion equipment comes in a variety of names, shapes and sizes including shell and tube heat exchanger, shell and coil heat exchanger, helical heat exchanger, plate heat exchanger, coils, boilers, furnaces, chillers, coolers, heaters, condensers, evaporators, converters, generators, etc. They are designed for a variety of fluid combinations. Some heat flow examples are:

From	To	Example
Refrigerant	Water	Water-Cooled Condenser
Refrigerant	Air	Air-Cooled Condenser
Water	Refrigerant	Water Cooler
Air	Refrigerant	DX Evaporator
Water	Water	Storage Tank
Air	Air	Heat Wheel
Air	Water	Cooling Coil
Water	Air	Heating Coil
Steam	Water	Storage Tank
Steam	Air	Heating Coil

In some cases only one fluid is involved, e.g., electricity and air or water, chemical and air or water.

Coil

A coil (Figure 8-12) is a common heat exchanger in commercial systems. They come in a variety of types, materials and sizes and are designed for various fluids such as water, brines, refrigerants or steam. Typically the fluid circulating through a coil either heats, cools or dehumidifies air passing over the coil. Heating or cooling coils have a frame and a core. The frame supports the core as it expands and contracts. The core is made of tubing, return bends and fins. The core tubing is usually copper (5/8 inch diameter) but tubing may be made from aluminum, carbon steel, stainless steel, brass and cupro-nickel. The number of tubes in the coil varies in both depth and height. Usually there are 1 to 12 rows in the direction of airflow (depth) and 4 to 36 tubes per row (height). Return bends are the U-shaped pieces of tubing at the end of the coil. The number and arrangement of the return bends will depend on the number of rows and the circuiting arrangement. The circuiting arrangement is the order in which the tubes are connected together in series and in parallel to produce the best heat transfer, capacity, fluid flow and pressure drop combination.

Tube fins are normally aluminum or copper. Fins on a coil increase the area of heat transfer surface to improve the efficiency and rate of transfer. They are generally spaced from 4 to 14 fins per inch (fpi). Aluminum is usually picked over copper for fin material to reduce cost. However, when cooling coils are sprayed with water, copper fins are needed to prevent electrolysis between dissimilar metals (copper tubes

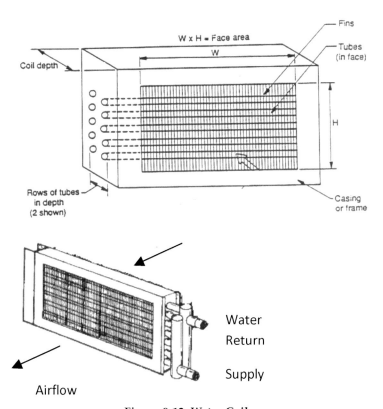

Figure 8-12: Water Coils.

and aluminum fins). Coils wetted only by condensation are seldom affected by electrolysis and are usually copper headers, copper tubes, aluminum fins and steel frames. For applications where the air stream may contain corrosives, various protective coatings can be applied to the fins and tubes.

Heat transfer (heating or cooling) increases with more rows, more tubes per row, and more fins per inch and, for chilled water coils, more dehumidification. Of course, with more tubes, rows, and fins, there is an increase in initial cost of the coil and ongoing energy cost increase also because more horsepower is required to overcome the increased resistance to airflow.

HVAC water coils can be piped either counter flow (Figures 8-3, 8-5 and 8-6) or parallel flow (Figure 8-13). System Efficiency: For the greatest heat transfer for a given set of conditions, water coils should be piped counter flow. Counter flow means that the flow of air and water are in opposite directions to each other. In other words, water enters on the same side of the coil that air leaves. For cooling coils, this would mean that the coldest water is entering the coil on the same side that the coldest air is leaving the coil. A coil that is piped parallel flow means that the flow of air and water are in the same direction to each other. Water and air enter on the same side. For parallel flow cooling coils, the coldest water enters the coil on the same side that the warmest air enters the coil. This means less heat transfer. System Operation: In some applications such as preheat coils, the coils are intentionally designed for parallel flow. For example, a preheat coil may be used to heat outside air in cold climates to prevent freezing of downstream coils. Therefore, the coil is piped parallel flow so the hot water or steam enters the coil on the same side that the cold air enters. Heat transfer, in this example, is critical, and getting the most heat to the coil as quickly as possible is what is important.

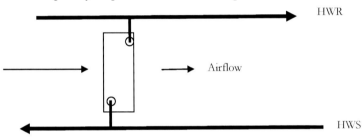

Figure 8-13. Coil piped for parallel flow.

In addition to being piped counter flow water coils should also be piped so that the inlet water is at the bottom of the coil and water flow is up through the coil and out the top. This will enable air entrained in the water to be pushed ahead of the water and accumulate in the top portion of the system where it can be easily vented. Note: Steam coils are piped stream inlet at the top and condensate at the bottom of the coil.

WATER CHILLER

The two categories of water chillers used in HVAC systems are absorption and mechanical. The absorption chiller does not have a mechanical compressor, but instead uses a generator (aka concentrator), an absorber, a condenser, an evaporator and control devices to chill water.

The absorption chiller will be outlined at the end of this chapter.

Mechanical vapor-compression water chiller systems (Figure 8-14) are the most common and will be discussed first. The mechanical system uses a centrifugal compressor or a positive displacement reciprocating or rotary mechanical compressor. The mechanical compressor is the heart of the vapor-compression air-conditioning cooling system. It pumps refrigerant around the system and compresses the vapor from a lower pressure to a higher pressure. The other components in the vapor-compression refrigeration mechanical system are a condenser, an evaporator, and various control devices.

CENTRIFUGAL WATER CHILLER

Willis Carrier, the "father of air conditioning," patented the centrifugal chiller in 1921. This was the first practical method of air conditioning large spaces. Before this, refrigeration machines used reciprocating

Figure 8-14. Centrifugal Mechanical Vapor-Compression Chiller. Foreground, left to right: Chilled Water Supply Pipe (white insulation), Evaporator (black insulation) with compressor (gray, top right, no insulation). Background: Return Water Pipe (no insulation), Condenser (gray, no insulation).

compressors to move refrigerant through the system. The heart of a centrifugal water chiller is the centrifugal compressor.

Centrifugal Compressor

Compression in a centrifugal compressor is done primarily by the centrifugal force produced as high-speed impellers rotate the refrigerant vapor. Centrifugal compressors are high capacity machines moving high volumes of vapor and can't be economically built for small capacity systems. The centrifugal compressors used in HVAC water chillers start about 80 tons of refrigeration and go to several thousand tons. The larger the capacity the more advantageous the centrifugal compressor becomes. I saw a centrifugal chiller from Carrier Corporation in New York state at one of its universities which was 10,000 tons.

Centrifugal compressors, centrifugal pumps and centrifugal fans are members of the same family of machines…the pumping force is based on impeller size (wheel size for fans) and rotating speed. The operating principles are also similar. In a centrifugal compressor, low pressure, low temperature and low velocity refrigerant vapor is drawn into the impeller housing near the center of the compressor, and then enters the inlet of the impeller. As the impeller spins, the vapor is discharged at high velocities and higher temperature and pressure to the outside of the housing. To maintain the centrifugal force, the impeller rotates at very high speeds (much higher than centrifugal fans or pumps). Speeds to 25,000 rpm are common. However, centrifugal compressors don't build up as much pressure as do positive displacement compressors. Therefore, several impellers are put in series to increase the pressure of the vapor. Commonly, centrifugal compressors will have two, three or four impellers. Each impeller is a stage of compression. After the vapor leaves an impeller, it's directed into another impeller or into the discharge line

Centrifugal compressors may be either open or hermetic. And, since they have no cylinders, valves or pistons there are fewer parts needing lubrication. Capacity control of centrifugals is usually done by varying the speed of the compressor or by variable inlet guide vanes. Both reducing impeller speed and closing inlet vanes reduces vapor flow and reduces refrigeration capacity. Variable inlet guide vanes or pre-rotation vanes (similar to vortex dampers or inlet guide vanes in centrifugal fans) are located directly ahead of the impeller inlet. These vanes change the direction, or rotation, of flow of the refrigerant vapor immediately before it enters the impeller. This change in direction results in a reduction in total flow.

Operation

Let's start the explanation of how a centrifugal water chiller works at the evaporator. Refrigerant vapor (gas) from the evaporator flows into the center (eye) of the centrifugal compressor impeller. Vanes in the eye of the impeller draw the gas into radial passages. As the impeller rotates, it increases the velocity of the gas, which then goes through diffuser passages, and into a space in the perimeter of the compressor housing where the gas is stored. Inset vanes installed ahead of the impeller stabilize the performance of the compressor over a range of load conditions. The inlet vanes adjust the gas quantity (gas flow rate) as well as the angle of the vapor as it enters the impeller, which creates a new compressor performance characteristic for each vane position.

The vapor goes through several stages of compression and then is discharged into the condenser. The hot vapor enters the condenser. The vapor is cooled by water from the cooling tower (Figure 8-15) circulating through the condenser. As the water passes through tubes in the condenser it picks up heat from the vapor. The water then goes back to the cooling tower to release into the outside air the heat that it has picked up in the condenser. The condenser pump moves the water around this

Figure 8-15. Cooling Tower. Top pipe condenser water return. Bottom pipe CWS.

circuit. Once the entire vapor refrigerant is condensed to a hot liquid, the refrigerant leaves the condenser through the liquid line.

The hot liquid refrigerant from the condenser is metered through an orifice system into a pressure chamber, which is called an economizer. The purpose of the economizer is to pre-flash the liquid refrigerant. When the liquid refrigerant enters the economizer the pressure on the refrigerant is reduced. This reduction in pressure reduces the boiling point of the liquid refrigerant. However, the temperature of the liquid refrigerant is still above the new boiling point. Because the liquid refrigerant is hotter than its boiling point a part of the liquid refrigerant begins to boil off. This boiling off of the liquid refrigerant is called flashing. The liquid refrigerant which is boiled off, or flashed, changes state to a vapor or gas. This gas is called flash gas. When a part of the liquid refrigerant is flashed, it removes heat. This flashing continues until the remaining liquid refrigerant is cooled down below the boiling point which corresponds to the pressure on the liquid.

Pre-flashing in the economizer reduces the volume of flash gas required to cool the refrigerant flowing through the metering device into the evaporator. This reduction in the volume of flash gas in the metering device means that more of the liquid refrigerant is available for use in the evaporator. This makes the chiller system more efficient. Also, there is less of a load on the first stage of the compressor therefore, a reduction in the power requirements on the compressor. The pre-flashed gas from the economizer is sent back to the compressor to be compressed. The liquid refrigerant in the economizer, which is now at an intermediate pressure, that is, a pressure somewhere between the higher pressure of the compressor and the lower pressure of the evaporator, continues through the liquid line into the metering device.

The condensed liquid refrigerant from the economizer goes through the metering device into the evaporator. The metering device is a system of orifices in the liquid line. The purpose of the metering system is to maintain the required refrigerant flow for each load condition. As the liquid refrigerant flows through the various orifices, its pressure and temperature is reduced. This causes a part of the liquid refrigerant to flash, reducing the temperature of the remaining liquid to the required evaporator temperature. This cooler, lower pressure liquid-vapor mixture now enters into the evaporator through the liquid line. As the refrigerant liquid-vapor mixture leaves the liquid line it goes into a liquid distributor which runs the length of the evaporator. The distributor helps to promote a more uniform heat transfer throughout the entire length of the evaporator. The temperature of the refrigerant liquid-vapor mixture is about 40F. The temperature of the water from the air handling units is about 55F. As the water travels through tubes in the evaporator it is cooled down about 10 degrees to approximately 45F.

Water Temperature and Pressure

One of the functions of the chilled water temperature control system is to modulate the centrifugal compressor's inlet vane position. By varying inlet vanes or compressor speed, the capacity of the compressor is modulated according to the system load. A typical sequence of operation would be that a rising return chilled water temperature would cause a sensor to send an increasing signal to a controller. A rising return water temperature indicates that the conditioned space is becoming warmer and the system load is increasing. The controller sends a signal to start the compressor motor. The signal also goes to a variable speed drive or to the compressor inlet vanes. The compressor inlet vanes are set normally closed (NC). As the pressure increases, the vanes start to open, allowing more refrigerant vapor into the compressor.

A load limiting relay (LLR), allows the control signal to pass on to the inlet vane operator as long as the compressor motor current is less than 98% of maximum. If the motor current exceeds 98% the LLR impedes the control signal which closes the inlet vanes. The refrigerant vapor flow to the compressor is reduced which reduces the load on the motor. The load limiting relay remains in control until the load on the compressor is reduced below the setting on the relay. At this point, the LLR once again allows the control signal to pass on to the vane operator and control of the chiller is returned to the chilled water temperature control system. In addition to the automatic load limiting relay a manual demand limiter device is installed in the control panel to set the limits of chiller operation. By adjusting the demand limiter the load on the compressor can be set to 40%, 60%, 80% or 100%.

The temperature and pressure in the condenser is a function of condenser water temperature, condenser water flow rate, the amount of non-condensable gases in the condenser and the cleanliness of the condenser tubes. Air is an example of a non-condensable gas, i.e., a gas that will not condense at the pressures and temperatures in this system. Condenser water temperature is normally about 95F leaving the condenser and going to the cooling tower and about 85F coming back from the tower into the condenser. The condenser water flow rate in gallons per minute (gpm) will vary depending upon the size of the system. The water flow should be tested, adjusted and balanced between plus or minus 10% of full flow.

The temperature and pressure in the evaporator is also function of water flow. If the temperature difference is lower while the pressure drop is higher from previously recorded readings the chilled system is over-pumping. If however, the temperature difference is higher, while the pressure is lower, the system is probably low on water. As with the condenser water flow rate the evaporator water flow should be balanced between plus or minus 10% of full flow.

RECIPROCATING AND SCREW COMPRESSOR WATER CHILLER

A positive displacement compressor such as a reciprocating or screw compressor is usually more economical below 100 tons of refrigeration capacity as centrifugal compressors are high-capacity machines moving large volumes of vapor and can't be economically built for small capacity systems. Reciprocating compressors (Figure 8-16) are among the most widely used compressors because of design and the range of capacity (approximately one ton up to hundreds of tons of refrigeration capacity in single unit). Reciprocating compressors are generally classified according to type of drive, motor accessibility, piston type, number

Figure 8-16. Reciprocating Compressor. Demonstration model showing valves and pistons left and motor right.

and arrangement of cylinders, valve construction, method of lubrication, and capacity control.

A screw compressor (Figure 8-17) uses two helically grooved rotors to compressor refrigerant vapor. The rotors intermesh to progressively reduce the space inside the cylinder and reduce the volume of refrigerant vapor and increase its pressure. As the rotors turn, vapor from an inlet port at the suction end of the screw cylinder enters the space between the rotors. The rotors continue to turn and close off the suction port. The screw action then forces the vapor to the discharge end and compresses it against a discharge plate. At a given point, the rotating screws uncover discharge ports in the discharge plate and the compressed vapor is forced out into the discharge line. Capacity control on the screw compressor is done by a slide valve in the housing wall underneath the rotors. The slide valve is hydraulically operated. When the system calls for a slowing of the refrigeration process the valve is opened allowing some vapor to recirculate in the cylinder without being compressed. A screw compressor is generally used on systems of 50 tons or larger.

ABSORPTION CHILLERS

Absorption chillers are similar to mechanical chillers. They use heat to vaporize a volatile refrigerant at low pressures in the evapora-

Figure 8-17. Large Screw Compressor.

tor. They take a low pressure refrigerant vapor from the evaporator and deliver a high pressure refrigerant vapor to the condenser. They condense the refrigerant vapor in the condenser. They recycle the refrigerant. One of the major differences is the absorption chiller does not use a mechanical compressor. Instead, heat from various sources is used to change the concentration of the refrigerant-absorbent solution. As the concentrations change the pressures in the various components will change. This pressure difference moves the refrigerant around the system.

A generator and absorber replace the mechanical compressor. An absorbent is used, usually water or lithium bromide. The refrigerant is typically either ammonia or water. The energy input comes from heat (steam or hot water) supplied directly to the generator. The heat may come from boilers or furnaces. In some cases, heat from other processes may be used. These include low pressure steam or hot water from industrial plant work, waste heat recovered from exhaust gases of gas engines or turbines, or low pressure steam from steam turbine exhaust. Generally, commercial HVAC absorption systems will have capacities from 25 to over 1,000 tons.

An absorption chiller has the following components: generator, condenser, evaporator, absorber, fluid pumps, purge unit and controls. The evaporator and the absorber are on the low pressure side of the system in a vacuum of about 0.12 pounds per square inch absolute (0.248 inches of mercury). The generator and the condenser are also in a vacuum on the high pressure side of the system. The pressure on the high side is 1.5 pounds per square inch absolute (3.06 inches of mercury) or approximately one-tenth of atmospheric pressure (14.7 pounds per square inch absolute). There are two working fluids in an absorption system—a refrigerant and an absorbent. In our example system the refrigerant is water and the absorbent is lithium bromide.

The refrigerant flow is from the condenser to the evaporator (liquid state); evaporator to the absorber (vapor); absorber to generator (liquid); generator to condenser (vapor). The absorbent flow is from absorber to generator (diluted to concentrated solution); generator to absorber (concentrated to diluted solution).

The operation of the absorption chiller depends on an absorbent that has a great attraction for the refrigerant. For use in large capacity absorption chillers, lithium bromide, a salt, is such an absorbent. Lithium bromide, when dry, is in a crystal form. The amount of lithium bromide dissolved in the water to form the lithium bromide solution is measured by weight and not by volume. The concentration of the lithium bromide solution is stated in terms of the amount of lithium bromide in the solution. For example, a 100 pound solution of lithium bromide and water may have 65 pounds of lithium bromide and 35 pounds of water. This would be called a 65% concentration because 65% of the total weight is lithium bromide. The evaporator is maintained in a vacuum of about 0.248 inches of mercury (0.12 psia). The temperature of the refrigerant corresponding to this pressure is approximately 40F. The refrigerant is water.

Let's start the cycle with "high" pressure liquid refrigerant (the pressure are relative, see paragraphs above). The high pressure liquid refrigerant (water in this example system) from the condenser passes through a metering device into the evaporator (Figure 8-18, top right in chiller shell). The pressure on the refrigerant is reduced lowering the temperature of the refrigerant to approximately 40F.

This refrigerant goes through a spray header over the evaporator tubes. The sprayers help keep the tubes wet at all times. They also break the liquid refrigerant into small droplets allowing it to vaporize more easily. This improves heat transfer and makes maximum use of the refrigerant. As the refrigerant comes in contact with the warmer evaporator tubes part of the refrigerant is vaporized and absorbs latent heat from the fluid being cooled. In this example, the fluid being cooled is chilled water return water from the air handling units.

The return water (CHWR) enters the evaporator tubes at 55F. The refrigerant cools the water down to a leaving temperature of 45F (CHWS). In addition to the refrigerant from the condenser any liquid refrigerant that is not vaporized collects under the tubes in the bottom of the evaporator. A pump re-circulates the liquid refrigerant collected back to the sprayers.

The low pressure refrigerant vapor passes through eliminators which remove any liquid refrigerant and flows from the evaporator to the absorber. This flow occurs because the vapor pressure of the lithium bromide solution in the absorber is lower than the vapor pressure of the refrigerant vapor in the evaporator.

The refrigerant vapor from the evaporator is drawn into the low pressure area created by controlling the temperature and concentration of the refrigerant-absorbent solution (i.e., the higher the concentration and the lower the temperature of the lithium bromide and refrigerant

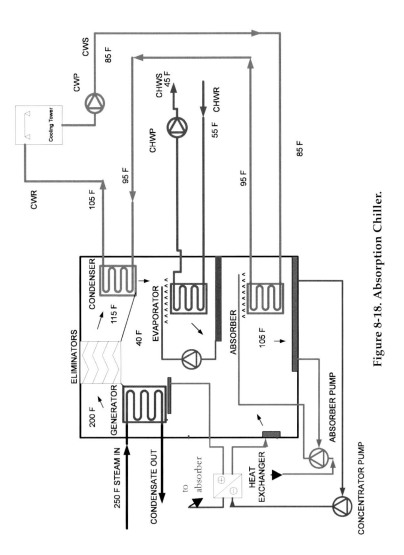

Figure 8-18. Absorption Chiller.

water solution the lower its vapor pressure).

An absorber pump circulates the concentrated solution of lithium bromide and water through a spray header over the absorber tubes. The lithium bromide solution from the sprayers mixes with and absorbs the refrigeration vapor from the evaporator. When the mixing takes place the refrigerant vapor condenses into a liquid to create a diluted solution of liquid refrigerant and lithium bromide. This is called the dilute solution. This dilute solution then drops into the bottom of the absorber shell and is pumped by the concentrator pump through a heat exchanger to the generator. Heat is generated during the mixing process. To maintain the temperature in the absorber, water from a cooling tower flows through the absorber tubes. The temperature of the water entering the absorber tubes is approximately 85F. The water in the tubes picks up the heat in the absorber and leaves at 95F. If the heat is not removed the temperature and pressure in the absorber will rise and the vapor flow from the evaporator will stop.

The dilute solution must be re-concentrated in order for the refrigeration process to continue. The dilute solution flows through the heat exchanger and picks up heat from the concentrated lithium bromide solution*. After passing through the heat exchanger the dilute solution flows over the generator tubes.

_____

*The heat exchanger transfers heat from the warmer concentrated lithium bromide solution leaving the generator (and going to the absorber) to the cooler dilute lithium bromide solution coming from the absorber. The efficiency of the absorption system is improved by using the heat exchanger. The temperature of the dilute solution is increased, which reduces the amount of heat needed in the generator. The temperature of the concentrated solution is reduced, resulting in less heat having to be removed in the absorber by the water from the cooling tower. The dilute solution pumped by the absorber pump from the bottom of the absorber mixes with the cooler concentrated solution from the generator and goes through the sprayers. This solution is now ready to absorb refrigerant vapor from the evaporator.

Low pressure steam flows through the generator tubes. The heat from the steam vaporizes a portion of the refrigerant (which reduces the water content of the lithium bromide solution) making it a concentrated solution. The vaporization of the refrigerant is possible because it has a lower boiling point than the absorbent and the generator is never at a temperature high enough for the absorbent to boil. The refrigerant vapor in the generator is about 200F. It passes through eliminators (which remove any entrained lithium bromide) and flows into the condenser.

Approximately 3.06 inches of mercury (1.5 pounds per square inch absolute is maintained in the high side which means that the refrigerant vapor will condense on the condenser tubes at approximately 115F. Water from the absorber is pumped through the condenser tubes and then into the cooling tower. The water enters the condenser at 95F and picks up the latent heat of condensation. The water leaves the condenser at about 105F. The water goes to the cooling tower where it is cooled down to 85F and is re-circulated back to the absorber. The condensed refrigerant liquid flows by pressure differential into the evaporator through the metering device and the operating cycle is repeated.

Non-condensable gases will tend to collect on the surface of the dilute lithium bromide solution in the bottom of the absorber. If the non-condensable gases are not removed by purging, they will increase the pressure in the absorber to a point where the flow of refrigerant vapor from the evaporator will stop. Control of the operating cycle is accomplished with an automatic control valve on the steam line. This control valve modulates the flow of steam to the generator tubes. The control valve operates from a temperature sensor in the chilled water line leaving the evaporator. For example, if the chilled water temperature is too cold, the control valve will close, reducing the quantity of steam supplied to the generator. The reduced amount of steam will boil a smaller quantity of refrigerant out of the generator and reduce the concentration of the lithium bromide solution being sprayed in the absorber. The solution will not be able to absorb as much refrigerant vapor and the cooling ability of the evaporator will be reduced. Water flow rate through the condenser will be controlled with a bypass loop in the condenser piping.

Chapter 9

Variable Volume Systems

In this chapter I will take you through variable volume air, refrigerant and water systems. On the air side, aka dry side, variable air volume (VAV) and variable refrigerant flow (VRF) systems and on the water side, aka wet side, variable water volume (VWV) systems. The primary reason for installing a variable volume system is energy savings and therefore, cost savings.

VARIABLE VOLUME AIR SYSTEM

Let's start off talking about a constant air volume (CAV) system. The design engineer may elect to design the system so that the air temperature varies with the changes in heating or cooling load. Now, if the Btuh (sensible cooling load) increases, the controls will decrease the temperature of the air into the space while the cfm remains constant. This will cool the space. Likewise, if the sensible cooling load decreases, the controls will increase the temperature of the air while the cfm into the space remains constant. We vary the temperature of the supply air to maintain the space temperature. This is a constant air volume system.

Remember our equation for a heating or cooling sensible load is Btuh = cfm x 1.08 x TD. Let's put some numbers in. For this example the air handling unit (AHU) is in the cooling mode. The system cfm is 2000 and the TD is 17 (75F room air temperature, RAT- 58F cooling coil leaving air temperature, LAT). At this moment the cooling load in the space is 36,720 Btuh. Now say the load in the space begins to decrease to 32,400 Btuh. The space thermostat on the wall senses the change and sends a signal to the water valve on the chilled water coil to allow less water flow through coil increasing the coil LAT to 60F and 32,400 Btuh (2000 cfm x 1.08 x 75-60) is going into the space. So, as the space cools down toward 32,400 Btuh the air temperature leaving the cooling coil goes up towards 60F. Voila! Constant volume (2000 cfm), variable temperature.

Now, the basic pressure independent variable air volume system is constant air temperature and variable air volume. Okay, same room (conditioned space) with a cooling load of 36,720 Btuh and the RAT is 75F. The air temperature leaving the coil and flowing into the room is a constant 55F. The cfm at this moment would be 1700 (36,720 Btuh/1.08 x 20 TD).

Same scenario as the CAV system above, the room cooling load begins to decrease to 32,400 Btuh. The volume of air going into the space must decrease to 1500 cfm in order to match the load (32,400 Btuh/1.08 x 20 TD). There you have it! Variable volume, constant temperature (55F).

Variable Volume, Constant and Variable Temperature

Figure 9-1 illustrates a pressure independent (PI), constant temperature, variable air volume (VAV) system. We'll designate it VAV-101. Starting with the fan we see the fan has a maximum output of 60,000 cfm. However, the total of all the terminal boxes or all the diffusers is 80,000 cfm. This system has a 25% diversity, or difference between the total output of the fan and the total of the diffusers (60,000/80,000). Therefore, if all the space thermostats called for full cooling and all the boxes opened to maximum simultaneously then the fan would not have enough air output to meet the demand. The boxes closest to the fan would get their required air while the boxes further downstream would "starve" for air.

Now, let's look at the boxes. All the boxes are set for a maximum flow of 1000 cfm. The interior boxes have a maximum shutoff which is set by the design engineer according to code. I've listed them as 75%. The minimum flow then is 250 cfm. All these boxes aka terminal boxes are variable volume, constant temperature.

Some of the exterior boxes have a reheat coil and have a maximum shutoff at 50 %, once again set by design and code. In other words, for theses boxes the minimum flow is 500 cfm to allow for proper heat transfer. That is, the required heat from the steam, hot water or electrical coil may not be transferred to the air if the airflow is too slow. Also, safety and fire may be a concern, especially with electric reheats (if you still have them) if not enough air is moving across the electric resistance coil to remove the heat. These boxes are variable volume and variable temperature to the conditioned space.

VAV Operation

Most VAV systems use a single duct, which supplies a constant air temperature, generally between 55F and 60F, to the VAV terminal box. Our system, VAV-101, has a 55F constant air temperature into the ter-

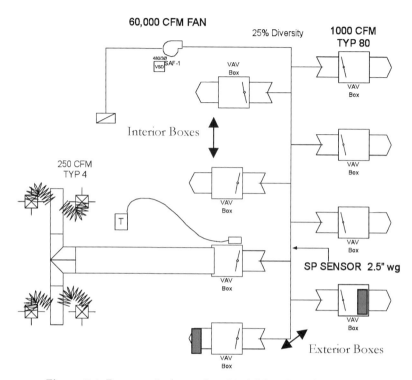

Figure 9-1. Pressure Independent Variable Air Volume System

minal box. The air volume through the terminal box and into the space is varied to maintain the space temperature. There are also dual duct systems that use two single duct supplies to the boxes. One duct carries heated air and the other chilled air (see Chapter 7).

As the cooling load increases in the conditioned space the cfm into the space also increases. In other words, as the temperature increases in the space (as sensed by the space thermostat) more air at 55F flows into the space. For example, in our system VAV-101, a given room thermostat is satisfied. The airflow into the room is 500 cfm. Circumstances change in the room and the temperature rises above the 75F thermostat setpoint (cooling load is increasing). The volume damper in the box opens to allow more cfm into the space. This will cool the space. Likewise, if the Btuh (sensible load) decreases the box will decrease the cfm into the space while the temperature of the air remains constant. This will allow the space to warm up (from the heat generated by the people, lights and equipment in the space).

VAV Terminal Box and Static Pressure Sensor Operation

As the VAV terminal box dampers or air valve throttle back to allow less airflow into the space (cooling load is reducing and the space thermostat is calling for less cfm) static pressure increases in the supply duct. A static pressure sensor (SPS, Figure 9-1) is installed in the system. It senses the increase or decrease in duct static pressure and sends a signal back to a control device. The controlled device regulates the supply fan air volume (cfm) and static pressure. For centrifugal fans the volume controlled device is a variable frequency drive aka variable speed drive, inlet dampers or outlet dampers (see VAV Controller). The SPS and the fan volume controls react to maintain a constant static pressure at the sensor location as the system's air volume fluctuates.

For example the SPS for VAV-101 is set at 2.5 inches water gauge. The cooling load is increasing on the southwest side of the building and the space thermostats are calling for more cfm. The terminal box volume dampers or air valves start opening to allow more airflow into the spaces. This decreases the static pressure in the supply duct. The static pressure sensor senses the decrease in duct static pressure and sends a signal back to the volume controlled device (a variable speed drive in this system) to increase the supply air volume to the boxes. If, on the other hand, the conditioned spaces are too cool the box volume dampers close which increases the duct static pressure. When the duct static pressure exceeds the 2.5" wg setpoint the controls slow the fan and reduce cfm and static pressure in the duct.

SP Sensor Location

The SPS is typically about two-thirds to three-quarters of the way from the fan to the end of the duct system. The location of the sensor is a compromise between energy efficiency and control of the system. For the maximum energy efficiency the sensor would be installed at the inlet to the "farthest" (static pressure-wise) terminal box from the fan and the static pressure set just high enough to operate that box and its associated low-pressure system. However, because the entire system may be continuously changing, i.e., some boxes are closing while others are opening, the "farthest" box may also be constantly changing. This means that at any given time other terminal boxes may be the "farthest" box and may need more static pressure than was required at the original box. Therefore, the sensor is located closer to the fan and set at a higher static pressure to accommodate any additional pressure losses. This gives control of the system but also increases the pressure that the fan must produce and some energy savings is lost.

VAV Controller and Controlled Device

A volume and static pressure controller receives the signal from the sensor. The controller sends a signal to a controlled device. The controlled device may be an electronic variable frequency drive (VFD) aka variable speed drive (VSD), which varies the frequency and speed of the motor and in turn varies fan speed. Another controlled device is a variable pitch propeller in an axial fan. Other controlled devices include: (1) static pressure dampers at the fan discharge which reduces air volume and static pressure output and (2) inlet guide vane dampers aka vortex dampers at the fan inlet which varies the spin of the air and reduces the open area into the fan which varies the volume and pressure output. All these methods save energy. Table 9-1 compares the most common. The exception to the energy-saving control schemes is a bypass type of system which uses relief dampers to bypass the air to the fan inlet as the terminal boxes or VAV ceiling diffusers close. This type of VAV system is variable only into the conditioned space and has no energy savings since the primary fan is constant air volume.

Table 9-1. VAV Fan Airflow Control and Approximate Horsepower.

Percent of Airflow	80%	70%	60%	50%
System Type	Horsepower Required			
BI (SPD)	90%	85%	80%	75%
FC (SPD)	80%	65%	52%	42%
AF (IGV)	75%	68%	55%	50%
FC (IGV)	70%	55%	42%	32%
FC, BI, AF (VFD)	64%	50%	40%	28%

FC—Forward Curved fan, BI—Backward Inclined fan, AF—AirFoil fan, SPD—Static Pressure Dampers, IGV—Inlet Guide Vanes, VFD—variable frequency drive. Example: If the airflow in a 10,000 cfm BI fan operating at 10 bhp is reduced to 8,000 (80%) using SPD the horsepower would be 9 bhp. Using VFD the horsepower would be 6.4 bhp.

Pressure Independent and Pressure Dependent VAV Systems

Variable air volume systems can be pressure independent or pressure dependent. Pressure independent (PI) systems have terminal boxes which work off the space thermostat signal as the master control. This signal operates a motor which in turn opens and closes the box's volume damper or air valve. A velocity (or volume) controller is used as a submaster control to maintain the maximum and minimum air volume to the space. The maximum to minimum airflow will be maintained when the static pressure at the inlet of the box is in compliance with the box manufacturer's published operational data. A pressure independent terminal box can be single duct, double duct, induction, or fan powered. Pressure independent boxes are independent of what is happening in the system duct or at any other box.

Pressure dependent (PD) terminal boxes do not have an automatic volume controller to regulate airflow as the inlet static pressure changes as do pressure independent boxes. What they have is an automatic inlet volume damper controlled by the space thermostat. These volume dampers may or may not have a minimum position limiter. The airflow delivered by the box is solely dependent on the inlet static pressure and therefore, changes as the inlet static changes. This type of system may have a manual balancing damper at the inlet of the box and balancing dampers in the branch lines. With pressure dependent systems when the system is in normal operation a change in a damper setting at one box may affect adjacent boxes. Pressure dependent systems can be single duct, fan powered, or bypass. For a simple explanation of how the two systems work look at Figure 9-2.

Typically, in office buildings there will be at least five zones served by any VAV system... north, south, east, west, and center or core zone. The PD system on the left has four VAV terminal boxes serving the North Zone. Boxes 3 and 4 (numbering from top to bottom) at the end of the line are 1000 cfm maximum. Scenario: In the conditioned space served by Box 3 some people leave the space and some lights are turned off. The cooling load decreases. The thermostat senses that there is a drop in temperature below setpoint. The room is being over-cooled with 1000 cfm at 55F air (variable volume, constant temperature). The thermostat sends a signal to reduce the cfm into the space from 1000 cfm to 500 cfm and allow the room to warm back to setpoint. The other 500 cfm (1000 – 500) "backs-up" in the branch duct with most of it going down to Box 4. At about the same time that Box 3 is closing down some people leave

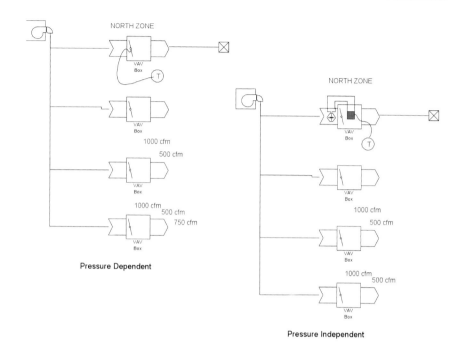

Figure 9-2. Pressure Dependent (left) and Pressure Independent (right) VAV Systems. Boxes numbered 1-4 (top to bottom). PI boxes have a thermostat (T) connected a controller (■) which is connected to a VAV damper (\) and a velocity pressure sensor. PD boxes have thermostat connected to damper.

the space served by Box 4 and some lights are turned off. The cooling load decreases. The thermostat senses that there is a drop in temperature below setpoint. The room is being over-cooled. The thermostat sends a signal to reduce the cfm into the space from 1000 cfm to 500 cfm and allow the room to warm back to setpoint. This happens at Box 4 but the air from Box 3 is now coming into Box 4 and then into its conditioned space. Instead of getting 500 cfm as the thermostat called for the room is getting 750 cfm (at 55F). The room is still being over-cooled. The room will continue to be over-cooled until the thermostat senses that it is still below setpoint and sends a signal to close the damper more, to 500 cfm in this example. This time lag can make the space uncomfortable.

The PI system on the right has four VAV terminal boxes serving the North Zone. Boxes 3 and 4 (numbering from top to bottom) at the end of the line are the same, 1000 cfm maximum. In the conditioned space served by Box 3 some people leave the space and some lights are turned off. The cooling load drops. The thermostat senses that there is a drop in temperature below setpoint. The room is being over-cooled. The thermostat sends a signal to reduce the cfm into the space from 1000 cfm to 500 cfm and allow the room to warm back to setpoint. The other 500 cfm "backs-up" in the branch duct with much of it going down to Box 4. At about the same time that Box 3 is closing down, some people leave the space served by Box 4 and some lights are turned off. The cooling load drops. The thermostat senses that there is a drop in temperature below setpoint. The room is being over-cooled. The thermostat sends a signal to reduce the cfm into the space from 1000 cfm to 500 cfm and allow the room to warm back to setpoint. This happens at Box 4 but the air from Box 3 is now coming into Box 4. At the inlet of Box 4 is a velocity sensor (shown at Box 1) which senses that more than 500 cfm is coming into the box. The sensor sends a signal to the controller to close the damper and reduce the cfm to 500. So the damper is repositioned by a signal at the box instead of the room thermostat resulting in minimal lag time. Also, as the boxes close, the static pressure sensor in the main duct will detect when the static pressure goes above setpoint. When it does, a signal is sent to the controller to reduce airflow and static pressure out to the system, which reduces the tendency for the air from one box to flow into another box.

VAV Volume and Static Pressure Controls

Figure 9-3 (top) shows an air handling unit (AHU) with automatic control dampers (ATCD) at the return air (RA), exhaust air (EA) and outside air (OA). This unit has filters, a cooling coil and a "draw-thru" (the fan is after the coil, if the fan is before the coil it is a "blow-thru") forward curved centrifugal supply fan. In the discharge duct is a set of static pressure dampers for system volume and static pressure control. A static pressure sensor (SP) is located in the supply duct, appropriately two-thirds to three-quarters of the way down the length of the main duct from the fan. The static pressure sensor is set for a static pressure to operate all the boxes that need to be served at any given time. If the static pressure sensed in the duct is above or below setpoint a signal is sent to the receiver-controller (RC). The RC then sends a signal to the damper motor (DM) to reposition the damper to bring the static pressure in the duct back to setpoint.

Figure 9-3 (bottom) shows a system with inlet guide vanes (IGV), aka vortex dampers (photo, Figure 9-4). The signal from the RC goes to

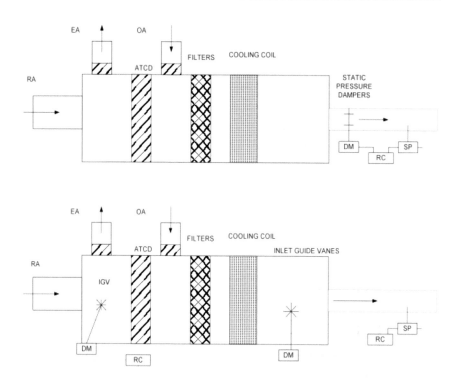

Figure 9-3. Top: Static Pressure Damper VAV System. Bottom: Inlet Guide Vane VAV System.

both the damper motors on the supply fan and the return fan. IGV are installed on variable air volume fan systems to control discharge pressure and air volume. The IGV gives the air a spinning motion as it enters the fan to provide reduced fan performance at reduced loads. Inlet guide vanes are a restriction to airflow and even with the vanes (blades) wide open; they reduce the fan volume output by approximately 5%. In this system both the supply air volume and the return air volume is controlled by the supply duct static pressure sensor.

If this system is not working properly, for example the return fan sometimes leads or lags the supply fan too much (e.g., the fans are pulsating or "hunting") it may be necessary to put in another static pressure sensor(s) and controller (or cfm air monitor and controller) to control the return fan separately with an appropriate static pressure differential (or an appropriate cfm differential) to allow the return fan to positively track the supply fan. Additionally, in this example system or any VAV

Figure 9-4. Inlet guide vanes (vortex dampers) installed on backward inclined fan

system, as the SA fan slows to reduce SA cfm to the system as boxes close down to maintain condition space temperature the OA may drop below minimum. Therefore, another static pressure or air monitor controller may be required in the OA to open the outside air dampers above minimum position in order to maintain minimum OA cfm.

Figure 9-5 shows an AHU with the variable frequency drive (VFD), also known as a variable speed drive (VSD), and has a control scheme with both return and supply fans controlled by a single supply duct sensor. As an example of how the unit controls volume and static pressure out of the fans let's say that a number of terminals in the system close down to minimum cfm. The dampers or air valves in the terminal boxes are closing on a signal from their respective space thermostats indicating that the temperature in the space is below setpoint. As the boxes close to reduce airflow into the space, static pressure begins to build up in the main supply duct. When the static pressure sensor in the main duct senses that the static pressure is above setpoint it sends a signal to the VFD. The VFD varies the frequency of the motor, which in turn varies the motor speed; in this case, to slow the motor down and slow the fan down. Slowing the fan reduces the air volume and static pressure out to the system.

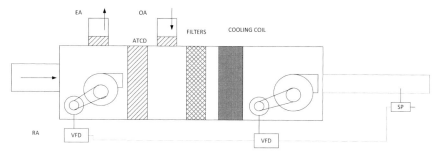

Figure 9-5. VAV Volume and Static Pressure Control: Variable Frequency Drive (VFD) aka Variable Speed Drive (VSD).

Sometime later a number of terminals in the system open to maximum cfm. The dampers or air valves in the terminal boxes are opening on a signal from their respective space thermostats indicating that the temperature in the space is above setpoint. As the boxes open to increase airflow into the space static pressure begins to fall in the main supply duct. When the static pressure sensor in the main duct senses that the static pressure is below setpoint it sends a signal to the VFD to increase the motor and fan speed. Speeding up the fan increases the air volume and static pressure out to the system.

VAV Fan Powered Terminal Boxes

Fan powered boxes change the basic VAV system from variable volume, constant temperature, to constant volume, variable temperature and then variable volume and variable temperature. See Chapter 7 for additional information on fan powered VAV boxes.

System 1: A single duct pressure independent series fan powered VAV box. The primary fan to the AHU is variable volume and the secondary fan in the box is constant volume (Figure 9-6). The primary fan is designed to overcome the pressure losses in the main and branch duct system and into the terminal box. So it is designed and set up to have the static pressure to operate any given box in the system and supply maximum cfm to said box. The secondary fan is energized along with the primary fan, and operates continuously. Here is an example of how this box operates. This box operates at 1000 cfm. On a call for cooling the thermostat sends a signal to open the VAV damper or air valve. 1000 cfm (at 55F) from the primary fan comes into the box. The constant volume secondary fan is set up to move 1000 cfm (maximum airflow). It picks up the air after the VAV damper and blows it into the conditioned space. If the space over-cools the thermostat sends a signal to close the damper. In this example, the damper closes to allow only 500 cfm of 55F into the box. However, the secondary fan is set up to move 1000 cfm. At this condition the dampers in the RA inlet open and the secondary fan pulls 500 cfm from the return air ceiling plenum. The temperature of the air in the ceiling plenum is 75F. Therefore, the secondary fan will blow 1000 cfm of air at 65F into the conditioned space. The calculation is: 500 cfm @ 55F + 500 cfm @ 75F = 1000 cfm @ 65F. This fan powered box maintains a constant volume into the space at a varying temperature. This box also has a hot water reheat coil. If the box closes to minimum and the thermostat is still calling for heat, the reheat coil's two-way valve will open and hot water will transfer heat into the coil. Now, 1000 cfm will be blown across the heating coil picking up additional heat and flowing into the space.

System 2, Figure 9-6: A single duct pressure independent parallel fan powered VAV box. The primary fan for this is designed to overcome the pressure losses in the main and branch duct system, into and through the terminal box, and through the low pressure duct from the box and through the outlet just as a basic VAV SA fan does.

Each type of box contains a constant air volume fan and a return air opening from the ceiling space. When the room thermostat is calling for cooling, the box operates as would the standard cooling VAV box. However, on a call for heat the fan draws warm (secondary) air from the ceiling plenum. Unlike the secondary fan in the series box, this secondary fan is intermittent, that is, it only comes on when there is a call for heat or low volume of primary air. Varying amounts of cool primary air from the main system are introduced into the box and mix with the secondary air. A system of dampers, backdraft or motorized, controls the airflow and mixing of the air streams. As the room thermostat continues to call for heat the primary air damper closes down towards minimum and more secondary air is drawn into the box. The volume and temperature sequence is similar to the series fan powered scenario above. If more heat is needed reheat coils may be installed in the boxes or discharge ductwork.

So the boxes have changed the variable volume, constant tem-

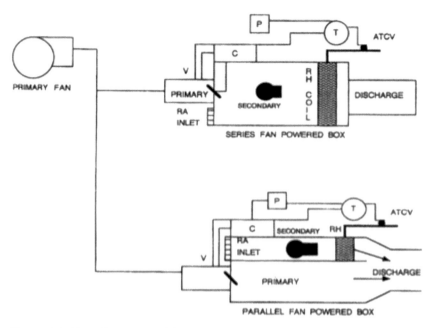

Figure 9-6. Top: System 1 with single duct pressure independent series fan powered VAV boxes. Bottom: System 2 with single duct pressure independent parallel fan powered VAV boxes.

perature primary system to constant volume, variable temperature secondary system into the conditioned space. If the secondary fan is VAV then the secondary system becomes variable volume, variable temperature.

MAXIMIZING VAV PERFORMANCE

How do you maximize the performance of VAV systems? Well, that probably depends on what stage of the system you're working on. If your system is already installed and it's not working properly you'll probably want to look at controls, system maintenance, and an understanding of the operation of the system by the O&M staff. The next step might be to do a Verification of System Performance (my term when I did test and balance and energy management consulting, now most likely termed commissioning and specifically existing building commissioning, (re-commissioning or retro commissioning) to find out what the system is doing or check the air balance of the system. While you're doing the VOSP or having the system air balanced the next step may be to look at the installation. Finally, is the system properly designed? You may have a commissioning report which should provide information on installation and design.

Let's start with the first step: the understanding and maintenance of the system. The operations and maintenance staff will need to have training on the systems and their components. In addition, as with any other HVAC system, regularly scheduled inspections and maintenance should be conducted. A leak test may need to be done. You may have heard that the three most important things in real estate are location, location, location? In VAV systems, and other variable flow systems, its controls, controls, controls.

Diffuser Dumps Cold Air: Airflow too low (velocity too slow). Check to determine if box is reducing too far. Evaluate box minimum setting. Diffuser is too large, check installation.

Conditioned Space is Too Warm: Supply air temperature setting is too warm. Not enough supply air. Refrigeration system is not operating properly. Fan-coil evaporator is iced over because of low airflow. Temperature sensor is located or set incorrectly or needs calibration. Low pressure duct is leaking. Low pressure duct not insulated. Cold air from diffuser isn't mixing properly with room air. Increase air volume or velocity change into space or retrofit diffuser as needed.

Noise: Too much air in low pressure duct; check box maximum setting. Static pressure in the system is too high. Diffuser is too small. Diffuser is dampered at face (always damper at takeoff). Pattern devices loose, tighten or remove.

Not Enough Air: Fan rotating backward, or fan wheel installed backward, cutoff plates not installed or installed incorrectly, fan speed not correct, system effect from poor design or installation of duct into or out of fan, fan improperly installed in AHU. Box not operating properly, check minimum setting, reset as necessary. Not enough static pressure at box inlet for proper operation. Damper or valve in VAV box is closed, may be loose on shaft or frozen. Low pressure duct damper closed. Restrictions in low pressure duct. Remove pattern devices in diffusers. Low pressure duct is leaking. Flex duct is disconnected or twisted. Install fan powered boxes.

Box Not Operating Properly: There's not enough static pressure at the box inlet. There's too much static pressure at the box inlet. Box pressure sensor is defective, clogged, or located incorrectly. Pressure or voltage setting at controller is incorrect. Controller needs calibrating. Flow coefficient is incorrect. Fan speed is not correct. Inlet duct leaking or disconnected. Box is leaking. Main ductwork improperly designed. Not enough straight ductwork into the inlet of the box. Diversity is incorrect. Box is wrong size or wrong nameplate. Damper is loose on shaft. Air valve is not functioning properly. Linkage from actuator to damper is incorrect or binding. Actuator is defective. Controls are defective, need calibration or are set incorrectly. The volume controller is not set properly. Controls are incorrectly set for normally open or normally closed operation. Damper linked incorrectly, NO for NC operation or vice versa. The controls are wired incorrectly. The computer board is defective. If you have a pneumatic system: tubing to controller is piped incorrectly, leaking, or pinched, restrictor in tubing is missing, broken, wrong size, placed incorrectly or clogged. There's oil or water in pneumatic lines. There is no pneumatic or electric power to the controls.

Fan Not Operating Properly: Inlet vanes on centrifugal fans are not operating properly. Pitch on vane axial fans not adjusted correctly. VFD is not operating or not set correctly. Fan speed is not correct, check drives and rpm. Fan is rotating backwards. Return air fan not tracking with supply fan. Fan allowing air recirculation, check cutoff plate. Parallel fans not getting required cfm at inlet or backdraft damper not correct. Wheel installed backwards. Wheel needs dynamic balancing or is dirty, clogged. Check static pressure sensors, move, clean or calibrate. Check airflow measuring stations, move, clean or calibrate.

Negative Pressure in the Building:

Check for stack effect. Check for improper return air control. Seal building properly. Balance the return system. Install manual balancing dampers needed to control OA, RA and EA at the unit. Get return fan to track with supply fan. Consider replacing return fan with relief (exhaust air) fan. Check that static pressure sensors are properly located and working. Install pressure controlled return air dampers in return air shafts. Supply fan is reducing air volume. Not enough outside air for the constant volume exhaust fans and exfiltration. Increase minimum outside air by opening manual volume damper. Increase outside air duct size. Control OA from supply fan (as fan slows, outside air damper opens). Control OA damper from flow monitor in OA duct. Maintain a constant minimum OA volume.

VARIABLE VOLUME REFRIGERANT SYSTEM

A conventional constant volume DX refrigerant system regulates refrigerant flow either on or off (that is, 100% flow or no flow) or is staged so that some stages (compressors) are full flow while others are 0% flow or everything is full on or off.

A variable volume refrigerant system is basically a traditional vapor-compression DX system with electronic metering devices (thermal expansion valves, aka TXV) and electronic inverters to vary motor and compressor speed and therefore directly vary refrigerant flow in response to load variations as rooms require more or less cooling or heating. In some systems multiple scroll compressors of varying capacity are used to vary refrigerant flow. Varying flow makes the variable volume refrigerant system more energy efficient during part-load conditions than on/off or staged systems.

There are several names for variable volume refrigerant systems. I'll use the generic term variable refrigerant flow (VRF) which refers to the ability of the system to control the amount of refrigerant flowing between indoor units (evaporators) and outdoor units (condensers, Figure 9-7). In other words, in a VRF system varying amounts of refrigerant is piped to an indoor unit to heat or cool the conditioned space and then back to a condensing unit. VRF systems are either two- or three-pipes.

A single outdoor condensing unit can be connected to multiple indoor units of varying capacity and configurations. Several outdoor units can handle 100 or more indoor units. Indoor units may be exposed or concealed and include: ceiling mount, wall-mount, floor mount, or heat pump units. See Chapters 5 for operation of vapor-compression systems and Chapter 11 for conventional heat pumps.

The basic VRF system provides single zone comfort control while more sophisticated systems provide simultaneous heating and cooling in different zones. In these systems heat extracted from zones requiring cooling (a heat recovery unit, VRF-HR) is put to use in the zones requiring heating. That is, the coil in a "currently heating" indoor unit is functioning as a condenser and providing cooled liquid to a "currently cooling" indoor unit which is functioning as an evaporator. The heat recovery system has a greater initial cost but it should provide better zone temperature control and operating efficiencies.

Figure 9-7. Outdoor Units. Typical air cooled condensers used for split or VRF systems.

VRF Benefits

Occupants in a given conditioned space set thermostat controls and the VRF system automatically adjusts the refrigerant flow to the temperature requirements to suit the comfort of those occupants while providing energy savings. Another benefit of the VRF system is that ductwork is only required for the ventilating system and it will be smaller and less than the ducting in standard HVAC cooling and heating systems. This will reduce sheet metal construction, installation, insulation, test and balance, and leak-sealing costs. Also eliminated is the water piping required for cooling and heating coils in a conventional air and water FCU or AHU, again reducing construction costs, etc. Additionally, from an architectural and financial prospective in systems with smaller ductwork, no water pipes, and relatively small refrigerant pipes there is the probability of having more saleable or usable conditioned space.

VARIABLE WATER VOLUME SYSTEM

A variable water volume system is used to achieve either full- or part-load heating or cooling conditions while reducing the energy consumed by the pump. Variable water volume systems typically use two-way automatic control valves with either a constant speed or a variable speed system pump, Figure 9-8. Both types of systems reduce flow to maintain a constant water temperature difference across the terminal. The equation is Q = gpm x 500 x delta T. As the heat load (Q in Btu/hr) in the conditioned space changes, the flow (gpm) changes to maintain constant temperature (delta T) across the terminal.

Variable flow systems that use a constant speed pump and two-way valves do not reduce the energy consumed by the pump nearly as much as variable speed systems. In a typical constant speed system, a temperature control device in the conditioned space sends a signal to a terminal's two-way valve to reduce or increase flow. If the valve closes to reduce flow there is an increase in resistance in the system. This increased resistance "backs the pump up on its curve," reducing water flow through the pump and decreasing the horsepower and the energy required by the pump. For example, a pump that uses 54.5 brake horse-

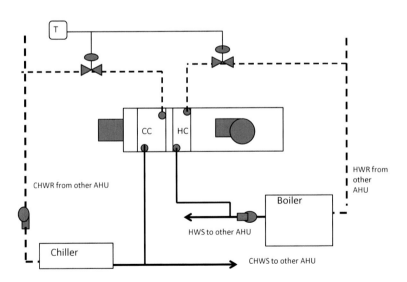

Figure 9-8. A four-pipe system with variable flow 2-way valves controlled from room thermostats.

power when circulating 1,250 gallons per minute of water at 138 feet of head will require only 49 bhp when pumping 1,000 gpm. The flow and the horsepower are reduced because the two-way valves in the system partially close to increase the resistance to 146 feet of head.

Systems that use variable speed pumping to control the water flow rate use an electronic device called a "variable frequency drive" to vary the speed of the pump motor and the pump. This type of system will also reduce pumping horsepower but to a greater extent than just using two-way valves. In the example above, if the system had variable speed pumping, the horsepower would be reduced to approximately 28 bhp when the flow is 1,000 gpm.

A differential pressure (DP) device is installed in between the supply and return piping to control the speed of the variable speed pump, Figure 9-9. In a typical commercial system, as the temperature controls are satisfied, the automatic control valves for the terminals be-

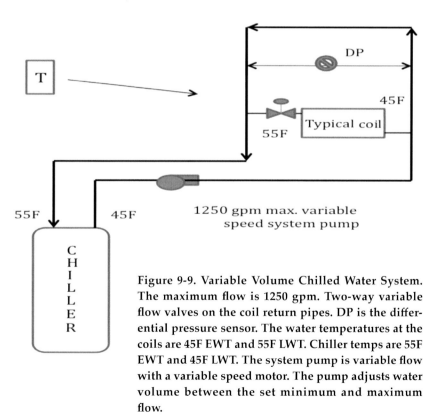

Figure 9-9. Variable Volume Chilled Water System. The maximum flow is 1250 gpm. Two-way variable flow valves on the coil return pipes. DP is the differential pressure sensor. The water temperatures at the coils are 45F EWT and 55F LWT. Chiller temps are 55F EWT and 45F LWT. The system pump is variable flow with a variable speed motor. The pump adjusts water volume between the set minimum and maximum flow.

gin to close. The differential pressure devices sense the increase in system pressure and the pump speed is reduced to meet actual system demand. To set the differential pressure on the DP device, the pump and the system are first set for full flow. The DP is then set to maintain the required differential pressure at that point in the system. For instance, if the device is placed in the ends of the supply and return mains, it is set for the required pressure drop across the last terminal in the system. This pressure drop includes the terminal's associated valves and piping. When the differential pressure device is located at the end of the system, the pump will operate at maximum energy savings. If the DP device is placed closer to the pump, some energy savings will be lost because the pump will have to operate at a higher speed to overcome the pressure losses in the piping and terminals that are past the device. In some systems the DP may have to be installed closer to the pump to maintain control of the system. The DP is installed in the correct location when it will control the variable frequency drive so that the pump operates at the lowest speed to maintain required demand to all terminals.

Chapter 10

Controls

The purpose of HVAC automatic control systems is to start, stop or regulate the flow of air, water or steam and to provide stable operation of the system by maintaining the desired temperature, humidity and pressure. The automatic control system is a group of components, each with a definite function designed to interact with the other components so that the system is self-regulating. HVAC control systems are classified according to the source of power used for the operation of the various components. The classifications and power sources are:

Direct Digital Control (DDC) electronic systems
 Low amperage, 4 to 20 milliamps dc (direct current)
 Low voltage, 0 to 15 volts dc (vdc)
Electrical Controls
 Low voltage, 24 volts ac (alternating current)
 Line voltage, 110 to 220 volts ac (vac)
 Lighting, 120, 177 vac
 Motor, 208, 220-240, 440-480 vac
Pneumatic Controls using compressed air
Interface Controls using electronic, electrical and pneumatic components
Sensors, Controllers, Relays and Switches, Actuators or Motors, and
 Controlled Devices

DIRECT DIGITAL CONTROL SYSTEMS OPERATION

Direct digital control (DDC) is the automated control of a condition or process by a digital computer. A comparison can be made between conventional pneumatic controls and DDC. A typical HVAC pneumatic control system may consist of a pneumatic temperature sensor, pneumatic controller, and a heating or cooling valve. In the pneumatic system the sensor provides a signal to the controller and the controller provides an output to the valve to position it to provide the correct temperature of supply air. A DDC system replaces this local control loop with an electronic temperature sensor and a microprocessor to replace the controller. The output from the microprocessor is converted to a pressure signal to position the same pneumatic heating and cooling valve as in the pneumatic control system. However, the DDC system is not limited to utilizing pneumatic control devices but may also interface with electric or electronic actuators.

Electric or pneumatic devices can be used to provide the control power to the final control elements (the controlled device), but the DDC system provides the signal to that device. In a true DDC system there is no conventional controller. The controller has been replaced by the microprocessor. A common application of DDC includes the control of the heating valve, cooling valve, mixed air damper, outside air damper, return air damper, and economizer cycles to maintain the desired supply air temperature. Other systems commonly controlled by DDC include: chilled water temperature, hot water temperature, and variable air volume and variable water volume capacity.

The DDC system uses a combination of software algorithms (mathematical equations) and hardware components to maintain the controlled variable according to the desires of the system operator. The controlled variables may be temperature, pressure, relative humidity, etc. In the past, the maintenance and operations personnel had to calibrate the local loop controller at the controller's location. Now, with a DDC system, the system's operator may tune the control loop by changing the software variables in the computer using the operator's keyboard. So, instead of calibrating the hardware controller the control sequence and setpoint are input to the computer by a software program and modified by a proper password and the appropriate command keyboard entry.

The DDC system monitors the controlled variable and compares its value to the desired value stored in the computer. If the measured value is less than or greater than the desired value, the system output is modified to provide the correct value. Because the microprocessor is a digital device, there must be some feature in the DDC panel to convert the digital signal to an output signal which the controlled device can use. Pneumatic actuators can be used to position the controlled device. If this is the case, there must be a component or translator incorporated to provide a digital-to-pneumatic conversion. This is done with a digital-to-voltage converter and voltage-to-pressure converter (aka electric-to-pneumatic transducer). It is the development of these transducers and the development of the computer hardware and software that have made DDC systems cost effective. If the measured value is less than or greater than the desired value the computer circuitry outputs a series of digital impulses that are converted to a modulating signal to the actuator by way of a transducer (electrical-to-pneumatic or electrical-to-electrical). The transducer maintains the computer output signal until readjusted by the computer. Other DDC systems may change the control signal by a series of on-off or open-closed signals to bleed air out of, or put air into the actuator. In all cases there is some interfacing signal device required to isolate the computer output circuitry from the control signal circuitry.

The DDC system can utilize many forms of logic to control the output from a given input. The input signal can be modified considerably by various logic statements as desired, thereby providing a great amount of flexibility in establishing the sequence of operation. With a practical understanding of the HVAC system, the system operator is able to fine tune the control system to provide the most efficient operation possible.

An example DDC controller's output signals (in volts direct current) for both direct acting and reverse acting operation are:

Temperature	D/A Output	R/A Output
72F	6vdc	9vdc
75F	7.5 vdc	7.5 vdc
78F	9vdc	6vdc

BENEFITS OF DIRECT DIGITAL CONTROL

The benefits of DDC include simplicity of operation (with one system providing control and energy management), tighter control, greater reliability, greater flexibility, and substantial energy and cost savings.

Direct Control

Because all the setpoints are now programmed within the microprocessor of the DDC system, the owner, energy manager, or system operator has direct control over the environment within the building by dictating the temperature, pressure and humidity setpoints. This prevents the occupants from constantly adjusting the setpoints of a wall thermostat up and down to their individual wishes which causes significant energy waste. An environmental control system can now be provided that is more attuned to the needs of the majority of the occupants and not the individual desires of a select few controlling the room thermostats.

Precise Control

The DDC systems provide the ability to control the setpoint much more accurately than traditional pneumatic systems. One of the inherent flaws of a pneumatic system is that it cannot provide a precise and repeatable setpoint. Pneumatic systems are only modulating control. (Figure 10-1). There is always an offset from the setpoint under minimum and maximum load conditions. The DDC system, because it can be programmed to provide proportional, integral, and derivative (PID) control, can provide absolute control of the setpoint under all load conditions. Therefore, if the setpoint is 72F, it will maintain that setpoint regardless of the load on the HVAC system. This provides considerable energy savings because the controlled variable (temperature, pressure or relative humidity) can be precisely maintained. The digital computer can be programmed to maintain the control point (the actual temperature, pressure or humidity which the controller is sensing) equal to the setpoint using proportional (modulating) control, and adding integral (reset) control. Derivative (rate) control is added for some control sequences (PID). A floating point (moving the controlled device only when the controlled variable reaches an upper or lower limit) may be added as well.

Deadband and Control Sequence

Based on the response time of any particular controlled device a small deadband (above and below the setpoint) can be established to maintain stability. These deadbands, as well as rate of change of the signal to the actuator and minimum length of time between control signal changes, are individually changeable by an authorized operator. The control sequence can be modified by changing the program algorithms, usually without any change in hardware. The ease of making the changes varies with the system design.

Schedule Changes

Direct digital control and energy management systems provide easy changing of schedules and therefore can reduce the energy waste caused by being on the wrong time and HVAC operation schedule. Day-

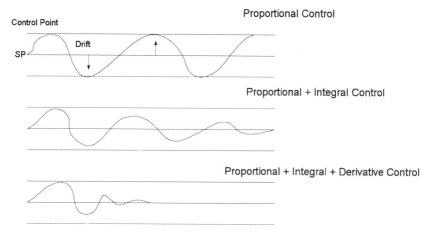

Figure 10-1. Proportional (P, modulating), Proportional and Integral (PI, modulating and reset. It has less drift each cycle.), Proportional, integral and Derivative (PID, It controls modulation, reset and rate, i.e., less drift each cycle and faster response to get to setpoint.) SP: Setpoint, Drift or offset: The distance from setpoint to control point.

night schedules, monthly schedules, seasonal schedules, winter-summer schedules, yearly schedules, holiday schedules, etc., can all be quickly changed with simple keyboard entries. For example, in a given facility to change the time clocks from Standard Time to Daylight Saving Time or vice versa might previously have literally required days or longer, whereas with DDC a knowledgeable operator can change the schedule in a few minutes.

Flexibility

A direct digital control system provides greater flexibility in determining how the control loop is to function. The owner-operator-manager has access to software programs which change settings as desired. The system operator can now optimize the control system and provide the most economical operation under all conditions. This is especially important in continually changing conditions within the building or conditioned space such as: number of people, schedule changes, work load changes, and environment changes (lighting, computer, and other heat-generating equipment).

ENERGY MANAGEMENT CONTROL SYSTEMS

Most DDC systems, in addition to providing local loop control, provide energy management functions that are usually associated with supervisory type energy management systems. Historically, energy management systems (EMS) also now known as Building Automated Systems (BAS), or Building Management Systems (BMS) and various other names, were installed separate from, and in addition to, the local loop controls to provide these functions. This resulted in a local pneumatic control system which was interfaced to a computer system to provide the energy management functions. Now a DDC system can provide both these functions in one system. The energy management functions provided by these systems include cooling demand control, hot and cold deck reset, chilled water reset, deadband, duty cycling, optimized start! stop, etc.

Cooling Demand Control

The DDC system can automatically reduce the fan speed and/or increase the cooling temperature to unload the refrigeration compressor(s). This provides a percentage of load reduction as opposed to the simple on/off function.

Hot and Cold Deck Reset

Because the local loop control is being set by the DDC system, the hot/cold deck temperatures can be controlled directly, which allows the hot and cold valves to be positioned independently of each other. The heating valve can be commanded, for example, to be closed during the cooling season so there is no overlap of the heating and cooling functions.

Chilled Water Reset

By directly controlling the capacity of the chiller, the water temperature can be set at any value desired. This can be a function of outside air temperature, building load, or a combination of both. It assures the most efficient operation of the chiller no matter what the load or outside temperature may be.

Deadband

By direct control of the setpoints of the various systems, a deadband can be programmed into the control algorithm to provide a separation of the heating and cooling setpoints, i.e., the heating setpoint may be set at 70F and the cooling set for 75F. Between these two temperatures, the system keeps the heating valve and cooling valve closed.

CONTROL COMPONENTS

The typical components of an electric or electronic system are: control wiring paths, sensors, controllers, relays and switches, actuators and controlled devices.

Sensor

A sensing element, either internal or remote of the controller, measures the controlled variable (temperature, humidity and pressure) and sends a signal back to the controller.

A temperature sensor senses a change in temperature. There are two general types of electronic temperature sensing elements, thermistor and thermocouple. A thermistor or resistance temperature detector (RTD) senses a change in temperature with a change in electrical resistance. A thermocouple senses a change in temperature with a change in voltage.

A humidity sensor (hygrometer) is used to measure relative humidity or dew point. An electronic hygrometer senses changes in humidity from either changes in capacitance or resistance in the electronic circuitry.

An electronic pressure sensor senses changes in pressures using mechanical devices and then converts this signal to produce current or voltage.

Control Wiring

The control wiring conveys the electricity to the various controllers. For electrical control systems the voltage is 24 volts ac (low voltage), or 110/220 volts ac (line voltage). Electronic control systems use a voltage of 0 to 15 volts dc, or a current of 4 to 20 milliamps dc.

Controller

A controller is a device designed to control a controlled device such as an air damper or a water valve to maintain temperature (thermostat), humidity (humidistat), or pressure (pressurestat). A controller (Figure

10-2) may be direct acting (D/A) or reverse acting (R/A). A direct acting controller increases its outgoing/output signal (electric) or branch line pressure (pneumatic) as the condition it is sensing increases. It decreases its signal as the condition it is sensing decreases. A reverse acting controller decreases its signal as the condition it is sensing increases. It decreases its signal as the condition it is sensing increases.

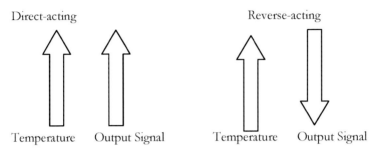

Figure 10-2. Direct-acting and Reverse-acting Controllers. Temperature is the controlled variable.

Some other important terms you'll need to know are throttling range and setpoint. Throttling range (TR) is the change in the controlled condition necessary for the controller output to change over a certain range. For example, if a DDC controller (a thermostat in this example, either direct acting or reverse acting) has a 6 degree throttling range it means the electrical signal output will vary from 6 to 9 volts direct current (vdc) over a 6 degree change in temperature. Another system is pneumatic. One of its controllers (a thermostat, either direct acting or reverse acting) has a 4 degree throttling range. In this example the thermostat's branch line output will vary from 3 to 15 psig over a 4 degree change in temperature.

Setpoint (SP) is the point at which the controller is set and the degree of temperature, or percent relative humidity, or pressure to be maintained.

A thermostat (T-stat) may be either direct or reverse acting. The standard thermostat has a temperature range of 55F to 85F. Generally, the throttling range is between 2 and 12 degrees. As an example, a direct acting thermostat with a setpoint of 72F and a 4 degree throttling range (70F to 74F). A direct acting thermostat means that a rise in space temperature causes a rise in the output of the thermostat. Therefore, when the room temperature is at or below 70F the thermostat signal output is at its minimum. As the space temperature goes above 70F the output signal will continue to increase until the setpoint temperature is reached or the controller output reaches maximum. Another example would be a room that has a reverse acting thermostat with a 6 degree throttling range and a setpoint of 72F. The control sequence is: at 75F, the signal output would be minimum (a rise in space temperature is a decrease in output) and at 69F the signal output would be maximum.

A humidistat measures humidity in the air and may be either direct acting or reverse acting. For example, reverse acting humidistat controls a normally closed two-way steam valve. As the relative humidity drops in a condition space the output signal to the valve is increased, opening the valve and allowing steam to enter the humidifier.

A pressurestat senses pressure and sends a signal to a controlled device. A static pressure sensor (SPS) in a VAV system is an example. As the static pressure increases in the duct where the SPS is located it senses the increase in pressure and sends a signal to the VFD to slow the fan speed. Pressurestats are also used to make or break an electrical, electronic or pneumatic control signal.

Electric controllers can control flow using the following methods: proportional (modulating), two-position, two-position timed, or floating control. They may be single-pole, double-throw (SPDT), or single-pole, single-throw (SPST). Proportional control uses a reversible motor with a feedback potentiometer. Two-position control is used simply to start or stop a device or control a spring-return motor. Two-position timed control uses SPDT circuits to actuate unidirectional motors. Floating control uses a SPDT circuit with a reversible motor.

An electronic direct digital controller (DDC) uses digital computers to receive electronic signals from sensors and converts the signals to numbers. The digital computer (microprocessor or microcomputer) compares the numbers to design conditions. Based on this comparison, the controller then sends out an electronic or pneumatic signal to the actuator.

Direct Digital Control differs from pneumatic or electric- electronic control in that the controller's algorithm (sequence of operation) is stored as a set of program instructions in a software memory bank. The controller itself calculates the proper control signal digitally, rather than using an analog circuit or mechanical change. Interface hardware allows the digital computer to receive input signals from sensors. The computer then takes the input data, and in conjunction with the stored algorithms calculates the changes required. It then sends output signals to relays or actuators to position the controlled devices.

Direct digital controllers are classified as preprogrammed or operator-programmable control. Preprogrammed control restricts the number of parameters, setpoints and limits that can be changed by the operator. Operator-programmable control allows the algorithms to be changed by the operator. Either hand-held or console type terminals allow the operator to communicate with and, where applicable, change the controller's programming.

Controlled Device and Actuator

A controlled device is the final component in the control system. It may be a damper for air control or a valve for water or steam control. Attached to the damper or valve is an actuator. The actuator receives the signal from the controller and positions the controlled device. Actuators are also known as motors or operators.

Some important terms to understanding the workings of controlled devices are: normally open, normally closed and actuator spring range. The terms normally closed (NC) and normally open (NO) refer to the position of a controlled device when the power source, compressed air in a pneumatic system, or electricity in an electrical, electronic or DDC system, is removed. A controlled device that moves toward the closed position as the output pressure from the controller decreases is normally closed. A controlled device that moves toward the open position as the output pressure decreases is normally open. The spring range of an actuator restricts the movement of the controlled device within set limits. Automatic dampers used in HVAC systems may be either single blade or multiple blade. Multiple blade dampers are either parallel blade or opposed blade. Parallel blade dampers are diverting and opposed blade dampers are non-diverting. See Figures 7-5, 7-6 and 7-7 in Chapter 7. Parallel blade dampers are typically used in the return air duct and in the outside air duct where the two ducts run parallel to each other into the mixed air plenum. The dampers are placed in the ducts so that when they open the return air and the outside air is directed into each other in a mixing application. Opposed blade dampers are used for mixing and volume control applications. The dampers may be installed either normally open or normally closed.

Damper actuators position dampers according to the signal from the controller. A pneumatic systems compressed air pressure may operate the actuator in either a two-position or proportioning manner. Inside the actuator the control air pressure expands the diaphragm around the piston and forces the piston outward against the spring, driving the pushrod out. As air pressure is increased the pushrod is forced to the maximum of the spring range. As air is removed from the actuator, the spring's tension drives the piston towards its normal position. For example, a damper with a proportioning actuator has a spring range of 3 to 7 psig. The actuator is in its normal position when the air pressure is 3 psig or less. Between 3 and 7 psig the stroke of the pushrod is proportional to the air pressure, for instance, 5 psig would mean that the pushrod is half-way extended. At or above 7 psig the maximum stroke is achieved.

Dampers can be sequenced by selecting actuators with different spring ranges. For example, two normally closed dampers operating from the same controller control the air to a conditioned space. The top damper operates from 3 to 7 psig. The bottom damper operates between 8 and 13 psig. Both dampers are closed at 3 psig. At 7 psig the top damper is full open and the bottom damper is closed. At 8 psig the top damper is full open and the bottom damper is starting to open. At 13 psig both dampers are full open. Damper actuators may be directly or remotely connected to the damper. The damper position, normally open or normally closed, is determined by the way the damper is connected to the actuator. In other words, if the damper closes when the actuator is at minimum stroke, the damper is normally closed. If, on the other hand, the damper opens when the actuator is at minimum stroke, the damper will be normally open.

Control valves are classified according to (1) their flow characteristics, such as quick opening, linear or equal percentage; (2) their control action, such as normally open or normally closed; and (3) the design of the valve body such as two-way, three-way, single seated or double seated. Flow characteristic refers to the relationship between the length of the valve stem travel expressed as a percent and the flow through the valve expressed as percent of full flow. For example, the quick opening valve has a flat plug, which gives maximum flow as soon as the stem starts up. This type of valve might deliver 90% of the flow when it's open only 10%. Therefore, a typical application for a quick opening valve might be on a stream preheat coil where it is important to have a lot of fluid flow as quickly as possible. By comparison, in a linear valve the percent of stem travel and percent of flow are proportional. For instance, if the stem travel is 50% the flow is 50%. In the equal percentage valve each equal increment of stem travel increases the flow by an equal percentage. For example, for each 10% of stem travel for a particular equal percentage valve the flow is increased by 50%. In other words, at 30% open the flow is 8%, when the stem travels to 40% open the flow is 12% (8% + 4%) and at 50% travel the flow is 18% (12% + 6%), etc. At 90% stem travel the flow is 91.125%.

There are two general categories of electric actuators. One is the solenoid type of actuator. It consists of a magnetic coil operating a moveable plunger. It is limited to the operation of smaller controlled devices. Most solenoid actuators are two-position. The other category of actuators is known as a motor. Motors are further classified as unidirectional, spring-return or reversible.

PNEUMATIC CONTROL SYSTEM

The power source for a pneumatic control system is compressed air. The typical components of a single or dual pressure pneumatic system are: Air compressor, Receiver Tank and Drain, Filters, Refrigerated Air Dryer, Pressure Relief Valve, Pressure Reducing Valve, and Control Piping (tubing). The following pneumatic components: sensors, controllers, relays and switches, actuators or motors, and controlled devices are similar or identical, except for the control power, to their electric/electronic counterparts. In an attempt to not be too redundant I have written about these components in either the electrical section or the following pneumatic section. As needed, for this "fundamentals" book, I have tried to make any differences clear.

Main Air

The source of the compressed air in a pneumatic system is an electrically driven, reciprocating, positive displacement air compressor generally sized at 25 horsepower or less. Air compressors are normally sized so that they do not operate more than one-third of the time. This extends compressor life and allows sufficient time to cool the air in the receiver tank. The receiver tank receives and stores the compressed air from the compressor for use throughout the system. In order for the pneumatic components to function correctly the compressed air used to operate the system must be kept clean, dry and oil-free. Therefore, a number of devices are installed in the system to dry the air and remove oil, vapors, dirt and other contaminants.

The first device is an air filter (top left, Figure 10-3) installed in the compressor's air intake to keep dirt and oil vapors from entering and being passed through the compressor and condensing into droplets in the air lines. As the air goes through the compressor its pressure is increased, generally to 60 to 100 pounds per square inch. Heat is also added to the air during the compression phase. As the air cools, moisture in the air is released. Therefore, an automatic or manual drain is installed in the receiver tank to remove any accumulated water, oil, dirt or scale which has settled to the bottom of the tank. A filter is installed in the supply line to collect any oil vapor or particles of dirt to ensure that the lines are oil and dirt free. To remove any moisture which may have been carried over, a refrigerated air dryer equipped with an automatic drain is placed downstream of the receiver tank. A manual bypass is installed around the refrigerated air dryer so it can be serviced without interrupting the system operation.

A pressure switch is installed on the compressor tank assembly to start and stop the compressor at predetermined setpoints. For example, the switch may be set to start the compressor when the pressure in the receiver tank falls to 60 psig and stop the compressor when the pressure in the tank reaches 100 psig. A pressure reducing valve downstream of the refrigerated air dryer and filter maintains the system pressure at 18 to 20 psig. A high pressure gauge is installed in the main supply line before the pressure reducing valve to indicate the pressure of the air stored in the receiver tank. A low pressure gauge is installed in the main supply line after the pressure reducing valve to show the pressure of the main air.

In addition to the devices installed to start and stop the compressor and to keep the system clean there must also be safety devices to protect the equipment. Generally, there are two safety relief valves installed in the system. A high pressure relief valve is installed on the receiver tank and a second relief valve is installed in the supply line downstream of the pressure reducing valve. The pressure relief valve at the receiver tank protects the tank from excessive pressures while the relief valve downstream of the pressure reducing valve protects the system if the pressure reducing valve fails. The relief valve in the supply line is normally set for 30 psig since this is the maximum safe operating pressure for most pneumatic devices.

The air lines coming from the compressor receiver tank assembly

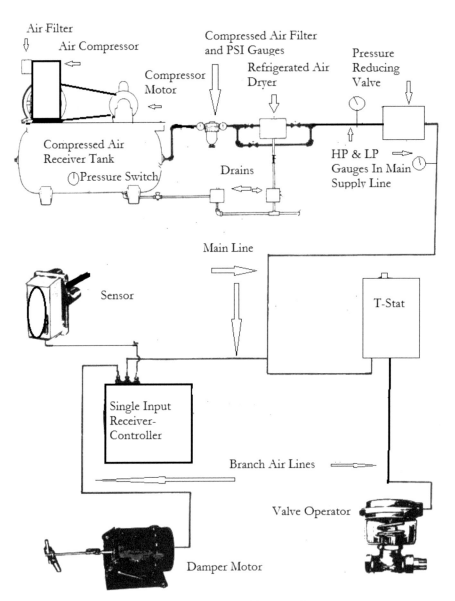

Figure 10-3. Pneumatic Control System Components.

and going to the controlling devices (aka controllers) such as thermostats, humidistats, etc., are called "mains." The air lines leading from controllers to the actuator (aka operator or motor) of controlled devices such as dampers or valves are called "branches." Air lines are generally made of either copper or polyethylene (plastic) tubing.

Types of Pneumatic Systems

Pneumatic systems are divided into single pressure and dual pressure systems. A single pressure system requires only one main air pressure. In a dual pressure system there are two different applications, summer/winter or day/night, which require two different main air pressures. The summer/winter system provides for the seasonal requirements of either cooling or heating. In other words, depending on the season, either chilled water or hot water is supplied to the water coil in the air handling unit. The day/night system allows for setting and controlling space temperature at different setpoints for the day and the night.

The configuration of a dual pressure system is the same as the single pressure system described before up to the pressure reducing valve. Since two different pressures are required, two pressure reducing valves are needed. One pressure reducing valve reduces pressures to about 13 to 16 psig and the other one is set for 18 to 25 psig. The higher-pressure reducing valve supplies air to the controlling device only when the device is on the winter or night setting. The lower pressure is supplied to the controller for summer or day operation. Downstream of the pressure reducing valves is a three-way air valve and a two-position manual or automatic switch. The function of this switch is to change the ports on the three-way valve. If the switch is set for summer or day then the normally open (NO) port is open, and the normally closed (NC) port is closed. Air from the lower pressure reducing valve is allowed to flow through the NO port into the common (C) port to the controllers. Air from the higher pressure, pressure reducing valve is blocked. When the switch is set for winter or night operation, the NO port closes and the NC port opens. This allows air from the higher pressure, pressure reducing valve to flow through the NC port into the common (C) port to the controllers. Air from the lower pressure, pressure reducing valve is blocked.

Single Pressure Thermostat Controller

A single-pressure thermostat may be a one-pipe, bleed-type or a two pipe, relay-type controller. The bleed-type thermostat has only one pipe connection. The main air is introduced through a restrictor into the branch line between the thermostat and the controlled device. The two-pipe thermostat has two connections, branch and main, and receives main air directly. The two-pipe thermostat will provide a greater volume of air to the controlled device which produces a faster response to a change in temperature.   Thermostats may be either direct or reverse acting. The standard thermostat has a temperature range of 55F to 85F and a 3 to 15 psig output range. Here's an example of a reverse acting room thermostat with a 6 degree throttling range (69F to 75F) and a setpoint of 72F. The control sequence is: at 75F, the branch output would be 3 psig (a rise in space temperature is a decrease in pressure output), at 72F the output pressure would be 9 psig and at 69F the branch output would be 15 psig.

Dual Pressure Thermostat (Summer/Winter or Day/Night) Controller

A summer/winter control system provides for the seasonal requirements of either cooling or heating and, depending on the season, either chilled water or hot water is supplied to the water coil in the air handing unit. Since the valve controlling the flow of hot water or chilled water remains the same, either normally open or normally closed, but not both, the system must have a thermostat which can be both direct acting and reverse acting. The bimetal strip in the thermostat is changed from direct acting to reverse acting by a change in the main air pressure. For example, a dual pressure thermostat controls a normally open two-way valve. When the system is in the winter condition the higher main air pressure (18-25 psig) is sent to the thermostat. This makes the thermostat direct acting. As the space temperature rises an increased branch pressure is sent to the valve causing it to close, allowing less hot water into the coil. With a reduced supply of hot water the space will begin to cool. When the system is switched to summer conditions the lower main air pressure (13-16 psig) is sent to the thermostat, which changes it over to reverse acting. Now on a rise in space temperature, a decreasing pressure is sent to the valve causing it to open, supplying chilled water to the coil and the space will begin to cool.

The day/night dual pressure thermostat lets you set and control space temperature at different points for the day and night or for varying load conditions. A day/night thermostat is essentially the same as a summer/winter thermostat except that the day/night thermostat has two bimetal strips and both are either direct acting or reverse acting. The two bimetal strips have separate setpoints. When the higher main air pressure (18 to 25 psig) is sent to the thermostat, the night bimetal strip is in control. When the lower pressure (13 to 16 psig) is sent the day bimetal is in control. For example, a direct acting day/night thermostat controls a two-way, normally open heating valve. During the day the lower main air pressure is sent to the thermostat. The thermostat's day setpoint is 71F. Any temperature above 71F will send an increasing branch pressure to the valve causing it to close. At 71F or below the thermostat will send a decreasing branch pressure to the valve causing it to open allowing more hot water into the coil. At night the main pressure is switched sending the higher main air pressure to the thermostat. The thermostat will now modulate the branch pressure based on the night bimetal strip which is set for 60F. At 60F or below the thermostat will send a decreasing branch pressure to the valve causing it to open allowing more hot water into the coil. Any temperature above 60F will send an increasing branch pressure to the valve causing it to close.

Deadband Thermostat Controller

A deadband thermostat is a two-pipe controller that operates in the same manner as a single pressure, single temperature thermostat. It's used for energy conservation when a temperature span or "deadband" is required between the heating and cooling setpoints. The deadband pressure is the output pressure at which neither heating nor cooling takes place. This type of thermostat uses two bimetal strips. One bimetal strip for heating and one for cooling. The heating bimetal modulates the output pressure between zero and the deadband pressure. The cooling bimetal modulates the output pressure between the deadband pressure and branch air pressure. For example, a deadband thermostat allows for heating below 70F and cooling above 76F. The deadband pressure is 8 psig and the temperature span is 6 degrees. Therefore, when the space temperature is 70F or below the branch output will be between 0 and 7 psig and heating will occur. However, when the space temperature is 76F or above the output pressure will be between 9 and 15 psig and cooling will occur. There will be no heating or cooling at 8 psig when the space temperature is between 70F and 76F Deadband thermostats are adjustable within the limits depending on the heating and cooling setpoints selected.

Humidistat Controller

Humidistats are similar in appearance to thermostats; how- ever, instead of using a bimetal strip as the sensor, humidity is sensed by a hygroscopic (water absorbing) material such as human hair, nylon, silk, wood or leather. Nylon is generally used. Humidistats may be direct acting or reverse acting. For example, a reverse acting humidistat controls a normally closed two-way steam valve. As the room's relative humidity drops the pressure to the valve is increased, opening the valve and allowing steam to enter the humidifier.

Master/Submaster Controller

A master controller is one which transmits its output signal to another controller. The second controller is called the sub-master. The submaster setpoint will change as the signal from the master controller changes. This is a reset type of control. The master controller sends its branch output pressure to the reset port on a second controller—the submaster controller—instead of a controlled device. Both master and submaster are piped with main air. The submaster controller setpoint changes as the signal pressure from the master controller changes. There are two types of reset—direct and reverse. When the change is direct-acting it is called a "direct reset" type of control. When the change is reverse-acting it is called a "reverse reset" control.

Direct Reset Controller Example: The master controller in this example is a space thermostat. The submaster controller is a thermostat with a remote sensing element located in the discharge air duct. The branch output from the submaster is piped to a two-way, normally open, heating water valve. As the space thermostat (master controller) senses an increase in room temperature an increased pressure is sent to the submaster controller to reset its setpoint lower. The submaster then senses discharge air temperature. If the discharge air temperature is above the submaster setpoint, the submaster controller sends a signal

Space Temp (F)	Master Control Output (psig)	Submaster Set Point (F)	Discharge Air Temp (F)	Heating Water Valve Position
70	3	100	100	Open
73	9	80	80	Modulating
76	15	60	60	Closed

to the heating water valve to close.

Reverse Reset Controller Example: Another application of a master/submaster controller is to reset heating water (submaster controller) from outside air temperature (master controller). This is a reverse reset. In this example (Figure 10-4), the sensor for the outside air master controller is set for 70F. The submaster controller operates the boiler heating water valve. In this example the setpoint on the sub-master is also 70F water temperature but of course both settings could be any appropriate design temperature. As the outside air temperature falls, the submaster setpoint is reset upwards and the heating water temperature is increased. The relationship between the master and submaster in this example is 1:1.5. That is, for or every degree the out-side air temperature drops the water temperature is reset upwards by 1.5 degrees.

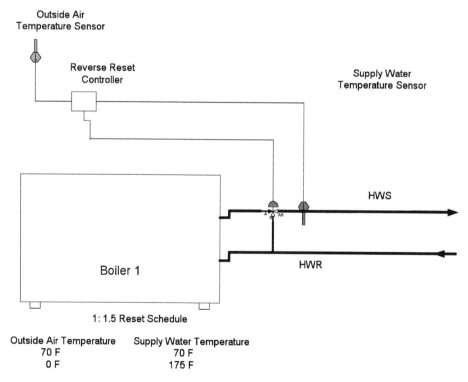

Figure 10-4. Reverse Reset Controller for Boiler Supply Temperature.

Outside Air Temperature (F)	Master Control Output (psig)	Submaster Set point (F)	Heating Water Temp (F)	Heating Water Valve Position
70	3	70	70	Closed
35	9	122.5	122.5	Modulating
0	15	175	175	Open

Receiver-Controller and Transmitter (Sensor)

A receiver-control consists of two main parts—the controller and the transmitter (sensor). The controller part of the receiver-controller, like the other controllers, receives a signal from a sensing element and then varies and "transmits" its branch output pressure to a controlled device or another receiver-controller. A single input receiver-controller functions the same as the controllers described before. Dual input pressure receiver-controllers are used in master/submaster reset applications. Typical transmitters and information below:

Transmitter Type	Transmitter Range	Transmitter Span	Pressure Output Span (psi)
Temperature	0 to 100F	100 degrees	12
Temperature	-25 to 125F	150 degrees	12
Humidity	30% to 80% rh	50% rh	12
Pressure	0 to 7 in. w. g.	7 in. w.g.	12

Relays and Switches

The number of applications for pneumatic (and electric/electronic) relays and switches is virtually unlimited. The following is a list and basic applications of ten common relays and switches.

Air Motion Relay: This relay senses air movement across a fan or coil to verify air flow. It is used as a safety device.

Amplifying or Retarding Relay: Changes the output start point. It is also know as a bias start relay or ratio relay. A typical application is when a retarding relay is installed to eliminate the simultaneous heating and cooling that occurs when a heating valve and cooling valve are operating from the same controller and have an overlapping spring range. Example: A normally open heating valve has a spring range of 3 to 7 psi. The normally closed cooling valve's spring range is 7 to 11 psi. A retarding relay is installed between the valves. The sequence is for the output pressure from the thermostat to go to the heating valve actuator and then to the relay. The relay then sends a signal to the cooling valve actuator. If the thermostat output pressure is 7 psi, the heating valve actuator senses 7 psi, as does the input to the relay. If this relay is set for a 2 psi retard bias, the input to the relay is 7 psi, but the output of the relay to the cooling valve actuator is 5 psi. Therefore, the cooling valve would not start to open until the output from the thermostat was 9 psi (the cooling valve gets 7 psi).

Averaging Relay: Resets a controller or operates a controlled device from the average signal of two or more controllers. For example, two direct-acting thermostats each send a signal to an averaging relay. One signal is 4 psi and the other is 8 psi. The output signal from the relay to the controlled device is 6 psi (4 + 8 divided by 2).

Electric-Pneumatic (E-P) Relay or Switch: Electrically operates a solenoid three-way air diverting valve. It is used to control a pneumatically operated device from an electric circuit. Example 1: A typical application is outside air dampers interlocked with the operation of the fan. The sequence is when the fan is turned on the E-P relay, which is wired to the fan, is energized. This allows control air piped at the normally closed (NC) port to connect to the common (C) port and go on to the damper actuator, opening the dampers. When the fan is turned off, an internal plunger blocks the NC port and connects the C port to the normally open (NO) port. This allows air in the actuator to bleed off through the NO port, closing the dampers. Example 2: A high limit diverting relay in the outside air is set at 72F. This allows the mixed air controller to control the outside air (OA) and return air (RA) dampers up to 72F. The output from the mixed air controller is piped into the normally open (NO) port of the diverting relay. As long as the outside air temperature is below 72F this signal is passed along to the dampers through the common (C) port to the damper actuators. The outside dampers are open and the return air dampers are closed. When the OA temperature reaches 72F the diverting relay switches and blocks the NO port and connects the C port to the normally closed (NC) exhaust port to allow the air pressure to be exhausted from the damper actuators. This closes the outside air dampers and opens the return air dampers.

Diverting Relay: Switches air signals. A diverting relay is a three-way air valve used primarily to convert a signal, at a predetermined setpoint, into a signal for a controlled device. A typical application for a diverting relay is to use it as either a high limit or low limit control in an economizer application (Figure 10-5). Example: A high limit diverting relay in the outside air is set at 72F. This allows the mixed air controller to control the outside air (OA) and return air (RA) dampers up to 72F. The output from the mixed air controller is piped into the normally open (NO) port of the diverting relay. As long as the outside air temperature is below 72F this signal is passed along to the dampers through the common (C) port to the damper actuators. The outside dampers are open and the return air dampers are closed. When the OA temperature reaches 72F, the diverting relay switches and blocks the NO port and connects the C port to the normally closed (NC) exhaust port to allow the air pressure to be exhausted from the damper actuators. This closes the outside air dampers and opens the return air dampers. Note—these OA dampers are the economizer dampers. They may or may not be the minimum OA dampers. There may be a separate set of dampers for minimum OA. If these dampers are both economizer and minimum OA then the dampers do not completely close when the AHU fan is on. They are always partially open for minimum OA. See minimum position switch below.

Gradual Switch: A manually operated device which selects a branch air pressure (0 to 20 psig) to be delivered to a controller or controlled device. A typical application is for the gradual switch to receive a pneumatic signal from a transmitter. The output pressure from the gradual switch can then be manually increased or decreased to raise or lower the setpoint on a receiver-controller.

Minimum Position Switch: A minimum position switch (MPS) is a gradual switch with a built-in high pressure selector relay. A typical application is to use a minimum position switch to position the outside air dampers on an economizer system. The MPS is piped with main air. When the supply fan is energized, an electric signal is sent to the E-P relay. The E-P relay is energized and allows control air to pass to the mixed air controller. The mixed air controller sends a branch signal to the minimum position switch, which sends the signal on to the outside air

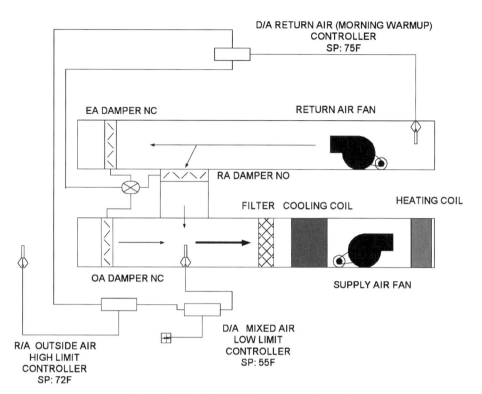

Figure 10-5. Air-Side Economizer Control

dampers. The MPS pressure has been manually set (in this example set-point is 5 psig) to maintain the outside air dampers open to the required minimum position whenever the fan is operating. When the output pressure from the mixed air controller is less than the MPS setpoint the switch will provide minimum pressure (5 psig) to the outside air damper actuator. This allows for the minimum volume of outside air required by local ventilation codes. When the fan is stopped control air to the mixed air controller is blocked by the E-P and branch air is exhausted from the controller, MPS and outside air damper actuator.

Pneumatic-Electric (P-E) Relay or Switch: Start or stop fans, pumps or other electrically driven equipment. It is an air actuated device to make or break electrical contacts and can be wired either normally open or normally closed. Note—when using electrical terms, normally open means the circuit is de-energized and normally closed means that the circuit is energized.

Reversing Relay: Reverses a signal from a controller. Example: A direct-acting space thermostat is controlling a normally open heating valve and a normally open cooling valve. A reversing relay is installed between the heating valve and the cooling valve. The branch pressure from the controller is piped into the heating valve and then to the reversing relay. The signal from the reversing relay is piped to the cooling valve. On an increase in pressure from the thermostat the heating valve receives this increase while the cooling valve gets a decreasing pressure. This sequence closes the heating valve and opens the cooling valve.

Selector Relay: Compares, selects and transmits control signals. The relay may be either a low select, high select or a high-low select. The relay receives two or more signals, compares them and selects and transmits either the lowest signal, the highest signal or both the lowest and the highest signals. It is also called a discriminating relay. Example: A high-low select on a multizone air handling unit receives the input from 7 direct-acting zone thermostats. The highest pressure (12 psig) is sent to the cooling valve to allow only enough chilled water into the coil to cool the zone (zone 6) with the greatest requirement. The lowest pressure (4 psig) is sent to the heating valve to allow only enough hot water into the coil to heat the zone (zone 4) with the greatest requirement.

Zone Number	1	2	3	4	5	6	7
Pressure (psig)	5	7	9	4	11	12	8

Actuator

Control air pressure from a controller is used to position an actuator. The actuator, sometimes called an operator or motor, consists of a cylindrical housing, air connection port, rubber diaphragm, piston, spring and a connector rod to the controlled device. The operation of the actuator is such that as the control air pressure is introduced into the actuator the diaphragm begins to expand. As the diaphragm expands it forces the piston outward against the spring driving the connector rod out. As air pressure is increased the connector rod is forced to the maximum of the spring range. When air is removed from the actuator the spring tension returns the piston to its normal position. Example: An actuator has a spring range of 3 to 7 psig. The actuator connector rod is in its normal position when the air pressure is 3 psig or less. Between 3 and 7 psig the stroke of the connector rod is proportional to the air pressure (5 psig would mean that the rod is halfway extended). Above 7 psig the maximum stroke is achieved.

Controlled Device

The automatic controlled devices used in HVAC systems are dampers (or air valves) for airflow and temperature control and water (or steam) valves for temperature and water or steam flow control. At this point I'll use the term damper for both damper and air valve.

Damper: The damper position is determined by the way the damper is connected to the actuator, either normally open or normally closed. If the damper opens when the actuator is at minimum stroke, the damper is normally open. If, on the other hand, the damper closes when the actuator is at minimum stroke, the damper will be normally closed. Damper actuators may be directly or remotely connected to the damper. Some systems will use multiple blade automatic face and bypass dampers for temperature control of coils. Most large air conditioning systems have multiple blade automatic temperature control dampers (see Chapter 7) to regulate the mixture of outside air and return air. The operation of these dampers is controlled by temperature requirements of the system, not by airflow requirements. These multiple blade dampers are either opposed blade or parallel blade. The terms "opposed" and "parallel" refer to the movement of the adjacent blades. The opposed blade damper has a linkage which cause the adjacent blades to rotate in opposite directions resulting in a series of slots that become increasingly narrow as the damper closes. This type of blade action results in a straight, relatively uniform airflow pattern sometimes called "non-diverting." The parallel blade damper is linked so all the blades rotate parallel to each other so it has a "diverting" pattern, because when closing, the damper blades have a tendency to divert the air sideways. This type of pattern is beneficial when properly used to mix incoming outside air and return air. However, a diverted flow pattern may adversely affect coil or fan performance if the damper is located too closely upstream. All temperature control dampers should be tight shut-off.

Water Valve: Automatic temperature control water valves are used to control flow rate or to mix or divert water streams. Valves are classified according to body design, control action and flow characteristics.

Body Design: Valves are constructed for either two-way or three-way operation.

Control Action: The control action (or valve position) is either normally open or normally closed and is determined by the way the valve is connected to the actuator. If the valve closes when the actuator is at minimum stroke the valve is normally closed. If, on the other hand, the valve opens when the actuator is at minimum stroke the valve will be normally open. Generally, normally open valves with low operating ranges are used in heating applications. Normally closed valves with higher operating ranges are used in cooling applications. This will allow for sequencing valves without simultaneous heating and cooling and for the system to fail safe to heating (see Figure 10-6).

Flow Characteristics: A valve's flow characteristic refers to the relationship between the percent of plug lift to the percent of flow. Control valves have three basic types of seat plug configurations: equal percentage, linear and quick opening. The equal percentage has each equal increment of plug lift and the flow increases by an equal percentage. This will provides a better relationship between lift and capacity and is used in heating or hot water systems when a small amount of flow results in a large heating capacity. A linear valve is used in chilled water systems

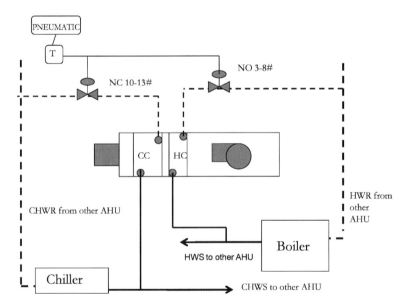

Figure 10-6. Four-Pipe Heating and Cooling System. Direct acting thermostat. Deadband is between 8 and 10 psig.

because the percent of plug lift and percent of flow are proportional. For instance, if the lift is 30%, the flow is approximately 30%. The quick opening valve has a flat plug which delivers nearly maximum flow at about 20% lift. A typical application for a quick opening valve is on a stream or water preheat coil where it is important to have maximum fluid flow as quick as possible.

EXAMPLE CONTROL SUBSYSTEMS

Air-side Economizer (Figure 10-5)

Let's start at the return air controller (aka morning warm up controller). When the AHU and controls are energized the return air sensor, in this example setpoint (SP) at 75F, will sense the return air temperature. The RA, EA, and OA dampers are in their normal position, that is, return air full open, exhaust air closed and outside air full closed or at minimum position (depending upon if the dampers are only economizer or they are minimum OA and economizer).

When the return air temperature goes above setpoint the direct acting (D/A) controller will send the control signal to the damper motors. However, before the signal goes to the damper motors it must get the "ok" from the MA and OA controller.

The D/A mixed air controller, in this example, has a 55F setpoint. Its sensor is in the mixed air plenum sensing the temperature of the mixture of the return air and outside air. When the temperature in the mixed air plenum goes above setpoint the signal from the RA controller will be passed on to the OA controller. In other words the MA controller has indicated that the MA temperature is above setpoint so the objective of this control subsystem is to open the outside air dampers so that more outside air can be brought in to be used to cool the conditioned space without energizing the refrigeration system fully or partially. This is sometimes called "free cooling."

The increasing signal will go from the mixed air controller to the damper motors to close the RA dampers and open the OA dampers (and EA dampers as appropriate to maintain the proper pressure in the controlled space). But wait! There's more.

Next is the outside air controller which is set for 72F outside air temperature. The OA controller is reverse acting (R/A). This allows the mixed air controller to modulate the outside air (OA) and return air (RA) dampers up to 72F. As long as the outside air temperature is below 72F the mixed air controller signal is passed along to the damper motors. Remember, R/A means an increase in temperature is a decrease in output signal pressure (Figure 10-1). The outside dampers are now modulating towards full open and the return air dampers are closing.

When the OA temperature reaches 72F the OA controller will close the outside air dampers to minimum (or full closed, see first paragraph above) and open fully the return air dampers. As appropriate, as the outside air dampers are closing the mechanical refrigeration system will be energized. In this example the OA controller is a dry bulb controller sensing sensible temperature. In geographic areas with high humidities a wet bulb controller sensing latent temperatures may be more appropriate.

Four-pipe Heating and Cooling System

Figure 10-6 shows an air handling unit one of several in this pneumatically controlled four-pipe (two supply pipes and two return pipes) heating and cooling system. The heating and cooling coils have a two-way variable volume control valve in the return pipe. The conditioned space thermostat is direct-acting (D/A). The heating valve is normally open (NO) operating from three to eight pounds (psig). The cooling valve is normally closed (NC) operating from 10 pounds to 13 pounds (psig).

As the conditioned space thermostat (T) senses a temperature above setpoint it sends an increasing signal to the valves. At 3 pounds the heating valve starts to close and continues to close is on the thermostat senses a temperature above setpoint. At 8 psig the heating valve is full closed. If the thermostat is still sensing a room temperature above setpoint the pneumatic air pressure continues to rise. There is a 2 psig deadband between the heating valve and the cooling valve when no hot water is going to the heating coil and no chilled water is going to the cooling coil. As room temperature continues to rise the cooling valve receives a pressure of 10 pounds and starts to open. It will continue to open if the thermostat is not satisfied until it reaches 13 pounds. At this time the cooling valve is full open.

When the space starts to cool below thermometer setpoint the pneumatic air pressure begins to drop from 13 pounds towards 10 pounds. At 10 pounds the chilled water valve is full closed. There is no chilled water or hot water going to their respective valves between 10 and 8 pounds. If the space is still too cold the pressure continues to drop and at 8 pounds the heating valve starts to open and is full open at 3 pounds and receiving full flow of hot water.

Single Zone Heating and Cooling Four Pipe System

This is a single zone system. This means that there is one thermostat in the conditioned space. If this space is changed by putting in ceiling-to-floor walls or partitions it will make it a multizone system with single zone control. That is, it's not going to work because whomever's in the "room" with the thermostat is going to control the temperature of everybody else's "new" room. That's going to be a problem, in all probability, because each of the "new" rooms is likely to have a different temperature requirement than the controlled "room."

The heating and cooling coils in this AHU (Figure 10-7) are controlled by three-way valves mixing valves in a bypass application. The valve ports are labeled "A," "B," and "AB." The "A" port is water coming from the coil. The "B" port is the water coming through the bypass. The "AB" port is water going to the boiler or chiller. It is the common port and might be labeled "C." Most systems fail (aka fail-safe) to the heating mode. Meaning that if control power, either pneumatic or electric, is lost the valves are going to go to their normal position so that we have hot water going to the heating coils and chilled water bypassing the cooling coils.

In this example the conditioned space thermostat is direct-acting (D/A). So for the cooling coil "A" is NC and "B" is NO. For the heating coil "A" is NO and "B" is NC. Authors note: Sorry about that the changes with "A" and "B" if you have a problem with how the piping is drawn.

As the conditioned space thermostat (T) senses a temperature above setpoint it sends an increasing signal to the valves. The "A" port on the heating valve starts to close and the "B" valve starts to open. So some hot water is going to the heating coil and some is bypassing the heating coil. Let's say 80% is going to the coil and 20% is bypassing. As the space temperature continues to rise the control pressure continues to increase and the normally closed port continues to open, while the normally open port continues to close. At some point the normally open "A" port is full closed and the normally closed "B" port is full open. So no new hot water is going to the coil. All the hot water is being bypassed back to the boiler.

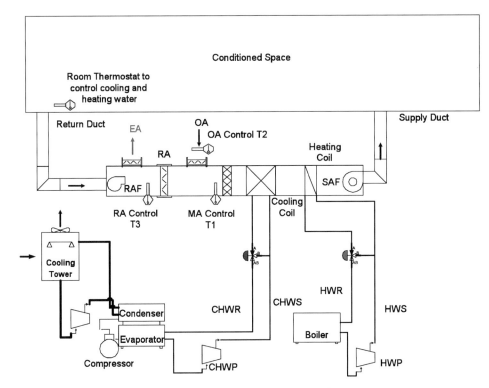

Figure 10-7. Single Zone Heating and Cooling System.

With the heating valve closed to new water and its bypass full open the chilled water valve normally closed port will start to open to new chilled water as the space temperature continues to increase. Let's say 20% chilled water going to the coil and 80% bypassing. As the temperature in the space continues to rise and control pressure continues to rise the flow will be 50-50 (coil and bypass) and at some point there will be 100% chilled water going to the coil and zero bypassing.

This drawing also has the sensors for the full economizer control at T1 for mixed air control, T2 outside air control and T3 return air control.

Reheat System

Figure 10-8 is a multizone system with central cooling and conditioned space reheat. There is a thermostat for each room which makes it multizone. This central cooling coil has a three-way valve and a sensor in the supply discharge duct. The "A" port is water coming from the coil.

The "B" port is the water coming through the bypass. The "AB" port is water going to the chiller. The "A" port is NC and "B" port is NO. The discharge duct temperature controller is direct-acting (D/A). With a rise in air temperature in the discharge duct an increasing signal will be sent to the chilled water valve to close the bypass "B" port and open the "A" port sending new chilled water through the coil.

The cool air will be discharged into each room (or zone). Let's say the rooms are numbered 1, 2 and 3 left to right. Rooms 1 and 2 need only cooled air. Room 3, however, is overcooled. The direct-acting thermostat senses the drop in temperature in that room and sends a decreasing signal to the three-way valve in the heating pipe to open the NO "A" port and close the NC "B" port. The system would fail to the heating mode. Hot or warm water flows through the reheat coil heating the supply air into the space. These systems can be made variable volume by a designing or retrofitting to two-way valves.

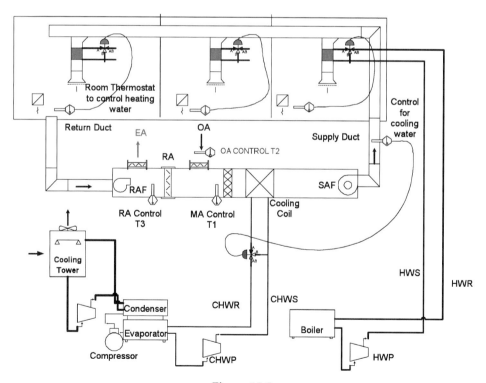

Figure 10-8.
Multizone System with Central Cooling and Conditioned Space Reheat.

Multizone Mixing Damper System

A mixing damper multizone system is shown in Figures 10-9 and 10-10. Figure 10-9 in the air handling unit. On the left is a fan. Next is a hot water heating coil at the top and chilled water coil on the bottom. The chilled water coil is bigger than the hot water coil. The hot water coil is controlled by a two-way valve in the return line. The temperature sensor in the hot deck. The chilled water coil is controlled by a three-way valve with the sensor in the cold deck. Next, are the hot and cold deck dampers. The coils the decks in the dampers are separate from each other. The hot deck dampers in the cold deck dampers move opposite of each other. In other words, when the hot deck dampers close the cold deck dampers open to maintain a constant flow of air for this constant air volume system. The controller for these dampers is a thermostat for each zone (room) thermostat. It is common to have a multizone system with between 5 and 12 zones. The ductwork is attached to the unit at the mixing dampers. We are looking at zone 1 as it comes out of the unit and elbows down. There would be a set of manual volume dampers after the

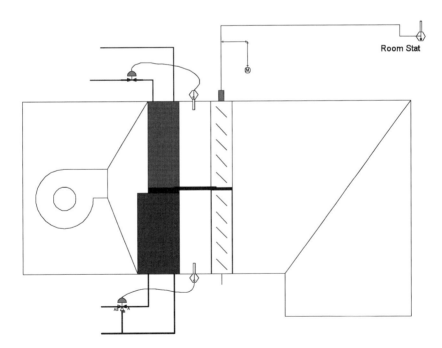

Figure 10-9. Mixing Damper Multizone System.

automatic temperature control mixing dampers to balance total airflow for each zone.

Figure 10-10 is a continuation of the ductwork as it goes to the various zones. This system has five zones. Typically the zones are fairly equal in size and cfm. One zone may span one hot and cold deck damper blade, another zone may span two damper blades, and still another zone might be three damper blades wide, etc.

Depending upon what the zone thermostat is calling for one zone could be cooling only, another zone could be heating only, and a third zone could be a mixture of heating and cooling air.

This system could also be made variable air volume. It could also be retrofitted or designed with dampers and controls to reduce or optimize simultaneous heating and cooling.

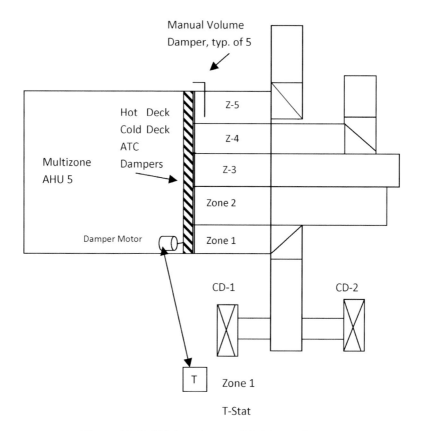

Figure 10-10. Mixing Damper Multizone System.

Chapter 11

System Selection and Optimization

SELECTION

The purpose of an HVAC (heating, ventilating, and air conditioning) system is to provide and maintain environmental conditions within an area called the "conditioned space." The type of system selected is determined by the mechanical designer's knowledge of systems and the building owner's financial and functional goals. The commercial system selected for a particular application endeavors to provide the optimum environment for employee comfort and productivity, process function, and good indoor air quality with energy efficiency and cost savings. Different systems will satisfy each of these objectives with different degrees of success. It is up to the designer and the building owner to make the correct assessments. Note: When I say the "designer and the "owner" I'm including their respective representatives.

In most applications, there are several choices for the type of system to use. The selection of the type of HVAC system by the designer and the building owner is a critical decision. It is the designer's responsibility to consider the various systems and select the one that will provide the best combination of initial cost, operating cost, performance, and reliability based on his understanding of the owner's needs and goals. In the selection process all factors must be analyzed, but initial cost and operation and maintenance cost are usually foremost.

Another cost concern that may be overlooked by the designer is the cost associated with equipment failure and equipment replacement cost. For example, how often might a selected system or component be expected to fail and what is the cost in loss of product and production? How long will the system be down? How will the comfort, safety and productivity of the occupants be affected by such a failure and what is the cost?

Depending on the owner's goals each of these concerns has a different priority. Most owners may not have detailed knowledge or understanding of the advantages of the different types of systems. Therefore, it is normally up to the designer to make the equipment selection. On the other hand, the designer may not have a complete understanding of the owner's financial and functional goals. For these reasons the best situation is when both the designer and owner are involved in the HVAC selection process.

The first step in the selection process is for the designer to ascertain and document from the owner the desired environmental conditions for the building or conditioned space. The designer must also learn and document the restrictions placed on the system design. For example, what is the required equipment space for a particular system versus what is available? Unfortunately, it is the nature of the business that very few projects allow as much detailed evaluation of all conditions and alternatives as some would like. Therefore, the designer must also rely on common sense and subjective experience to narrow the choice of systems.

Step two in the selection process is determining the building's heating and cooling loads. For example is the cooling load mostly sensible or latent? Is the load relatively high or low per square foot of conditioned area as compared to other similar buildings? Is the load uniformly distributed throughout the conditioned space? Is it relatively constant or does it vary greatly? How does the load vary with time and operating conditions? Determining the heating and cooling loads establishes the system's capacity requirements. Cooling loads and humidity requirements are used to size air conditioning (comfort and process cooling) system. In other systems, heating or ventilation may be the critical factors in sizing and selection. For example, a building may require a large air handling unit and duct system to provide huge quantities of outside air for ventilation or as make-up air to replace air exhausted from the building. In other buildings, in colder climates for instance, heating may be the determining factor on equipment size. The physical size of the equipment can be estimated from the heating and cooling load information alone. This information can help in the choice of systems to those that will fit the space available.

There are also choices to be made depending on whether the system is to be installed in a new building or an existing building. In existing buildings, for example, the HVAC system was designed for the loads when the building was built. This means if new systems are to be integrated with existing ones (in order to keep costs down or for other reasons) the new or retrofitted systems must be adaptable to existing equipment, ductwork and piping, and new equipment or systems must fit into existing spaces. If new systems are to perform properly when tied in with existing systems the old and the new must be looked at carefully and in its entirety. The designer will need to determine how a change to one part of a system will affect another part and how a change in one system will affect another system. The number of choices is narrowed further to those systems that will work well on projects of a given application and size and are compatible with the building architecture.

SELECTION GUIDELINES

Each of the following issues should be taken into consideration each time an HVAC system is selected.

Financial Factors
> Initial cost
> Operating cost
> Maintenance and repair cost
> Equipment replacement or upgrading cost Equipment failure cost
> Return on investment (ROI)
> Energy cost

Building Conditions
> New or existing building or space Location
> Orientation
> Architecture
> Climate and shading
> Configuration
> Construction
> Codes and standards

Building Use
> Occupancy
> Process equipment

Energy
> Types
> Availability
> Reliability

Control Scheme
> Zone control
> Individual control

TYPES OF HVAC SYSTEMS

The basic types of HVAC systems used in commercial buildings are built-up, unitary and split. A built-up heating, ventilating and air conditioning system is one that is custom designed for the building in which it is installed. It typically consists of a central plant with equipment and components that generate and distribute heated or cooled air or water energy to the conditioned spaces within the building. A unitary system is an air conditioning unit that provides all or part of the air conditioning functions (heating, cooling, fan, filter, compressor, condenser, controls) in one or a few assemblies, in essence in one package. A split system has an indoor section and matching outdoor section connected by refrigerant tubing. The indoor section typically consists of a fan, an indoor cooling coil, a heating function, and filters while the outdoor section houses the compressor and condenser. The unitary system is all-in-one and the split system is split into two sections, an outdoor section and an indoor section. Systems are further classified as air and water, all-air, or all-water (see Chapter 5). Note: Water systems are also called hydronic systems.

All-air systems provide heated or cooled air to the conditioned space through a ductwork system. The basic types of all-air duct systems are: single-zone, multizone, dual or double duct, terminal reheat, constant air volume, variable air volume, and combination systems. In the typical system, cooling and heating is accomplished by the mixed air (a combination of the return and outside air) passing over a refrigerant coil (cooling) or a heat exchanger (heating).

The basic air-water system (aka air-hydronic) is a central system similar to the all-air system with chilled water coils instead of refrigerant coils for cooling (with an air-cooled condenser) and hot water coils for heating. A variation of this system is the water-air (hydronic-air) system with refrigerant coils for cooling and a water-cooled condenser.

All-water (all-hydronic) systems accomplish space cooling by circulating chilled water from a central refrigeration system through cooling coils in air handling units (aka terminal units or fan-coil units). The units are located in the building's conditioned spaces. Heating is accomplished by circulating hot water through the same (cooling/heating) coil or through a separate heating coil. When one coil is used for cooling only, heating only, or heating and cooling at various times a two-pipe water distribution system is used. When two coils are used, one for heating and one for cooling, a four-pipe water distribution system is used.

Heating may also be accomplished using electricity or steam. Straight water heating systems will commonly use convectors, baseboard radiation, fin-tube radiation, standard fan-coil units, and unit heaters.

UNITARY SYSTEM

A unitary system is an air-conditioning unit that provides all or part of the air-conditioning functions. The components: fans, filters, controls, and the cooling apparatus (refrigerant coil, refrigerant piping, compressor, and condenser) are factory assembled into an integrated package. Unitary systems are used in a wide range of applications. Cooling capacity can range from fractional tonnage for window-type units to 100 tons of refrigeration or more for package units.

Components are matched and assembled at the factory to achieve specific performance objectives in accordance with industry-established increments of capacity (such as cfm of air per ton of refrigeration). These performance objectives are set by trade associations that have developed standards by which manufacturers may test and rate their equipment. These performance parameters and standards allow for the manufacture of quality-controlled, factory-tested systems. Types of unitary systems include package units, rooftop units, window-mounted air conditioners and heat pumps, through-the-wall air conditioners and heat pumps, and packaged terminal air conditioners and heat pumps.

Package Unit

Commercial grade unitary systems are known as package units. Some package units also have heating apparatus (e.g., natural gas heat exchanger, electric elements, steam or hot water coils) and humidifiers, Figure 11-1. A package unit designed to be placed on the roof is known as a rooftop unit, Figure 11-2 and 11-3. Package units are used in almost all types of building applications, especially in applications where performance requirements are less demanding and relatively low initial cost and simplified installation are important. Applications include hotels, manufacturing plants, medical facilities, motels, multi-occupancy dwellings, nursing homes, office buildings, schools, and shopping centers. However, package units are also used in applications where dedicated high performance levels are required such as computer rooms, laboratories and cleanrooms.

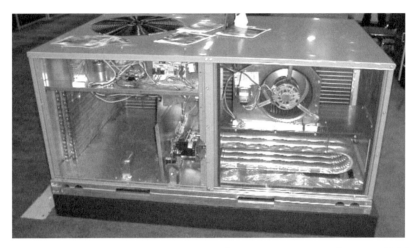

Figure 11-1. Cut-away of package unit. Left compartment: condenser coil (aka outdoor coil), condenser fan (aka blower) and compressor. Right compartment: Indoor fan, heating coil and cooling coil.

Figure 11-2. Package roof-top units. Unit in foreground, left to right: electrical disconnect box, indoor fan compartment, outside air inlet to natural gas heater and gas pipe, outdoor fan compartment and air-cooled condenser coil. Far right unit has condenser coil on left corner and outside air inlet into indoor fan on the right.

Figure 11-3. Close-up of RTU with left to right: indoor fan, natural gas pipe into heater compartment, condenser coil.

Window Unit

Window-mounted air conditioners and heat pumps cool or heat individual conditioned spaces. They have a low initial cost and are quick and easy to install. They are also used to supplement a central heating or cooling system or to condition selected spaces when the central system shuts down. When used with a central system the units usually serve only part of the spaces conditioned by the central system. In such applications, both the central system and the window units are sized to cool the particular conditioned space adequately without the other operating. In other applications where window units are added to supplement an inadequate existing system they are selected and sized to meet the required capacity when both systems operate. Window units require outside air and cannot be used for interior rooms. Window units are factory-assembled with individual controls. However, when several units are used in a single space, the controls may be interlocked to prevent simultaneous heating and cooling. For energy management in hotels, motels or other hospitality applications a central on/off control system may be used to de-energize units in unoccupied rooms or use of occupancy sensors or sensors on a sliding glass door or window which opens to the outside. Another factor to consider when selecting window unit systems is that window units are built to appliance standards rather than building equipment standards so they may have a relatively short life and high energy use.

Wall Unit

Through-the-wall air conditioners incorporate a complete self-contained air-cooled, direct expansion (DX) cooling system, controls, and fan in an individual package. Package terminal air conditioners (PTAC) add a heating system, typically natural gas, electric or hot water. Package terminal heat pumps (PTHP) and heat pumps (HP) heat by switching the evaporator and condensing coils. These unitary systems are designed to cool or heat individual spaces. Each space is an individual occupant-controlled zone into which cooled or heated air is discharged in response to thermostatic control to meet space requirements. Units range from appliance grade to heavy-duty commercial grade and are installed in homes, apartments, assisted-living facilities, hospitals, hotels, motels, office buildings, and schools.

Heat Pump

A unitary system that uses the refrigeration system as the primary heating source is a heat pump. Figure 11-4, shows a heat pump in the cooling mode (top) and heating mode (bottom). When a heat pump is in the cooling mode it is operating as would a traditional vapor compression air conditioning unit. The components listed are: TXV (thermal expansion valve, liquid refrigerant metering device), SB (sensing bulb to control the opening of the TXV), SG (sight glass shows if vapor or water is in liquid line), RV (reversing valve, changes direction of refrigerant flow), LLD (liquid line dryer, removes moisture from liquid refrigerant), SLFD (suction line filter dryer, removes moisture and particles from vapor), CV (check valve, limits flow in one direction), OD (outdoor coil and fan), ID (indoor coil and fan), Comp (compressor), Accumulator (suction line accumulator, removes liquid refrigerant from the suction line).

The right side of heat pump is facing into the conditioned space. The large arrows show the direction of airflow from the indoor fan (aka indoor blower) through the indoor coil into the conditioned space. The indoor coil is functioning as the evaporator. The direction of refrigerant flow is shown with the small arrows. Starting at the TXV (on the top righthand side) liq-

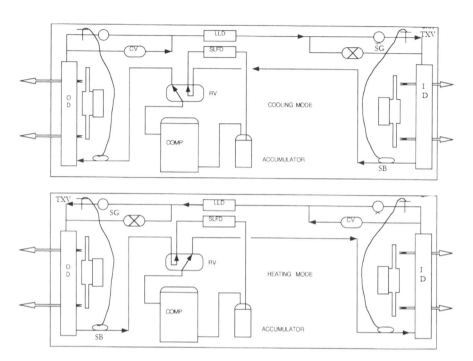

Figure 11-4. Air to Air Heat Pump. Components: TXV (thermal expansion valve), SB (sensing bulb for TXV), SG (sight glass), RV (reversing valve), LLD (liquid line dryer), SLFD (suction line filter dryer), CV (check valve), OD (outdoor coil and fan), ID (indoor coil and fan), Comp (compressor). Please note reversing valve and the check valves for the indoor and outdoor coils and how they change when the cooling mode changes to the heating mode and vice versa.

uid refrigerant flows into the evaporator coil where it is boiled. The boiling action removes heat from the conditioned space return air. The chilled supply air is blown into the conditioned space. The refrigerant leaves the evaporator as a gas and goes through the reversing valve, up through the suction line filter dryer, down through the accumulator and into the compressor where it's compressed into a high temperature, high pressure gas. Leaving the compressor the high temperature, high pressure gas goes through the reversing valve and into the outdoor coil where the relatively cooler air from the outside is being blown across the outdoor coil which is functioning as the condenser. The heat from the hot gas is blown into the outside air. Removing heat from the vapor causes the refrigerant to condense back to a liquid. The liquid refrigerant flows through the check valve (top left side) into the liquid line dryer and through the refrigerant piping. The refrigerant goes through the sight glass and through the TXV and the cooling cycle continues.

Now let's go through the heating cycle. When the thermostat in the conditioned space requires heat the controls switch the direction of refrigerant flow through the reversing valve and in essence makes the evaporator coil into a condenser coil and the condenser coil into an evaporator coil. Starting at the TXV on the top left side of the bottom "heating mode" drawing and following the arrows we see that liquid refrigerant goes through the TXV and into the outdoor coil. The relatively warmer outside air (even in the winter if the temperature is let's say 40F or higher) is blown through the outdoor coil and boils the liquid refrigerant. As you can see the outdoor coil is now functioning as an evaporator. The refrigerant gas leaves the outdoor coil and flows up through the reversing valve, which is now switched over, and flows through the suction line filter dryer down to the accumulator and into the compressor where it's compressed into a high temperature, high pressure gas on up through the reversing valve and over to the indoor coil. Return air from the conditioned space is blown through the indoor fan and through the indoor coil, which is now a condensing coil, where it picks up heat from the high temperature gas. The heated air goes into the conditioned space warming it. As heat is removed from the vapor it condenses into a liquid and goes through the check valve. If you look at the top drawing you'll see that this check valve in the cooling mode was not allowing fluid to go through it, remember, a check valve only allows flow in one direction. The liquid refrigerant flows through the check valve into the liquid line dryer over to the sight glass and into the TXV and the cycle starts over.

Selection Guidelines

Unitary systems are selected when it is decided that a central HVAC system is too large or too expensive for a particular project or a combination system (central and unitary) is needed for certain areas or zones to supplement the central system. For example, unitary systems are frequently used for perimeter spaces in combination with a central all-air system that serves interior building spaces. This combination will usually provide greater temperature and humidity control, air quality, and air (conditioned air and ventilation air) distribution patterns, than is possible with central or unitary units alone. As with any HVAC system both the advantages and the disadvantages of unitary systems should be carefully examined to ensure that the system selected will perform as intended for the particular application.

A solid understanding of the various types of commercial HVAC systems and their selection is important if you are the energy manager or facilities engineer. This position often calls for being the owner's representative, working with others to ensure that the owner gets the environmental system that will best fit his needs. Following are some of the advantages and disadvantages to consider when selecting unitary systems.

Temperature Control and Airflow

Individual room control (on/off and temperature) is simple and inexpensive. However, because temperature control is usually two-position there can be swings in room temperature. Also, the room occupant has limited adjustment on air distribution and airflow quantity which are fixed by design. Ventilation airflow quantity is also fixed by design as are the sizes of the cooling and condenser coils. On the plus side, ventilation air is provided whenever the unit operates.

Humidity Control

Unitary systems can provide heating and cooling capability at all times independent of other spaces in the building but basic systems do not provide close humidity control. However, close humidity control is not needed for most applications. But, if needed, in computer room applications or the like, close humidity control can be accomplished by selecting special purpose packaged units.

Manufacturing

Manufacturer-matched components have certified ratings and performance data and factory assembly provides improved quality control and reliability. There are a number of manufacturers so units are readily available but equipment life may be short (10-15 years) as compared to larger equipment which may have life expectancies of 20-25 years. Manufacturers' instructions and multiple-unit arrangements simplify the installation through repetition of tasks. However, from an architectural or esthetic point of view appearance may be unappealing.

Maintenance and Operation

An advantage of unitary systems is only the one unit and one temperature zone is affected if a unit malfunctions. Also, less mechanical and electrical space is required than with central systems. And, in general, trained operators are not required as might be with more sophisticated central systems. However, maintaining the units is more difficult because of the many pieces of equipment and their location such as on rooftops or in occupied spaces. Also, condensate can be a problem if proper removal or drainage is not provided. Other disadvantages are that air filtration options may be limited and the operating sound levels can be high.

Cost and Energy Efficiency

Initial cost is usually low but operating cost may be higher than for central systems. This will be the case when the unitary equipment efficiency is less than that of the central system components. Also, energy use may be greater because fixed unit size increments require over-sizing for some applications and outdoor air economizers are not always available to provide low cost cooling. However, for leased space applications such as offices, retail spaces, R&D labs, computer rooms, etc., energy use can be metered directly to each tenant. Also, units can be installed to condition just one space at a time as a building is completed, remodeled, or as individual areas are leased and occupied. Another energy management opportunity with unitary systems is that units serving unoccupied spaces can be turned off locally or from a central point without affecting occupied spaces.

SPLIT SYSTEM

The guidelines for selection of split and variable refrigerant flow (VRF) systems are similar to the information provided above for unitary systems. However, two other concerns that need to be addressed are length of refrigeration piping and ventilation air. Length of piping is not the concern it once was as newer systems have steadily increased maximum refrigerant pipe length, in some cases, to many hundreds of feet, however, meeting ANSI/ASHRAE ventilation code requirements is a still a concern as split systems and VRF systems do not inherently provide ventilation so a separate ventilation system may be necessary.

OPTIMIZATION

In the 1950s, air conditioning systems dramatically changed the way we live in the United States. As HVAC systems became more reliable, efficient and controllable, we were no longer dependent on the weather for work or leisure. We made the environment adapt to our needs. In fact, we started cooling to temperatures lower than the temperatures to which we had previously heated. Today, indoor climate control has become so reliable and affordable it is common in industry and homes alike as almost all commercial buildings have HVAC systems and many U.S. households have air conditioning. The goal of HVAC systems continues to be to provide a high degree of occupancy comfort and indoor air quality and to maintain environmental conditions for work processes while holding operating costs to a minimum.

Evaluate System Efficiency

There are many ways to evaluate whether or not the system is efficient. For instance, general appearance, equipment down-time, maintenance records, maintenance costs and occupant complaints are items that need to be taken into consideration. However, one of the best ways to evaluate how the system is performing, if energy usage costs are a major concern, is by comparing energy used over several years. To make this comparison, develop a Building Energy Use Number, Table 11-1.

Table 11-1. Example BEUN. LPG: liquid petroleum gas, kWh: kilowatt hour, mcf: 1000 cubic feet, gal: gallon, mlb: 1000 pounds, *Consult energy supplier for conversion factor.

Commodity	Unit	Units Consumed per year	Btu Conversion	Btu/year
Electricity	kWh	1000	3413 Btu/kWh	3,413,000
Natural Gas	therm	10	100,000 Btu/therm	1,000,000
Natural Gas	mcf		*	
LPG	gal		*	
Fuel Oil	gal		*	
Coal	ton		*	
Steam	mlb		Steam table	
Other				
Total Btu				4,413,000
Building sf				10,000
Btu/sf/year				441.3

To develop the BEUN gather the building's utility records for the past 24 months. Obtain the rates charged by energy suppliers including commodity rates, demand rates, discounts, taxes, on- and off-peak rates, power factor rates, ratchet charges, etc. Determine the total energy used by the building's HVAC systems for one year. This will include electricity, natural gas and oil and any other commodity (such as liquid petroleum gas) used to operate the HVAC system. Convert the energy used to Btu per year. Divide the Btu per year by the square feet of the building's conditioned space. This is the BEUN for the base year in Btu per year per square foot.

Compare the BEUN with other similarly constructed and used buildings in the area or across the country. Information on BEUN for similar buildings can be obtained from Federal, state or local government agencies or the utility company. Additional useful information for evaluating the HVAC system and energy efficiency would include occupancy data (hours the building is occupied, type of work performed and number of people per shift), weather data for the base year and present year, data on how the building is constructed, and HVAC operation and maintenance logs and manuals.

Establish Efficiency Goals

Organize a team consisting of staff, maintenance personnel, consultants, contractors and energy suppliers. Gather all information needed to develop the BEUN and evaluate HVAC systems. Identify type of HVAC systems and their interaction with each other. Determine system performance. Evaluate system maintenance. State system problems and opportunities and set efficiency goals. Determine what resources will be needed to achieve goals, including staffing and money. Assign responsibilities for achieving goals and set a time schedule to reach goals. Monitor results. Review and revise goals as needed.

Energy Systems

It's generally assumed that the HVAC systems and lighting systems account for most of a building's energy use. HVAC energy consumption is affected in part by the common practice of specifying oversized heating and cooling equipment to compensate for the energy inefficiency in a building's design and construction. The following are energy conservation opportunities (ECO) for HVAC systems and subsystems.

HVAC SYSTEM ECO

- Compare field measurements (air, water, steam and electrical) with the air or water balance report, commissioning report, and fan, pump, and motor curves to determine if the correct amount of air and water is flowing.

- Use nameplate data to prepare an up-to-date list of motors for fans, compressors, pumps, etc., and list routine maintenance to be performed on each.

- Routinely check time clocks and other control equipment for proper operation, correct time and day, and proper programming of on-off setpoints.

- Reduce or turn off heating and cooling systems during the last hour of occupancy.

- Close interior blinds and shades to reduce night heat loss in the winter or night and solar heat gain in the summer or day. Repair or replace damaged or missing shading devices.

- Inspect room supply air outlets and return and exhaust air inlets, diffusers, grilles and registers.

- Clean ducts. Open access doors to check for possible obstructions such as loose insulation in lined ducts, loose turning vanes and closed volume or fire dampers. Adjust, repair or replace these items as necessary.

- Reduce outdoor air intake quantity to the minimum allowed under codes by adjusting outdoor air dampers. Maintain a rate of 15-25 cubic feet per minute (cfm) of air per person. Maintain outside air dampers.

- List automatic and gravity dampers and routinely check that they open and close properly. Adjust linkage or replace dampers if the blades do not close tightly.

- Replace unsatisfactory automatic dampers with higher quality opposed blade or parallel blade dampers with seals at edges and ends to reduce air leaks. Readjust position indicators as needed to accurately show the position of all dampers.

- Regularly clean or replace dirty or ineffective filters.

- Clean coils and other heat exchangers.

- Ensure that all fans rotate in the proper direction.

- Check fan, pump, or compressor motor voltage and current.

- Adjust fan speed, inlet guide vanes, or VFD (variable frequency drive) for proper airflow.

- Measure total static pressure across fans and total dynamic head across pumps.

- Maintain correct belt tension on fan-motor drives.

- Check drives for misalignment.

- Discontinue use of unneeded exhaust fans.

- Rewire toilet exhaust fans to only operate when lights are on or there's a signal from an occupancy sensor.

- Check pump suction and discharge pressures and plot differential pressure on the pump curve.

- Close the discharge valve if the pump circulation is more than 10 percent greater than required flow.

- Reduce pump impeller size for greater energy savings.

- Adjust pump speed, impeller, or VFD (variable frequency drive) for proper water flow.

- Properly adjust and balance air and water systems.

- Adjust controls.

- Install a time clock or automated energy management system that will reduce heating and cooling.

- Close some air conditioning supply and return ducts for HVAC systems operating in lobbies, corridors, vestibules, public areas, unoccupied areas or little-used areas. Disconnect electrical or natural gas heating units to these areas.

AIRSIDE ECO

- Design or retrofit system for lowest pressure needed

- Design or retrofit system for lower airflow

- Install balancing dampers

- Air balance system

- Avoid installing restrictive ductwork on inlets and discharges of fans

- Install variable volume systems where applicable

- Reduce equipment on-time

- Use economizers

- Maintain systems including controls

- Verify fans are rotating in correct direction

- Clean fan blades

- Clean filters and coils

- Repair leaks in duct system

- Insulate duct

- Reduce resistance in system

- When reducing airflow change fan speed instead of closing main dampers

Filters

How often filters need changing will depend on type of HVAC used and the cleanliness of the environment in which the system is located. Install differential pressure gauges on either side of filters to provide a pressure drop across the filters. When the pressure drop reaches an established point the filters need to be changed. In order for filters to function properly they must be installed correctly. Most filters will have an arrow showing direction of airflow. Inspect filter frame. It should be tight against all sides of the filter housing so that no air bypasses the filters. Turn off fan when changing filters. The job is easier and the coil is protected from dirt and debris which might be drawn in.

Coils

Coils need to be kept clean. A coil is a heat exchanger. If a coil is dirty or has debris impinged upon it the coil loses some of its heat transferring surface and becomes less efficient. This means not only a loss in efficiency, but also an increase in energy as the coil tries to maintain a normal operating condition. Also check fins on the coil. If the fins are mashed together the surface area of the coil is reduced, which reduces the effectiveness of the coil. If the fins have been pushed in they need to be straightened.

Fans

Momentarily start and stop the motor to "bump" the fan to determine the direction of rotation. If the rotation is incorrect reverse the rotation on a three-phase motor by changing any two of the three power leads at the motor control center or disconnect. On single-phase motors change rotation, as applicable, by switching the internal motor leads within the electrical terminal box.

Fan performance can be reduced if the space between the fan inlet and the fan plenum wall is too close. There should be at least a one half fan wheel diameters between the plenum walls and the fan inlet. There should be at least one wheel diameter between inlets of fans in parallel. Check the installation and operation of parallel fans. If one of the fans has a restricted inlet it may handle less air than the non-restricted fan. This can result in fan pulsations which may reduce fan performance. The fan pulsations can also cause noise and vibration problems. The vibrations, if excessive, can damage the fan or the ductwork. Inspect fan cutoff plate for integrity and proper positioning. Keep fan wheel clean, an accumulation of dirt on the wheel can reduce the performance characteristics of the fan. It can also put the fan wheel in an unbalanced condition, causing noise or vibration problems.

If fans are not operating at the proper speed the conditioned space will not be able to maintain the correct temperatures and air changes per hour. Fan speed must be correct (plus or minus 10% of design or determined rpm) to have the proper heat transfer across heat exchangers. For example, if the fan is turning too slow, the evaporator (refrigerant) coil in a fan-coil unit may ice over. Icing of the coil can occur when the moisture in the air condenses on the coil and there is not enough warm air moving across the coil. The coil temperature drops to freezing and ice begins to form on the coil. Icing can also occur if fan belt breaks or is thrown off or removed. If the fan is turning too fast there could be damage to the fan wheel or the shaft and bearings. Calculate fan blade tip speed and compare with the manufacturer's maximum tip speed chart. Lubricate fan bearings only to manufacturer specifications.

Drive Belts and Sheaves

At least quarterly inspect drive belts. Check for wear and proper tension. Belts should not be too tight or too loose. The correct operating tension is defined as the lowest tension at which the belts will perform without slipping under peak load conditions. The tension can be checked by either using a tension gauge or using light finger pressure (depressing each belt at its middle). A belt tension checker is available from some belt manufacturers. The proper tension, as determined by finger pressure, is when the belt will depress approximately one half to three quarters of an inch. An indication that a belt is too loose is if it squeals when the motor is started. Another indication that a belt is too loose is when belt dust is found on the inside of the belt guard. If a belt is too loose, not only will this mean premature failure of the belt, but the fan being driven will not operate at the proper speed. If a belt is too tight (excessive tension) the belt will wear faster and cause excessive wear on motor and fan shaft bearings and possibly overload the motor.

Inspect drives (sheaves and belts) for integrity, alignment and proper size. Verify drive components are correct. If a different size component is needed calculate the size of the new drive component and install. Replace worn or broken drive components with new components of the same size.

Motors

Maintenance on motors is generally limited to proper lubrication of bearings and protection from moisture, dirt, and debris blocking airflow in and around motors or restricting motor movement.

Motor amperage measured on any phase should not exceed the motor nameplate amperage. However, if operating amperage reading is over nameplate amperage take one of the following steps to correct the problem: If reading is over nameplate amperage but within service factor and voltage limits reduce fan speed or close main discharge damper until amperage reading is down to nameplate or below. If reading is over nameplate amperage and outside service factor limit immediately turn off the fan and inform the person responsible for the fan. An exception is if fan has been running and it is serving a critical area (such as a cleanroom or surgical room). If this is the case leave fan operating and immediately notify the person responsible for the fan.

Fan System Effect (FSE)

Fan system effect is any condition in the fan plenum, inlet, outlet or distribution system that adversely affects the aerodynamic characteristics of the fan and reduces fan performance. Inspect fan, fan cabinet and the ductwork attached to (or around) the fan for any of the following inlet or outlet conditions which may cause system effect.

FSE, Duct, Fan Inlet

The inlet duct or fan inlet should have a smooth, rounded entry to reduce inlet losses. Square edges causes turbulence. A converging taper entry or a flat flange on the end of the duct will also reduce inlet losses. Inlet conditions can be improved by installing splitters or airflow straighteners in the duct.

FSE, Duct, Fan Outlet

The length of straight duct on the fan outlet should be at least one duct diameter for each 1,000 feet per minute of outlet velocity, with a minimum length of two and a half duct diameters. For example, a fan with an outlet velocity of 1,500 fpm would need a straight duct of 2.5 duct diameters while a fan with an outlet velocity of 3,000 fpm would need a straight length of duct of 3 duct diameters.

FSE, Duct Size

The size of the outlet (or inlet duct) should be within plus or minus 10% of the fan outlet (or inlet) area.

FSE, Duct Slope

The slope of an outlet or inlet transition should not be greater than 15% for converging transitions or greater than 7% for diverging transitions.

Fse, Elbow

If an elbow must be installed closer than one duct diameter for each 1,000 fpm of outlet velocity the ratio of the elbow radius to duct diameter should be at least 1.5 to 1. A 90 degree elbow without turning vanes installed at the fan inlet will create uneven airflow distribution into the fan. Losses can be reduced if a straight length of duct and turning vanes are added to this type of elbow.

FSE, Inlet Cone
Inspect the inlet cone for integrity and proper positioning.

Duct Configuration, Friction Loss, And Leakage
From an energy use standpoint round duct is more efficient than other duct configurations because it has less material in contact with the air and therefore has less friction loss. See Table 11-2, which compares a round duct to a square duct and a rectangular duct all with the same area of one square foot.

Materials used in the fabrication of duct also affect friction loss: Galvanized duct has less friction loss than fiberglass-lined galvanized duct or fiberglass duct. Flexible duct aka flex duct, the plastic-wrapped, fiberglass-insulated, wire duct, has the most friction loss of the ducts mentioned.

The aspect ratio of rectangular ducts is the ratio of the adjacent sides and is another consideration. Generally, for energy conservation, the aspect ratio should not exceed 3 to1. A duct that is 24 inches x 18 inches has an aspect ratio of 1.33:1 (24 ÷ 18). The rectangular duct in Table 11-2 has an aspect ratio of 4:1 (24 divided by 6).

For maximum energy conservation seal all ducts. Pressure test medium and high pressure ducts to determine percentage of leakage.

Table 11-2. Comparison of Round Duct to Square and Rectangular Duct.

Duct	Round	Square	Rectangular
Area	1.0 sq. ft.	1.0 sq. ft.	1.0 sq. ft.
Duct size	13.5"	12" x 12"	24" x 6"
Perimeter	42.4"	48"	60"
Friction Loss	0.12/100'	0.13/100'	0.16/100'

Volume Damper
Parallel blade dampers are recommended for mixing applications into the mixed air plenum when outside and return air ducts are parallel. Parallel blade dampers are not recommended as volume dampers entering and leaving the fan. Partially closed parallel blade dampers installed

at or near the fan inlet or outlet can divert the air to one side of the duct which can result in a non-uniform velocity profile beyond the damper. This can create airflow problems at takeoffs close to the damper.

Opposed blade dampers are recommended for volume control at the fan outlet or inlet. Opposed blade dampers are also recommended for mixing applications into the mixed air plenum when outside and return air ducts are perpendicular to each other.

Ventilation Control

Occupied commercial and industrial buildings require a specified quantity of outside air for ventilation. Depending on the usage of the building the outside air quantity will be approximately 15 to 40 cubic feet per minute (cfm) per person. In some buildings such as hospitals or chemical labs where there exists the possibility of hazardous airborne materials the HVAC system is supplied with 100% outside air.

Although the ventilation system may be ducted directly into the conditioned space most systems are designed to combine outside ventilation air with return air. This conserves the energy needed to condition the air entering the heating and cooling coils. The combination of return and outside air is known as mixed air (MA). Any return air not used in the mixed air is exhausted to the outside and is known as exhaust air (EA).

The control of the return air, outside air, exhaust air and mixed air is the "mixed air control" or "the economizer control." A mixed air, low limit thermostat (typically set for 55F) modulates the outside air, return air and exhaust air dampers to maintain the mixed air temperature. Other controls such as a minimum position switch, outside air high-limit and morning warm up low-limit may be added to make the mixed air economizer system function more economically with better temperature control.

When properly controlled the outside ventilation air can aid the heating, cooling and humidifying of the building spaces. It can also provide a positive static pressure in the conditioned spaces. This positive pressure reduces the amount of air infiltration. Commercial buildings will generally be pressurized at about 0.03 to 0.05 inches of water column static pressure.

Automatic dampers are used to control the amount of outside ventilation air entering the building. Typically, pneumatic actuators are used to position the dampers.

Ventilation Control Sequence

When the fan on-off switch is turned on the ventilation damper system is energized. The outside air (OA) dampers open to minimum position. Return air (RA) dampers close to maintain the same total amount of air going into the mixed air (MA) plenum (OA + RA = MA). When the fan switch goes to off the reverse happens…the OA dampers close and the RA dampers open. If freezing of water coils is a possibility a preheat coil is installed between the outside air dampers and the other water coils. The preheat coil (hot water or steam) is controlled by a sensor located in the outside air (typically set for 40F). When the outside air temperature drops to 40F the controller opens the hot water valve or steam valve to the preheat coil. A reverse-acting controller on the discharge of the preheat coil will close the valve when the temperature reaches 68F so the plenum does not overheat.

A freezestat is a ventilation safety control installed at the preheat coil or any downstream coil. Let's say it is set at 35F. If the preheat coil discharge air temperature drops to 35F the fan is automatically stopped and the outside air dampers are closed to prevent water freezing in the coil. If this happens the system must be inspected and the freezestat reset.

WATERSIDE ECO

- Design or retrofit system to lowest pressure needed
- Design or retrofit system for lower water flows
- Use primary-secondary circuits and variable flow systems where applicable
- Install flow meters and provide balancing valves
- Water balance system
- When reducing water flow change pump speed or trim impeller instead of closing main valves
- Avoid installing restrictive piping on the inlets and discharges of pumps
- Reduce equipment on-time
- Use economizers
- Ensure pumps have correct rotation
- Maintain systems including controls
- Clean water coils, condensers, evaporators and cooling towers

- Clean strainers and restricted valves
- Repair or replace leaking valves and pipes
- Re-pipe crossed-over piping
- Insulate pipe
- Reduce resistance in system
- Keep debris and other obstructions away from coils and towers
- Use two speed or variable speed fans on cooling towers
- Raise delta T on heat exchangers

Pump Rotation

Momentarily start and stop the motor to "bump" the pump just enough to determine the direction of rotation. If the rotation is incorrect, reverse the rotation on a three-phase motor by changing any two of the three power leads at the motor control center or disconnect. On single-phase motors change rotation, as applicable, by switching the internal motor leads within the terminal box.

Pump Cavitation

Cavitation can occur in a water pump when the pressure at the inlet of the impeller falls below the vapor pressure of the water causing the water to vaporize and form bubbles. The bubbles are entrained in the water and are carried through the pump impeller inlet to a zone of higher pressure where they implode with terrific force. Symptoms of a cavitating pump include: snapping and crackling noises at the pump inlet, severe vibration, drop in pressure, drop in brake horsepower, reduction in water flow or no water flow. Cavitation usually results in pitting and erosion of the impeller vane tips or inlet.

The problem of cavitation can be eliminated by maintaining a minimum suction pressure at the pump inlet to overcome the pump's internal losses. This minimum pressure is called net positive suction head. Generally, if pumping is limited to air conditioning chilled water closed circuit systems there will be enough pressure for good pump operation. Cavitation is not ordinarily a factor in open systems or hot water systems unless there's considerable friction loss in the pipe, or the water source is well below the pump and the suction lift is excessive. However, if the friction losses in the system are too great (meaning a lower than designed pressure at the pump suction), or the water temperature is high, the pressure at some point inside the pump may fall below the operating vapor pressure of the water allowing the water to vaporize.

If the pump is cavitating, or the measured net positive suction head available (NPSHA) is insufficient, check the suction line for undersized pipe, too many fittings, throttled valves or clogged strainers.

Pump Motor

Motor amperage measured on any phase should not exceed the motor nameplate amperage. If reading is over the nameplate amperage but within service factor and voltage limits close main discharge valve or slow the pump until the amperage reading is down to nameplate or below. If reading is over nameplate amperage and outside service factor limit immediately turn pump off and inform the person responsible for the pump. An exception is if the pump has been running and is a critical service then immediately notify the person responsible for the pump operation.

Economizer

The purpose of a water side economizer is, when practical, to reduce or eliminate the need for mechanical refrigeration. "Free cooling" is available with open cooling towers when the outside ambient air temperature is low enough that a water side economizer system can be used to directly or indirectly cool the space load. The "strainer cycle" and the "heat exchanger" are the two basic types of open cooling tower water side economizers.

A strainer cycle system, Figure 11-5, is a direct cooling system. It is conventional vapor-compression refrigeration system during warm weather conditions. In cooler weather, when the ambient outside air temperature is low enough, the cooling tower water is diverted from the condenser and circulated directly through the chilled water circuit. Special strainers and water treatments are incorporated in the piping to minimize corrosion and fouling in the chilled water circuit. This type of water side economizer system has the greatest potential for energy savings.

A heat exchanger water side economizer system, Figure 11-6, is an indirect cooling system. Like the strainer cycle this economizer system is also a conventional vapor-compression refrigeration system during warm weather conditions. In cooler weather, when the ambient outside air temperature is low enough, the cooling tower water is diverted from the condenser through a plate-type heat exchanger. The closed loop chilled water circuit is also diverted through the heat exchanger. There is an exchange of heat from the "warm" water in the chilled water loop to the "cooler" water in the condenser/cooling tower loop. The heat exchanger keeps the condenser water loop separate from the chilled water loop. Therefore, there is less potential for fouling problems than with the strainer cycle system, but there is also less potential for energy savings.

Water Chiller Optimization

To determine if the water chiller system can be optimized the first step is to conduct an energy survey and testing of the system's operating

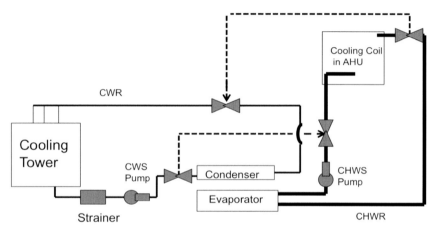

Figure 11-5. Strainer Cycle Water Side Economizer (dashed lines).

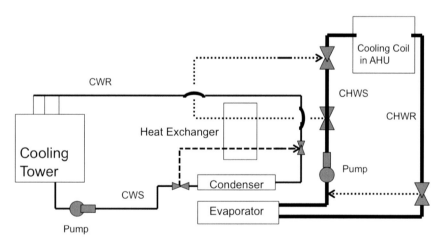

Figure 11-6. Heat Exchanger Cycle Water Side Economizer. Dashed lines are the condenser loop and dotted lines are the cooling coil loop.

performance. The second step is to consider the options. For example, is it more cost-effective to decrease chilled water supply temperature creating a higher delta T across the water cooler which will reduce pumping horsepower but increase compressor horsepower? Or is it better to increase chilled water supply temperature creating a lower delta T and increasing pumping horsepower but decreasing compressor horsepower? The third step to any proposed retrofit is to consider the consequences of the retrofit before starting the project. The following are some of the options to optimize the chilled water system.

Air Quantities and Heat Transfer Surfaces for Water Cooling Coils

Selecting the proper air quantities and heat transfer surfaces for cooling coils can substantially reduce circulating water qualities. Consider installing coils suited for a higher water temperature rise. For example a 12F rise (2 gpm per ton of refrigeration) rather than a 10F rise (2.4 gpm per ton of refrigeration) reduces circulated water quality by 17% and can reduce pump motor horsepower by 42%. However, a cooling coil with a higher delta T needs to be larger. There may also be an increase in air pressure drop through the coil that could increase the fan motor horsepower used. To compare the dollar savings use Equation 11-1. The cost of operation of a motor is equal to the horsepower times 0.746 kilowatts per horsepower times the hours of motor operation per year times the cost of kilowatt-hours divided by the motor efficiency.

Equation 11-1—Dollar Savings Per Year
$/yr = bhp x 0.746 kW/bhp x hours/year x $/kWh ÷ motor efficiency

Raise Evaporator Temperature

As the evaporator temperature and chilled water temperature increase the coefficient of performance (COP) of the system also increases and the power used by the compressor decreases (Equation 11-2). One way to increase evaporator temperature is to raise supply air temperature to follow the building's cooling load. Executing this task requires retrofitting the control system to raise temperature when the building cooling load permits and then raise chilled water temperature to meet the lighter cooling load. Raising the chilled water temperature increases the evaporator temperature. This control strategy monitors the chilled water valves. When the valves are closed or partially open (indicating that the water flow is reduced to match the light load condition) the chilled water temperature setpoint is raised. When one or more of the valves go to full open to match a heavier load the flow to the coils is increased and the supply water temperature is lowered.

Equation 11-2.—Horsepower Per Ton of Refrigeration.
$HP/Ton = 4.71 \div COP$ $(4.71 = 200\ Btu/min/ton \div 42.42\ Btu/min/hp)$

Lower Condenser Temperature
As the condenser temperature decreases the COP of the compressor increases and the power used decreases (Equation 11-2). One way to decrease condenser temperature is to decrease the temperature of the water entering the condenser. However, because there are practical limits to the lowest acceptable condensing temperature it is important to consult with the manufacturer for the recommended lower limit.

To lower condenser water temperature consider increasing the fan volume in the cooling tower by increasing fan speed or increasing the pitch of the blades on a propeller fan. Boosting fan volume increases the operating horsepower of the fan motor by the cube. Additionally, if the fan volume change is great enough it may require replacing existing fan motor with a larger size. Therefore, it is important with this or any other proposed retrofit to consider the consequences before commencing the project. In this case the trade-off between increased fan horsepower and decreased compressor horsepower needs to be calculated.

For chiller installations operating at constant condenser water temperature using on-off cycling of cooling tower fans consider modifying the controls to operate cooling tower fans continuously wherever there is a chiller on-line. Doing so will allow the condenser water temperature to drop until it reaches a predetermined low limit at which point the cooling tower fans can be cycled on and off to maintain the low limit. Also, consider using low temperature water from a well or other source for condenser water rather than cooling towers. Energy savings in compressor horsepower resulting from lower condenser water temperatures may make it a viable retrofit.

Condensing temperatures increase if the condenser heat exchanger is insulated. Scaling or fouling of the tubes in water-cooled condensers or restriction of airflow with air-cooled condenser are examples of unintentionally insulated heat exchangers. For water-cooler condensers consider installing automatic tube cleaners such as a cylindrical brush in each tube that is periodically forced from one end of the condenser tube bundle to the other by a reversal in the direction of water flow. This type of system can be effective in keeping condenser tubes clean by maintaining low fouling factors and reducing condensing temperatures. For air-cooled condensers keep coils clean and unobstructed. (Table 11-3.)

Table 11-3. Table shows increase in brake horsepower per ton of cooling with restriction (inside or outside) of heat exchangers (condenser and/or evaporator).

System Operating Condition	Evaporator Temperature	Condenser Temperature	Tons of Cooling	bhp	bhp per ton	Increased bhp per ton
Normal Operation	45	105	17	16	0.93	
Restricted Condenser	45	115	15.6	18	1.12	20%
Restricted Evaporator	35	105	13.8	15	1.10	18%
Both Restricted	35	115	12.7	16	1.29	39%

Optimize Chillers (Water Cooler) in Series

System efficiency is affected by the piping arrangement of the chillers. Chillers piped in parallel must each produce the coldest water required for the system. However, when chillers are piped in series the second (or third) chiller in the system operates at a higher suction pressure (higher evaporator temperature) and uses less energy for heavy cooling loads conditions. One drawback to the series arrangement is that at cooling lighter loads chilled water is still piped through the off-line chiller. Therefore, the pump must be selected to operate at the higher resistance created by two or more chillers in series.

To optimize this system add a bypass around the off-line chiller. Pumping horsepower will be reduced when the series chiller(s) go off-line (by the square of the pressure change). Pump horsepower can also be reduced (by the cube of the change in volume) if the chilled water volume can be reduced and the pump impeller trimmed or replaced by a smaller one. Increasing chiller and coil temperature differential (delta T) will reduce flow volume (gpm).

Optimize Chillers (Water Coolers) in Parallel

In some multiple-chiller operations, with chillers piped in parallel, chilled water is always circulated through all the chillers even when only one chiller is operating to meet a light cooling load demand. The result of this process results in wasted pump energy. The situation requires that the on-line chiller operate at a low evaporator temperature to produce chilled water at temperatures to meet the desired supply water temperature and offset the mixing effect of the water being circulated through the off-line chillers. COP is reduced and horsepower rises.

One option to optimize a parallel chiller system requires the installation of multiple constant volume pumps, one for each chiller. Each pump is selected and flow balanced for the gpm required and interlocked with its associated chiller. Isolating valves that close when a chiller goes off-line are installed on each chiller. Then, under lighter loads off-line chillers will be isolated. Only on-line chillers and pumps will be operating reducing pumping horsepower. The on-line chillers will operate at evaporator and water temperatures to meet the load. Pumping horsepower will drop by the cube if the chilled water volume through the chiller can be reduced after a water balance is completed and the pump impeller trimmed. A second option to consider is instead of multiple chilled water pumps install a multi-speed pump or variable frequency drive on the single present constant volume pump to make it variable volume if appropriate with condenser design flow limits.

Another option is installing variable volume system pumps sized for the critical load and a bypass (aka uncoupler loop or common loop) around the chillers. Each chiller has a dedicated constant volume pump. At lighter loads the chillers and the dedicated pumps go off-line and the system pumps draw through the bypass and from the on-line chillers to satisfy the cooling load. Example system operation follows in Figures 11-7 to 11-12.

REFRIGERATION SIDE ECO

• Adjust controls on multiple staging systems and inspect that staging functions properly. For example: second compressor doesn't energize until the first compressor can no longer satisfy the demand, etc., through all stages.

• Clean condenser coils on air-cooled systems. Remove debris restricting airflow. Also remove debris or restrictions (see icing next) from evaporator coils. Table 11-3.

• Defrost evaporator coils if iced. Determine the cause of icing and correct it (normally low air volume or low refrigerant charge).

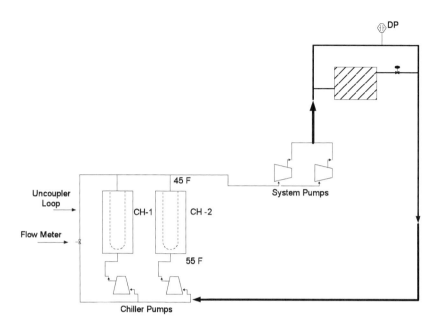

Figure 11-7. Parallel Chiller Optimization.
The system pumps are variable flow with variable frequency (speed) drives. In this example each pump circulates water from a maximum flow (1000 gpm) down to some set minimum flow. The water temperatures at the coils are 45F EWT and 55F LWT. Two-way variable flow valves on the coil return pipes are controlled by room T-stats. A differential pressure (DP) sensor and controller is installed in the system. Chiller pumps are constant speed and flow, i.e., when on, each pump moves 1000 gpm. Chiller temps are 55F EWT and 45F LWT.

- Clean scale build-up in water-cooled condensers.

- Record normal operating temperatures and pressures and check gauges frequently to ensure conditions are met.

- Check for proper refrigerant charge, superheat, and operation of the metering device.

- Repair leaking compressor valves.

- Repair leaking liquid line solenoid valves and clean liquid line strainers.

- Experiment with chilled water supply temperature while maintaining an acceptable comfort level.

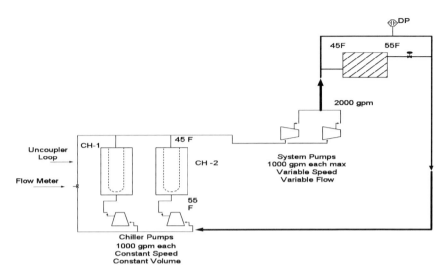

Figure 11-8. Maximum Flow (2000 gpm)

Chillers and their associated pumps are operating. Each chiller pump moves 1000 gpm, a total of 2000 gpm. The conditioned space thermostats in the system are calling for full cooling, full water flow 2000 gpm. Both system pumps are operating. The pumps are variable speed and volume. At maximum flow each pump is moving 1000 gpm, a total of 2000 gpm. Water leaves the chillers at 45F, goes through the coils where it picks up heat and returns to the chillers at 55F.

• Increase temperatures to reduce energy used by the compressor or decrease temperature to reduce water pump horsepower (see water chiller optimization above).

• To reduce condensing temperatures on air-cooled condensers consider (1) increasing the air volume through the condenser by increasing the fan speed (2) adding more air-cooled condensers in parallel to increase coil heat transfer surface area or (3) replace the existing condenser coil with one that has a larger surface area.

BOILER '

• Ensure the proper amount of air for combustion is available. Check that primary and secondary air can enter the boiler's combustion chamber in regulated quantities and at the correct place.

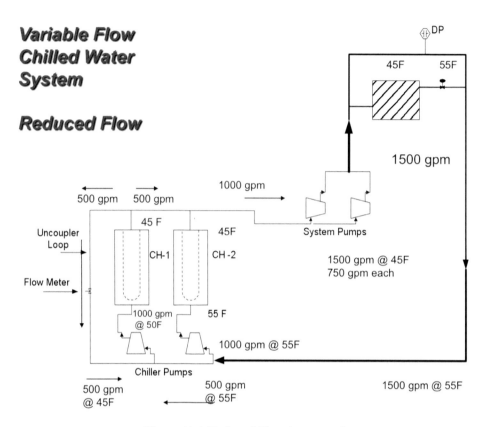

Figure 11-9. Reduced Flow (1500 gpm)

Chillers 1 and 2 and their associated pumps are operating. The thermostats in the system are calling for reduced flow (because of a reduced cooling load), 1500 gpm. The DP sensor and controller senses a rise in pressure in the system and sends a signal to slow the system pumps reducing water flow. Both system pumps are operating. At this reduced flow each pumps is moving 750 gpm, a total of 1500 gpm. Water leaves the chillers at 45F, goes through the coils where it picks up heat and returns to the chillers at 55F. The pump for CH-2 moves 1000 gpm of water at 55 F which goes into the chiller and is cooled down to 45F. The pump for CH-1 moves 500 gpm @ 55F. This is the remainder of the flow from the 1500 gpm back from the coils along with an additional 500 gpm of 45F water. This is water from the uncoupler loop. 1000 gpm of water at 50F (50% x 55F + 50% x 45F) goes into CH-1 and is cooled down to 45F. The chiller is not working to maximum capacity and therefore is using less energy. 1000 gpm leaves CH-1 of which 500 gpm goes to the system pumps and 500 gpm through the uncoupler loop.

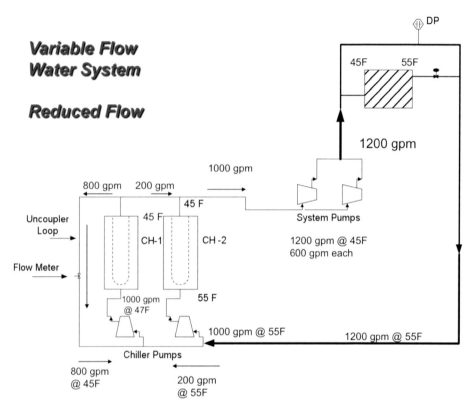

Figure 11-10. Reduced Flow (1200 gpm)

Both chillers and their associated pumps are operating. The thermostats in the system are calling for reduced flow, 1200 gpm. The DP controller senses a rise in pressure in the system and sends a signal to slow the system pumps reducing water flow. At this reduced flow each pump is moving 600 gpm. Water leaves the chillers at 45F goes through the coils where it picks up heat and returns to the chillers at 55F. CH-2 pump circulates 1000 gpm of water at 55F through the chiller and cools it to 45F. CH-1 pump circulates 200 gpm of water at 55F. This is the remainder of the flow from the 1200 gpm back from the coils. CH-1 pump also circulates an additional 800 gpm, 45F water from the uncoupler loop. Therefore, 1000 gpm of water at 47F goes into CH-1 and is cooled to 45F. The chiller is using less energy. The 1000 gpm leaves CH-1 with 200 gpm going to the system pumps and 800 gpm through the uncoupler loop. In this example system at this gpm or some lesser gpm CH-1 and its pump shuts down.

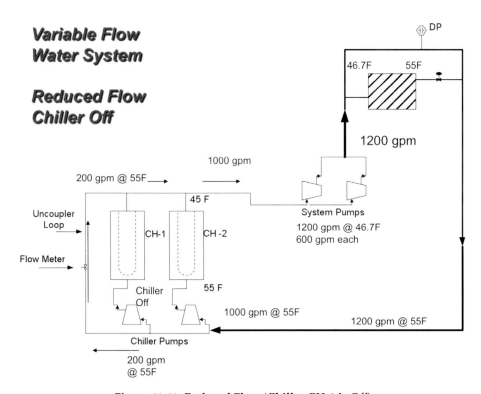

Figure 11-11. Reduced Flow (Chiller CH-1 is Off)
The thermostats in the system are calling for reduced flow (1200 gpm). Both system pumps are operating. Each pump is moving 600 gpm. CH-2 and associated pump are operating but CH-1 and its pump are off. CH-2 pump pumps 1000 gpm of water at 55F into the chiller and it is cooled down to 45F. CH-1 pump is off. 200 gpm of water at 55F (the remainder of the flow from the 1200 gpm from the coils) bypasses the chiller CH-1 pump and flows up through the uncoupler loop where it mixes with the water from chiller CH-2. The water temperature going out to the systems pumps is 46.7F.

- Inspect boiler gaskets, refractory, brickwork and castings for hot spots and air leaks. Defective gaskets, cracked brickwork and broken casings allow uncontrolled and varying amounts of air to enter the boiler and prevent accurate fuel-air ratio adjustment.

- Perform a flue-gas analysis. Take stack temperatures and oxygen readings routinely and inspect the boiler for leaks.

- Repair all defects before resetting the fuel-air ratio. Consider installing an oxygen analyzer with automatic trim for larger boilers.

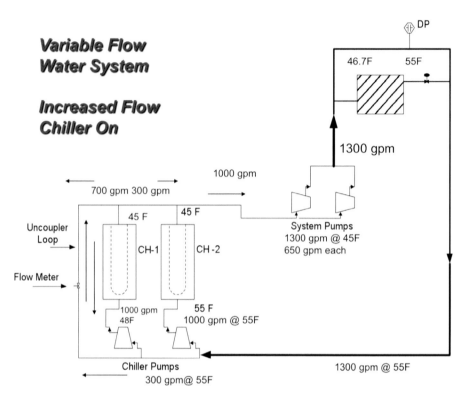

Figure 11-12. Increased Flow (1300 gpm), Chiller On

The thermostats in the system are calling for increased flow (because of increased cooling load), 1300 gpm. The DP (delta P) sensor and controller senses a drop in pressure in the system and sends a signal to speed up the system pumps thus increasing water flow. Chiller CH-2 and its associated pump are operating. CH-1 and its associated pump are off. Both system pumps are operating. At this point each system pump is moving 650 gpm, a total of 1300 gpm. At some designated low flow (a total flow less than 1000 gpm) one system pump would stop.

CH-2 pump circulates 1000 gpm of water at 55F through the chiller cooling it down to 45F. Chiller 1 is off. 300 gpm of water at 55F (the remainder of the flow from the 1300 gpm from the coils) bypasses the pump for CH-1 and flows up through the uncoupler loop mixing with the water from chiller 2. At this point, the flow meter in the uncoupler loop senses too much water (300 gpm) flowing up and sends a signal to start the pump for Ch-1. When CH-1 pumps starts it takes 700 gpm of water from the uncoupler loop and mixes it with 300 gpm of water back from the coils (the remainder of the flow from the 1300 gpm). 1000 gpm goes into chiller1 and is cooled down to 45F. Both chillers and their associated pumps are operating again.

This device continuously analyzes the fuel-air ratio and automatically adjusts it to meet the changing stack draft and load conditions.

- Check that controls are turning off boilers and pumps as outlined in the sequence of operations. Observe the fire when the boiler shuts down. If it does not cut off immediately check for a faulty solenoid valve and repair or replace it as needed.

- Adjust controls on multiple systems so a second boiler will not fire until the first boiler can no longer satisfy the demand. Make sure that reset controls work properly to schedule heating water temperature according to the outside air temperature.

- Experiment with hot water temperature reduction until reaching an acceptable comfort level.

- Install automatic blowdown controls. Pipe the blowdown water through a heat exchanger to recover and reuse waste heat.

- Inspect boiler nozzles for wear, dirt or incorrect spray angles. Clean fouled oil nozzles and dirty gas parts.

- Replace all oversized or undersized nozzles. Adjust nozzles as needed.

- Verify that fuel oil flows freely and oil pressure is correct. Watch for burner short-cycling.

- Inspect boiler and pipes for broken or missing insulation and repair or replace it as needed

- Clean the fire side and maintain it free from soot or other deposits.

- Clean the water side and maintain it free from scale deposits.

- Maintain the correct water treatment. Remove scale deposits and accumulation of sediment by scraping and/or treating chemically.

Documentation

A daily log of boiler operating pressures and temperatures and firing rate will detect variations in system performance. Any major variation in the recorded pressures or temperatures may indicate that a problem exists. Taking the time to investigate, analyze, and correct any developing problem will extend the life of the boiler and maintain high operating efficiency. Most commercial package boilers operate at a maximum efficiency of about 80 to 83%. However, this is only true when the burner is functioning properly and the tubes are clean. Maintaining a high efficiency in the boiler will decrease operating expenses and increase the life of the boiler.

Flue Gas Analysis

To determine the combustion efficiency of a boiler perform a flue gas analysis. Use an electronic instrument that measures temperatures and flue gases in the boiler stack. To make a test of boiler combustion efficiency drill a hole large enough to accommodate the instrument probe in the flue stack between the boiler shell and the stack damper. The hole should be at least 6 inches from the damper. Using the thermometer supplied with the instrument, or one that reads to 1000F, read the boiler stack temperature. Stack temperatures can vary 100 degrees within a few minutes during load changes therefore, note the firing rate when logging temperatures and pressures. While waiting for the stack temperature reading to stabilize use another thermometer to take the boiler room temperature. The difference between these readings is the net stack temperature.

Record firing rate, percent of carbon dioxide, percent of excess air, stack temperature, and the net stack temperature. These readings will determine the percent of stack loss and the combustion efficiency. The maximum combustion efficiency attainable for natural gas- and oil-fired boilers will be about 80% to 83%. A measurement of carbon dioxide (CO_2), oxygen (O_2), and carbon monoxide (CO) is a good indication of combustion efficiency and burner performance. Carbon dioxide should be good to excellent. Oxygen should be a maximum of 1-2% and there should not be any carbon monoxide. Table 11-4 provides boiler combus-

Table 11-4. Boiler Combustion Efficiency.

Fuel	Natural Gas	#2 Oil	#6 Oil
Rating	Percent Carbon Dioxide		
Excellent	10.0%	12.8%	13.8%
Good	9.0%	11.5%	13.0%
Fair	8.5%	10.0%	12.5%

tion efficiency rating based on the percent of carbon dioxide measured when burning either natural gas, number 2 oil, or number 6 oil.

Stack Temperatures

The boiler stack temperature should be no more than 150F above the steam or water temperature. If it is, the boiler is not working efficiently. The rule of thumb for stack temperature is that for each 100 degrees that the stack temperature can be lowered there is a 2.5% increase in efficiency. A high stack temperature means that there's poor combustion, the tubes are fouled, or there's too much combustion air being brought into the boiler and it's pushing the gases through the boiler without the proper heat exchange taking place. The stack temperature should be at least 320 degrees. If the stack temperature is too low the water vapor in the flue gas will start to condensate in the stack. This water mixes with the sulfur in the flue gas and creates sulfuric acid which will corrode the stack and the tubes. A minimum boiler water temperature of 170F should be maintained. This will mean a stack temperature of about 320F (150 degree delta T).

Excess Air

The amount of excess air, that is, the air needed for complete combustion plus some extra for a safety factor, should not exceed 10%. To take carbon dioxide and oxygen readings use the same hole in the stack insert the instrument probe and take the measurements. The instrument will read out directly in percent of carbon dioxide, oxygen and efficiency. The oxygen level should be at least 1 percent but should not exceed 2 percent. A rule of thumb says that there is approximately 5% excess air for each 1% of oxygen in the flue gas. The amount of carbon dioxide should be as high as possible. For maximum efficiency in natural gas boilers this will be about 10% while oil-fired boilers should have about 13 to 14% carbon dioxide.

Carbon Monoxide Test

Test for the presence of carbon monoxide. There should not be any. The existence of carbon monoxide indicates incomplete combustion. Carbon monoxide is a deadly gas. If its presence is found, the boiler should be shut down and the problem corrected. Either there is not enough air being brought into the boiler or there is a problem with the burner.

Smoke Test

In addition to the flue gas test a smoke test should be made on oil-fired boilers. Excessive smoke is evidence of incomplete combustion of the oil. This means that fuel is being wasted. It also can result in soot being deposited on the heat transfer surfaces which also means lower efficiencies. A one-eight inch thick soot deposit increases fuel consumption by approximately 10%

Boiler Scale

Scale acts as an insulator reducing boiler efficiency. But, it can also result in overheating of the firing chamber and tubes. This can cause cracking and eventually, leakage problems. Routinely do a visual check of the rear portion of your boiler. This is the area that's the most prone to scale buildup. Use a scraper or a small hammer to get some samples of any scale formation which may be present. Scale formation is the result of improper feedwater treatment or improper blowdown procedures. "Blowdown" is the process of draining off some of the boiler water to reduce the concentrations of minerals in the water. These minerals are brought in by the feedwater (water source from the city, a well, a river, etc.). When the water is boiled off concentrations of solids are left in the remaining boiler water. If these concentrations of solids are not reduced they will be deposited on the tube surfaces forming scale.

A buildup of 1/8" thickness of scale will result in a 15% loss of efficiency. When the heating surfaces become scaled heat transfer is reduced. Some of the excess heat goes out the stack but much of it overheats the boiler tubes causing corrosion, blistering and early tube failure. To combat losses in efficiency, excessive fuel costs and reduced life expectancy resulting from boiler scale, it probably will be necessary to call in a feedwater specialist to give you recommendations on blowdown procedures and frequency of blowdown to reduce the concentrations of scale producing elements. Blowdown is usually a percentage expressing the quantity of blowdown verses the quantity of feedwater. For instance, a 5% blowdown means that 5% of the water fed to the boiler is removed during the blowdown process. Give the scale samples removed from the boiler to the feedwater consultant. The consultant will analyze the water and then make recommendations on the proper chemical treatment to use to prevent scale formation on the boiler's heating surfaces.

Pitting and Sediment Accumulation

Pitting or oxygen corrosion of the tubes is another problem resulting from incorrect water treatment. A deaerating or oxygen-removing feedwater heater may also be needed. Check for pitting problems at least once a year. This will require draining the boiler and using a flashlight and a mirror and visually checking the tubes and shell for blisters, pock marks or any other type of erosion of the metal surfaces. Contact a feedwater specialist for help with diagnosing problems. Sometimes the water conditions, or even the chemicals used to treat the water, will cause an accumulation of sediment in the bottom of the boiler. This sediment or "mud" will be found during visual inspection. It must be removed. Use a high pressure hose to wash out the bottom of the boiler and then check by hand to make sure that all the sediment has been removed.

Soot Deposit

Soot deposits act as an insulator decreasing heat transfer and boiler efficiency. The boiler tubes should be checked frequently for evidence of soot deposits. To reduce the time needed for visual inspection of the tubes install a thermometer in the exhaust stack and keep a daily log. If the stack gas temperature rises above normal it means that the tubes are dirty and need cleaning. The length of time between cleaning varies with the type of burner, the type of fuel used and the burner adjustment. However, if there's heavy sooting within a short period of time after cleaning the tubes it probably means that the fuel-to-air ratio is incorrect. In other words, there is too much fuel and not enough air. If this is the case the burner needs adjustment.

Linkages, Tube Sheets, Gaskets, Refractory and Stack

Check the burner and air damper linkages for tightness. Watch as the linkage moves back and forth. If there's any jerking motion or slippage this will need to be corrected. Check for white streaks or deposits at the ends of the tube sheets. The white streaks mean that the tube ends are leaking and a re-rolling of the tubes may be needed. Check for any leaking gaskets around doors, handholds, or manholes. At least once a year wash down and check the refractory surfaces for loose, cracked, broken or missing tiles. Replace or repair as instructed by the boiler manufacturer. Inspect the stack. It should be free of haze. If not, it indicates that a burner adjustment is needed.

Water Level Control

On steam boilers the low water cutoff and water column should be blown daily to remove the solids. Additionally, the low water cutoff should be checked under operating conditions at least once a week. Turn off the feedwater pump and let the system operate as normal. Watch the gauge glass and mark the glass at the precise level where the low water cutoff shuts off the boiler. This is now a reference point. The cutoff control should shut down the boiler at the same water level each time. If it doesn't the controls may need to be replaced. On water boilers check the low water cutoff periodically by manually tripping the control.

High Fuel-to-Air Ratio and High Air-to-Fuel Ratio

A high fuel-to-air ratio causes sooting and lowers boiler efficiency. In certain conditions it may also be dangerous if there's not enough air for complete combustion and dilution of the fuel. A high fuel-to-air condition can be caused by an improperly adjusted burner, a blocked exhaust stack, the blower or dampers set incorrectly or any condition which results in a negative pressure in the boiler room. A negative pressure in the boiler room can be the result of one or a combination of conditions such as an exhaust fan pulling a negative pressure in the boiler room, a restricted combustion air louver into the room, or even adverse wind conditions.

High air-to-fuel ratios also reduce boiler efficiency. If too much air is brought in the hot gases are diluted too much and rapidly swept out of the tubes before proper heat transfer can occur. High air volumes are caused by improper blower or damper settings.

HEAT RECOVERY

The objective of heat recovery systems is to reduce the energy consumption and cost of operating a building by transferring heat between two fluids. Exhaust air and outside air is one example. In many cases, the proper application of heat recovery systems can result in reduced energy consumption and lower energy bills while adding little or no additional cost to building maintenance or operations. However, if it cannot be shown that the benefits of a heat recovery system outweigh the cost building owners will not be motivated to make a financial investment in such a system.

During the past 50 years building owners and other commercial energy end-users have had to find ways to cope with increasing uncertainty about the supply and economic volatility of nuclear, electrical and fossil fuels used to generate energy for their facilities. Indeed, weather, politics, and market forces play a significant role in determining the availability of energy and its cost. End-users need only to recall the power shortages in the last decade that plagued sections of the U.S. crimping supplies and sending energy prices soaring. Those with a greater sense of history are aware of the oil shortages of the mid-1970s. I was just starting in HVAC then and I remember that first oil embargo and the threat that gasoline could go as high as sixty cents ($0.60) per gallon.

The continuing unrest suggests that an oil crisis is not only possible but probable. It is just another wakeup call that energy created by fossil fuels will not always be readily accessible. It also serves notice about the need to reduce our reliance on such energy sources to better insulate business from forces beyond our control. Designers and facility managers of commercial buildings need to focus their efforts on energy conservation as well as maximizing occupant comfort and process function. While many of the conservation measures implemented are voluntary regulators will also continue to mandate energy conservation strategies.

One energy conservation measurement worth considering is heat recovery which captures waste heat from one fluid and transfers it to another cooler fluid to be used in a heating application. An example of a heat recovery system is heated exhaust air to wintertime outside air coming into an AHU. A heat recovery system reduces energy consumption by eliminating the need to generate new heat and in turn lowers building operating costs. Another example is capturing waste heat from the flue gas of a boiler and the reusing that heat to preheat the boiler input water and therefore the amount of heat required to generate steam or hot water is reduced. A third example is capturing waste heat from large ovens or other similar heat-generating equipment to be used for comfort heating.

Additionally, heat recovery systems (also known as heat energy or energy recovery systems) can be used to provide reserve heat (energy) capacity. The reason for having reserve capacity is that many times the implemented energy conservation measures substantially reduce the capacities of HVAC equipment. Therefore, the installed equipment capacities now closely match the design load and less reserve capacity is available for new projects that may substantially change the building HVAC system. For example, many installed systems may not have enough reserve capacity to make needed changes to accommodate increased outdoor air requirements in order to satisfy indoor air quality (IAQ) concerns. When more outside air is needed a heat recovery system can help to offset the increased energy cost to heat up or cool down the increased volume of outside air.

There are three basic types of heat recovery systems: comfort-to-comfort, process-to-comfort, and process-to-process. The types of heat exchangers for these systems include rotary wheel, fixed plate, heat pipe, and run-around coil. To obtain a better understanding of how to best make use of heat transfer systems it is important to first understand the components that make up heat recovery. Heat is a form of energy, which can be converted from-or-to other forms of energy such as mechanical energy or electrical energy (more in Chapter 2, Heat Flow).

Thermodynamics is the science of heat energy and the study of how heat energy can be changed from one form of energy to another. One of the laws of thermodynamics states that heat energy flows from a higher level to a lower level. When this law is applied to heat recovery systems it tells us that the waste heat in a fluid such as air, flue gas, steam, refrigerant, brine, or water from a heat-generating process can be captured and transferred to a cooler fluid for use in another process. The intent of heat recovery is to reduce energy costs by supplementing the energy required to fuel the process or comfort system. Conduction, convection, and radiation are the three means of heat transfer. Conduction is the transfer of heat from on substance to another when each substance is in direct physical contact with the other. An example of conduction is a human hand on a cold pipe. Warmth from the skin is transferred to the pipe. Convection is heat transfer by movement of a fluid over an object. Convection is demonstrated when heated air flows into a room and warms the occupants. Radiation is heat transfer by waves transmitted from the source of the heat to an object receiving the heat waves without heating the space. Examples are when the sun's rays heat a glass window or when a person is warmed by the heat waves from a fire or infrared heater.

The effectiveness of a heat exchanger (coil, plate heat exchanger, heat wheel, etc.) in a heat recovery system is dependent upon three factors: (1) the temperature difference of the fluids circulated through the exchanger; (2) the thermal conductivity (ability to conduct heat) of the material (copper, aluminum, steel, etc.) in the exchanger; and (3) the flow pattern (e.g., counter flow or parallel flow) of the fluids. Heat transfer is greatest in counter flow exchangers. Counter flow is when "Fluid A" enters on the same side of the exchanger that "Fluid B" is leaving (Figure 11-13). Parallel flow is when "Fluid A" enters on the same side of the exchanger that "Fluid B" is entering.

Figure 11-13. Counter Flow Heat Exchanger

Comfort-to-Comfort Heat Recovery System

Comfort-to-comfort systems are typically used in HVAC applications. These heat recovery systems capture a building's exhaust air and reuse the energy to precondition the outside air coming into the building. In comfort-to-comfort applications the energy recovery process is reversible, i.e., the enthalpy (total heat content) of the building supply air is lowered during warm weather and raised during cold weather. Air-to-air heat recovery systems for comfort-to-comfort applications fall into two general categories: sensible heat systems and total heat systems.

Sensible heat recovery systems transfer sensible heat between exhaust air leaving the building and make -up or supply air entering the building. Rotary wheel heat exchangers are used in comfort-to-comfort sensible heat recovery applications. To determine the amount of sensible heat transferred through the air use equation 11-3. Btuhs is sensible heat transfer, cfm is quantity of airflow, 1.08 is a constant, and TD is dry bulb temperature difference between the airstreams.

Equation 11-3.
Btuhs = cfm x 1.08 x TD x system efficiency

Example: The summer outside air temperature is 90Fdb and the building exhaust air temperature is 75Fdb. The recovery system operates at 20,000 cfm and is 73% efficient. The amount of heat transferred from the outside air into the exhaust air is 236,520 Btuhs (20,000 x 1.08 x 15 x 0.73. In the winter the OA temp is 30Fdb and the EA temp is 75Fdb. The amount of heat transferred from the EA into the OA is 709,560 Btuhs.

For a total heat recovery system use Equation 11-4 for air total heat transfer. Btuht is total heat transfer, cfm is quantity of airflow, 4.5 the constant, and Δh is total heat difference between the airstreams.

Equation 11-4.
Btuht = cfm x 4.5 x Δh x system efficiency

Example: The average summer outside air dry bulb and wet bulb temperatures are 90Fdb and 70Fwb. From the psychrometric chart the enthalpy at this condition is 33.8 Btu/lb. The average exhaust air dry bulb and wet bulb temperatures are 75Fdb and 60Fwb. The enthalpy at this condition is 26.3 Btu/lb. The total heat transferred from the outside air is 492,750 Btuh. In the winter the OA temperatures are 30Fdb and 20Fwb (7.1 Btu/lb) and the EA temperatures are 75Fdb and 60Fwb (26.3 Btu/lb). The heat recovered from the EA is 1,261,440 Btuh.

Process-to-Comfort Heat Recovery System
     Process-to-comfort systems are generally only sensible heat recovery used during the spring, fall, and winter months. When considering process-to-comfort heat recovery system the process effluent (the exhaust from equipment and the actual work process) must be evaluated for harmful material such as corrosives, condensables (moisture or water vapor), contaminants, and noxious or toxic substances.

Process-to-Process System Recovery System
     Process-to-process systems are typically only sensible heat recovery, usually full recovery, but in some cases, partial recovery can be performed. Determining when to use a process-to-process system for partial sensible heat recovery instead of full sensible recovery is based on the circumstances under which the system will operate. For example, when the exhaust stream contains condensables such as moisture or water vapor and possible overcooling of the exhaust air stream could occur with full recovery then a partial recovery system is more appropriate. As with process-to-comfort systems the process effluent must be evaluated for harmful substances such as corrosives, contaminants, and noxious or toxic materials.

Determining Value
     A heat recovery system, as with any other energy-conservation retrofit, must have resultant benefits that exceed the investment costs.

The reason for installing and the effective use of a heat recovery system depends on many factors. A heat recovery system is warranted when all the factors are analyzed and there are (1) annual energy savings, and (2) the payback period is reasonable. In some cases heat recovery systems are also considered when additional energy is currently unavailable or unattainable.

Considerations for Selecting a Heat Recovery System
>   Space requirements of the heat recovery system
>   Distance between airstreams
>   Temperature difference of the airstreams (sensible and latent heat)
>   Mass flow rates (pounds per hour) of the airstreams
>   Efficiency of the heat recovery system
>   Additional operating energy required for the heat recovery system
>   Quality of the indoor and outdoor air
>   Existing modifications to the HVAC systems
>   Mandatory energy conservation regulations and cost
>   Mandatory pollution regulations and cost
>   Pollution abatement savings
>   Construction cost of heat recovery system
>   Reduced capital cost of HVAC equipment due to reduced capacity
>   Maintenance cost of the heat recovery system
>   Reduced maintenance cost for existing equipment
>   Revenue from sales of recovered heat or energy
>   Increased production
>   Reduction in production labor cost
>   Availability of fuel
>   Forecast for the rising cost of fuels
>   Possibility of changeover to other fuels
>   Fuel savings

HEAT EXCHANGER

A heat exchanger is a device specifically designed to transfer heat between two physically separated fluids. The term heat exchanger can describe any heat transfer device such as a coil or a particular category of devices. Heat exchangers are made in various sizes and types. The basic types of heat exchangers are shell and tube, shell and coil, U-tube, helical, and plate. Typical HVAC heat exchangers are designed for a number of fluid combinations including steam to water (converter), water to steam (generator), refrigerant to water (condenser), water to refrigerant (chiller), water to water (heat exchanger), air to refrigerant (coil), and air to water or water to air (coil).

HEAT RECOVERY HEAT EXCHANGER

Double Bundle Condenser

Double bundle condensers contain two sets of water tubes bundled within the condenser shell. One set of tubes is the heating bundle with water from and back to a boiler and water storage heat exchanger. The other set is the cooling bundle with water from and back to a cooling tower. Heat is rejected from the refrigerant system as superheated gas from the refrigerant compressor flows into the condenser shell and is condensed back to liquid refrigerant. During the heating season water pumped through the "winter bundle" absorbs heat. This heat can be used for heating domestic water or heating the perimeter of the building, etc. During the cooling season water pumped through the "summer bundle" rejects heat to the cooling tower after any hot water needs are met.

Fixed Plate

Fixed plate exchangers have no moving parts. They have alternative layers of plates that are separated and sealed. Heat is transferred directly from the warm fluid through the separating plates into the cooler fluid. Some plate exchangers, called plate fin heat exchangers, have alternating layers of separate plates and interconnecting fins. Fixed plate heat exchangers use counter flow, parallel flow, and cross flow patterns, Figure 11-14.

Heat Jacket

Various heat exchangers including a water jacket placed on an engines in order to capture heat from the engine exhaust. A heat exchanger placed in the exhausts of reciprocating engines and gas turbines to capture heat for water heating systems or steam generation.

Heat Pipe

Heat pipe exchangers are similar in appearance to a fin-tube water coil except that the tubes are not interconnected and the exchanger is

Figure 11-14. Cross Flow Heat Exchanger. 1. Cool supply air into heat exchanger. 2. Warmed supply air out. 3. Warm exhaust air in. 4. Cooled exhaust air out.

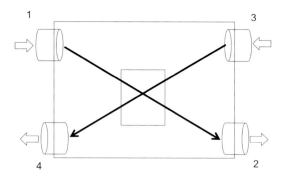

divided into evaporator and condenser sections by a partition plate. In HVAC systems the pipe is filled with a suitable refrigerant. On the evaporator side of the exchanger hot air flows past the evaporator boiling off the liquid refrigerant. This air is then cooled and the vapor refrigerant goes to the condenser side of the heat pipe. On the other side of the partition cold air passes over the condenser side. This air is heated. The vapor refrigerant is condensed to a liquid and flows to the evaporator side of the heat pipe. Heat pipes are essentially sensible heat transfer equipment but condensation of moisture (latent heat) from the hot air on the fins improves recovery performance. This type of heat exchanger uses a counter flow pattern. Other heat pipe exchangers use parallel flow.

Heat Pump

Rooms containing laundries and food preparation facilities are often extremely hot and uncomfortable. Heat from the air can be captured for heating water by using a dedicated heat pump that mechanically concentrates the heat contained in the air. See Hot Refrigerant Gas below and Chapters 9 and 11 for more information on heat pump and vapor-compression refrigerants operation.

Hot Flue Gas

Hot flue gases from boilers can provide a source of waste heat for a variety of uses. The most common use is pre-heating boiler feed water. Heat exchangers used in flues must be constructed to withstand the highly corrosive nature of cooled flue gases.

Hot Refrigerant Gas

The refrigeration cycle of air conditioners and heat pumps provides an opportunity to capture waste heat for heating domestic water.

HVAC compressors concentrate heat by compressing gaseous (vapor) refrigerant. The resultant superheated gas is normally pumped to a condenser for heat rejection. However, a hot gas-to-water heat exchanger may be placed into the refrigerant line between the compressor and the condenser coils to capture a portion of the rejected heat. In this system, water is looped between the water storage tank and the heat exchanger when the HVAC system is on. Heat pumps operating in the heating mode do not have waste heat because the hot gas is used for space heating. The heat pump, however, can still heat water more efficiently compared to electric resistance heating. Also commercial refrigerators and freezers can be installed with all the condensing units in one location. This will enhance the economic feasibility of capturing heat from refrigerant gases for heating.

Rotary Wheel

The rotary wheel, aka heat wheel, is a sensible heat air-to-air heat exchanger that has a large-surface revolving cylinder. The cylinder is filled with a gas-permeable material. As the cylinder revolves each of two air streams flows through approximately half of the wheel in a counter flow pattern. The two air streams (exhaust and outside air, for example) flow axially to the shaft. The heat from the warm air, for example, is absorbed into the porous materials and then released into the cooler airstream as the wheel rotates. Some rotary wheels are treated with a hygroscopic (water absorbing) material to enable them to transfer moisture (latent heat) from one airstream to another. The moist airstream is dehumidified and the drier airstream is humidified. A hygroscopic heat wheel is an example of a total heat rotary wheel because both sensible and latent heat is transferred simultaneously. Rotary wheel heat exchangers use counter flow and parallel patterns.

Run-Around

The heat exchangers in a run-around heat recovery system for HVAC comfort-to-comfort applications are fin-tube type connected by counter flow piping. A pump circulates water, glycol, or other liquids through the system. The coils are mounted in different airstreams connected either in series or parallel to provide the greatest heat recovery. This system is seasonally reversible, meaning that the exhaust air coil either preheats or precools the outside air, depending on the season.

SYSTEM AND COMPONENT LIST

Airside System Information and Condition (D) Design		(M) Measured
Air Handling Unit and System		
Cooling only Heating only	Heating and Cooling	Reheat
Constant Air Volume	Variable Air Volume	
Single duct	Double duct	
Single zone	Multizone	
Fan Wheel and Blade		
Rotation	Clockwise	Counter Clockwise
Clearance		
Fan Housing Condition		
Air Leakage	Extensive	Nominal/Negligible
Plenum Condition		
Air Leakage	Extensive	Nominal/Negligible
Flexible Connection Condition		
Air Leakage	Extensive	Nominal/Negligible
Supply Fan		
Forward Curved Airfoil	Backward Inclined	Backward Curved
Tube axial Vane axial	Radial Other	Inline Centrifugal
Manufacturer/Model Number		
Speed (rpm)	(D)	(M)
Static Pressure		
Suction	(D)	(M)
Discharge	(D)	(M)
Total Pressure		
Suction	(D)	(M)
Discharge	(D)	(M)
Total Static Pressure	(D)	(M)
External Static Pressure	(D)	(M)
Fan Static Pressure	(D)	(M)
Fan Total Pressure	(D)	(M)
Discharge Air Temperature		
Dry Bulb	(D)	(M)
Wet Bulb	(D)	(M)
Return/Relief Fan		
Forward Curved Airfoil	Backward Inclined	Backward Curved
Tube axial Vane axial	Radial Other	Inline Centrifugal
Static Pressure		

Suction	(D)	(M)
Discharge	(D)	(M)
Total Pressure		
Suction	(D)	(M)
Discharge	(D)	(M)
Total Static Pressure	(D)	(M)
External Static Pressure	(D)	(M)
Fan Static Pressure	(D)	(M)
Fan Total Pressure	(D)	(M)
Exhaust Fan		
Forward Curved Airfoil	Backward Inclined	Backward Curved
Tube axial Vane axial	Radial Other	Inline Centrifugal
Manufacturer/Model Number		
Speed (rpm)	(D)	(M)
Static Pressure		
Suction	(D)	(M)
Discharge	(D)	(M)
Total Pressure		
Suction	(D)	(M)
Discharge	(D)	(M)
Fan Total Pressure	(D)	(M)
Air Volume		
Outside Air Cubic Feet/Minute	(D)	(M)
Exhaust Air Cubic Feet/Minute	(D)	(M)
Return Air Cubic Feet/Minute	(D)	(M)
Supply Air Cubic Feet/Minute	(D)	(M)
Outside Air		
Outside Air Temperature		
Dry Bulb	(D)	(M)
Wet Bulb	(D)	(M)
Louver and Screen Condition	Clean Dirty	Clogged
Outside Air Damper		
Position	Minimum Full open	Modulating
Close Properly	Yes	No
Open Properly	Yes	No
Sealed All Sides	Yes	No
Return Air		
Return Air Temperature		

Dry Bulb	(D)	(M)
Wet Bulb	(D)	(M)
Return Air Damper		
Position	Minimum Full open	Modulating
Close Properly	Yes	No
Open Properly	Yes	No
Sealed All Sides	Yes	No
Exhaust/Relief Air		
Exhaust Air Temperature		
Dry Bulb	(D)	(M)
Wet Bulb	(D)	(M)
Exhaust Air Damper		
Position	Minimum Full open	Modulating
Close Properly	Yes	No
Open Properly	Yes	No
Sealed All Sides	Yes	No
Filters		
Fiber Type		
Viscous	Dry	HEPA
Bag	Continuous Roll	ULPA
Renewable Type		
Viscous	Dry	Electronic
Sealed All Sides	Yes	No
Filter Pressure		
Entering Air	(D)	(M)
Leaving Air	(D)	(M)
Drop	(D)	(M)
Fan Motor		
Type	Single phase	Three phase
Manufacturer/enclosure		
Nameplate Horsepower	(D)	(M)
Service Factor	(D)	(M)
Voltage	(D)	(M)
Amperage	(D)	(M)
Drives		
Belt		
Condition	Good Worn	Tight Loose
Number of Belts		

Motor Sheave, Drive	Fixed	Adjustable
Number of Grooves		
Manufacturer/Size		
Fan Sheave, Driven	Fixed	Adjustable
Number of Grooves		
Manufacturer/Size		
Air Distribution		
General Condition	Good	Fair Poor
Aspect ratios	Good	Fair Poor
Use of Fittings	Good	Fair Poor
Main Duct Pressure		
Classification		
High inches wg		
Low inches wg		
High inches wg		
Medium or High Pressure Duct		
Leak Tested	Yes	No
Leakage Class		
Leakage Rate		
Sealed	Yes	No
Air Leakage	Extensive	Nominal/ Negligible
Low Pressure Duct Condition		
Sealed	Yes	No
Air Leakage	Extensive	Nominal/ Negligible
Balancing Dampers		
Supply Outlets	Yes	No Need #
Return Inlets	Yes	No Need #
Exhaust Inlets	Yes	No Need #
Zones	Yes	No Need #
Building or Space Pressurization		
Positive inches wg		
Negative inches wg		
Waterside System Information and Condition (D) Design) (M) Measured (A) Actual		
Boiler		
Manufacturer/Model		
Water Pressure		
Entering Water	(D)	(M)

Leaving Water	(D)	(M)
Pressure Drop	(D)	(M)
Water Temperature		
Entering Water	(D)	(M)
Leaving Water	(D)	(M)
Rise	(D)	(M)
Water Flow		
Gallons per minute	(D)	(M)
Temperature Controls		
Cut-in	(D)	(A)
Cut-out	(D)	(A)
Combustion Analysis		
Percent Oxygen		
Percent Carbon Dioxide		
Percent Excess Oxygen		
Carbon Spot Test		
Flue Gas Temperature		
Room Temperature		
Boiler Efficiency, Calculated		
Combustion Air Fan Motor		
Type	Single phase	Three phase
Manufacturer/enclosure		
Nameplate		
Horsepower	(D)	(M)
Amperage	(D)	(M)
Voltage	(D)	(M)
Service Factor		
Drives		
Condition	Good Worn	Tight Loose
Number of Belts		
Number of Grooves		
Condition	Good	Worn
Manufacturer/Size		
Fan Sheave, Driven	Adjustable	Fixed
Number of Grooves		
Condition	Good	Worn
Manufacturer/Size		
Heated Water		

Piping	Counter flow	Parallel flow	
Supply	Top	Bottom	
Return	Top	Bottom	
Coil Condition	Dirty	Clean	
Sealed All Sides	Yes	No	
Coil Pressure			
Entering Air	(D)	(M)	
Leaving Air	(D)	(M)	
Drop	(D)	(M)	
Coil Bypass Factor			
Coil Face Velocity			
Feet Per Minute	(D)	(M)	
Entering Air Temperature			
Dry Bulb	(D)	(M)	
Wet Bulb	(D)	(M)	
Leaving Air Temperature			
Dry Bulb	(D)	(M)	
Wet Bulb	(D)	(M)	
Water Temperature			
Entering Water	(D)	(M)	
Leaving Water	(D)	(M)	
Rise	(D)	(M)	
Water Flow			
Gallons Per Minute	(D)	(M)	
Steam			
Piping	Counter flow	Parallel flow	
Supply	Top	Bottom	Center
Return	Top	Bottom	Center
Coil Condition	Dirty	Clean	Combed
Sealed All Sides	Yes	No	
Coil Pressure			
Entering Air	(D)	(M)	
Leaving Air	(D)	(M)	
Drop	(D)	(M)	
Coil Face Velocity			
Feet Per Minute	(D)	(M)	
Entering Air Temperature:			
Dry Bulb	(D)	(M)	

Wet Bulb	(D)	(M)
Leaving Air Temperature:		
Dry Bulb	(D)	(M)
Wet Bulb	(D)	(M)
Fluid Flow		
Condensate GPM	(D)	(M)
Steam Pounds/hour	(D)	(M)
Cooling Equipment		
Chiller Water Cooler-Evaporator		
Manufacturer/Model		
Water Pressure		
Entering Water	(D)	(M)
Leaving Water	(D)	(M)
Pressure Drop	(D)	(M)
Water Temperature		
Entering Water	(D)	(M)
Leaving Water	(D)	(M)
Drop	(D)	(M)
Water Flow		
Gallons per minute	(D)	(M)
Cooling Water Pump		
Type	Single suction	Double suction
Manufacturer/Model		
Speed (rpm)	(D)	(M)
Pump Static Head		
Suction	(D)	(M)
Discharge	(D)	(M)
TDH	(D)	(M)
Cooling Water Pump Motor		
Type	Single phase	Three phase
Manufacturer/enclosure		
Nameplate Horsepower	(D)	(M)
Amperage	(D)	(M)
Voltage	(D)	(M)
Service Factor		
Drives (Note: Most HVAC pumps are direct drive)		
Belt Condition	Good	Tight Loose Worn
Number of Belts		

Motor Sheave, Drive	Adjustable	Fixed
Number of Grooves		
Condition	Good	Worn
Fan Sheave, Driven	Adjustable	Fixed
Motor Sheave Manuf./Size		
Fan Sheave Manuf./Size		
Condenser		
Manufacturer/Model		
Water Pressure		
Entering Water	(D)	(M)
Leaving Water	(D)	(M)
Pressure Drop	(D)	(M)
Water Temperature		
Entering Water	(D)	(M)
Leaving Water	(D)	(M)
Drop	(D)	(M)
Water Flow		
Gallons per minute	(D)	(M)
Condenser Pump		
Type	Single suction	Double suction
Manufacturer/Model		
Speed (rpm)	(D)	(M)
Pump Static Head		
Suction	(D)	(M)
Discharge	(D)	(M)
TDH	(D)	(M)
Condenser Pump Motor		
Type	Single phase	Three phase
Manufacturer/enclosure		
Nameplate		
Horsepower	(D)	(M)
Amperage	(D)	(M)
Voltage	(D)	(M)
Service Factor		
Drives (Note: Most HVAC pumps are direct drive)		
Belt Condition	Good Worn	Tight Loose
Driver Belt Position	Correct	High Low

Number of Belts		
Motor Sheave, Drive	Adjustable	Fixed
Number of Grooves		
Condition	Good	Worn
Fan Sheave, Driven	Adjustable	Fixed
Number of Grooves		
Motor Sheave Manuf./Size		
Fan Sheave Manuf./Size		
Compressor		
Manufacturer/Model		
Refrigerant		
Charge		
Operating Pressure	Suction	Discharge
Type Metering Device		
Speed (rpm)		
Compressor Motor		
Type	Single phase	Three phase
Motor Manuf./enclosure		
Nameplate		
Horsepower	(D)	(M)
Amperage	(D)	(M)
Voltage	(D)	(M)
Service Factor		
Drives		
Belt Condition	Good	Tight Loose Worn
Driver Belt Position	Correct	High Low
Number of Belts		
Motor Sheave, Drive	Adjustable	Fixed
Number of Grooves		
Condition	Good	Worn
Fan Sheave, Driven	Adjustable	Fixed
Number of Grooves		
Motor Sheave Manuf./Size		
Fan Sheave Manuf./Size		
Cooling Tower		
Manufacturer/Model		
Range	(D)	(M)
Rise	(D)	(M)

Cooling Coil		
Type	Refrigeration	Chilled Water
Coil Size		
Height		
Width		
Depth		
Rows		
Fins Per Inch		
Chilled Water		
Piping	Counter flow	Parallel flow
Supply	Top	Bottom Center
Return	Top	Bottom Center
Coil Condition	Dirty	Clean Combed
Sealed All Sides	Yes	No
Coil Pressure		
Entering Air	(D)	(M)
Leaving Air	(D)	(M)
Drop	(D)	(M)
Coil Bypass Factor		
Coil Face Velocity		
Feet Per Minute	(D)	(M)
Entering Air Temperature		
Dry Bulb	(D)	(M)
Wet Bulb	(D)	(M)
Leaving Air Temperature		
Dry Bulb	(D)	(M)
Wet Bulb	(D)	(M)
Water Temperature		
Entering Water	(D)	(M)
Leaving Water	(D)	(M)
Rise	(D)	(M)
Water Flow		
Gallons Per Minute	(D)	(M)

Chapter 12

Lexicon

A is for Absolute

absolute: (Definition) Relating to or derived in the simplest manner from the fundamental units of length, mass, and time.

absolute pressure: (Physics) Pressure measured with respect to zero pressure. Total of the indicated gauge pressure plus atmospheric pressure. Aka true pressure. Units: (force per unit of area) pounds per square inch absolute.

absolute temperature: (Thermodynamics) Temperature measured with respect to zero temperature, Kelvin scale or Rankine scale. Kelvin is degrees Celsius plus 273 (rounded from 273.15). Rankine is degrees Fahrenheit plus 460 (rounded from 459.67). See table in Chapter 14.

absolute zero: (Thermodynamics) The theoretical absence of any thermal energy. It is minus 459.67 degrees Fahrenheit or minus 273.15 degrees Celsius. Also, zero degrees Rankine or zero degrees Kelvin.

absorb: (Definition) To take in or soak up—Also, to retain wholly, without reflection or transmission of that which is taken in.

absorbent: (Refrigeration) A substance which has the ability of taking in or absorbing another substance. Used in a refrigerant system as a dehydrator. The absorbent removes moisture by chemical action which converts the water to some other substance or compound. In some absorption (chiller) systems, lithium bromide is the absorbent. In other systems, water is the absorbent.

absorber: (Refrigeration) A vessel containing liquid for absorbing refrigerant vapor. See absorption chiller.

absorption chiller: (Refrigeration) A chiller (aka water cooler) in which a refrigerant is continuously evaporated from a refrigerant-absorbent solution then is condensed and allowed to evaporate (i.e., absorbing heat) and then reabsorbed. See Chapter 8.

acceleration: (Physics) The time rate of change of velocity (either an increase or decrease in velocity) expressed in feet (or meters) per second squared. It is a vector quantity having both magnitude and direction.

acceleration due to gravity: (Physics) The rate of increase in velocity of a body free falling in a vacuum. The speed of a body falling freely toward the earth through the action of gravity alone. Equal to 32.2 feet per second squared.

access door: (Airflow) Door installed in ductwork to allow access to fire dampers, controls, etc.

access opening: (Fume Hood) That part of the fume hood through which work is performed.

actuator: (Control) The control device which receives the signal from the controller and positions the controlled device. Aka motor or operator.

adiabatic process or change: (Thermodynamics) Without loss or gain of heat. A thermodynamic process when no heat is added to, or taken from, a substance—A process in which a gas is assumed to change its condition (pressure, volume or temperature) without the transfer of heat to or from the surroundings during the process—A change in volume and pressure of the contents of an enclosure without exchange of heat between the enclosure and its surroundings.

adjustable differential: (Control) The ability to change the difference between the cut-in and cut-out points.

adjustable pulley: (Mechanical, Drive) A pulley that has adjustable belt grooves. Pulleys are used on smaller equipment.

adjustable sheave: (Mechanical, Drive) A pulley that has adjustable belt grooves. Adjustable sheaves are used on larger equipment usually on the driver (motor) shaft but can be on the driven (fan, pump) shaft. The belt grooves can be adjusted within a limited range to increase or decrease fan (pump) speed. Aka variable pitch sheave or variable speed sheave.

adsorb, adsorption: (Definition) The taking up of one substance at the surface of another substance.

adsorbent: (Chemical) The substance on whose surface adsorption of another substance takes place.

after tax cash flow: (Energy Management, Financial) A measure of financial performance. The ability to generate operational cash flow. After Tax Cash Flow is Net income + Depreciation + Amortization + other non-cash charges.

air balancing: (Airflow) Adjusting airflow to provide the correct air velocity and air volume to a conditioned space in order to maintain comfort or working conditions with regards to temperature, pressure and humidity. Aka testing, adjusting and balancing.

air chamber: (Airflow) An air compartment. Aka plenum.

air changes per hour: (Airflow) A method of expressing the quantity of air exchanged per hour in terms of conditioned space volume.

air compressor: (Control) A machine which draws in air at atmospheric pressure, compresses it, and delivers it at higher pressure. In HVAC pneumatic control systems, an electrically driven, reciprocating positive displacement pump 25 horsepower or less. The operating schedule is typically 35% on-time and 65% off-time. See Chapter 10.

air conditioning: (HVAC) Treating or conditioning the temperature, humidity, and cleanliness of the air as well as providing the correct air volume and velocity to meet the design requirements of the conditioned space.

air conditioning unit: (HVAC) An assembly of components for the treatment and conditioning of air. Also known as air handling unit for larger systems, or fan coil unit for smaller systems.

air differential pressure gauge (Instrumentation) An instrument for measuring air pressures and differential pressures. These gauges contain no liquid, but instead work on a diaphragm-and-pointer-system which move within a certain pressure range. Although absolutely level mounting isn't necessary, if the position of the gauge is changed, resetting of the zero adjustment may be required for proper gauge reading as specified by the manufacturer. These gauges have two sets of tubing connector ports for different permanent mounting positions; however, only one set of ports is used for readings and the other set is capped off. The ports are stamped on the gauges as "high" pressure and "low" pressure and when used with a Pitot tube, static tip or other sensing device to measure total pressure, static pressure, or velocity pressure, the tubing is connected from the sensing device to the tubing connector ports in the same manner as the connection to the inclined-vertical manometer. These gauges are generally used more for static pressure readings than for velocity pressures readings.

air dryer: (Control) A refrigeration device for removing moisture from compressed air. To ensure that air lines are oil and dirt free, a refrigerated air dryer equipped with an automatic drain is placed downstream of the receiver tank to remove any moisture carryover.

A manual bypass valve is installed to service the refrigerated air dryer and any downstream filter without interrupting the system operation. See Chapter 10.

air duct: (Airflow) A passageway used to convey air for distribution throughout a building. Air duct is formed to fit the architecture and construction of the building. Duct may be round (the most efficient), rectangular, square, or flat-oval (a round duct that is spread to form essentially a rectangular duct with a semicircle at either side). Chapter 7.

air entrainment: (Fluid Flow) The induced flow of the secondary (room) air by the primary (supply) air, creating a mixed air path. Also, air captured in water.

air filter: (Airflow) A device made of various materials which removes solid particulates such as dust, debris, pollen, mold, and bacteria from air. Air filters are used where air quality is important such as building ventilation and cleanroom applications. See High Efficiency Particulate Air or Ultra Low Penetration Air filters. See Chapter 7.

air filter: (Control) A device installed in a pneumatic control system's compressor's air intake to keep dirt, oil vapors and other contaminants from entering the compressor and being passed through the compressor. Oil and moisture can condense into droplets, mixing with the dirt and clogging the air lines. Air filters are used in the intakes of compressed air systems are typically paper, foam, or cotton filters. See Chapter 10.

airflow pattern: (Airflow) Airflow patterns are important for proper mixing of supply and room air. Cooling only air systems usually distribute air from ceilings or high sidewall outlets with a horizontal pattern across the ceiling. Airflow for cooling only systems distributed from low sidewall or floor outlets have the outlets adjusted to direct the air up (e.g. computer rooms) Heating only air systems are typically distributed from low sidewall or floor outlets. If the heated air is distributed from ceilings or high sidewall outlets then the airflow pattern for is directed down. Most HVAC systems handle both heated and cooled air as the seasons change, therefore a compromise is made which usually favors the cooling season and ceiling or high sidewall outlets are specified. Generally, these outlets are adjusted for a horizontal pattern. However, if the ceiling is very high or velocities are very low, such as when variable air volume boxes close off, it may be necessary to use a vertical pattern to force the air down to the occupied zone.

airfoil: (Fume Hood) The curved or angular component at the entrance to a fume hood. Airfoils are used to counteract the effect of turbulence generated at the opening of plain entrance hoods.

air handling unit: (HVAC) See air conditioning unit. Aka AHU.

air horsepower: (Mechanical) The theoretical horsepower required to drive a fan if the fan were 100% efficient. Cubic feet per minute times pressure divided by 6356.

air intake: (Airflow) Any opening introducing air into the return air duct, outside air duct, exhaust air duct of an exhaust air system or suction side of the fan. Also see air compressor.

air lines: (Control) The pneumatic air lines (aka tubing or piping) going to controlling devices such as thermostats, humidistats, or pressurestats are mains. The pneumatic air lines leading from controlling devices to the actuator of controlled devices, such as dampers or valves, are branch lines. Pneumatic air lines are generally either copper or polyethylene (plastic) tubing. See Chapter 10.

air lock: (Environmentally Controlled Area) An anteroom with airtight doors between a controlled and uncontrolled space or between controlled (different pressure) spaces.

air meter: (Instrumentation) A portable or stationary apparatus used to measure the rate of airflow.

air monitor: (Instrumentation) Typically, a stationary air meter. See air meter. Also, a measuring instrument used for monitoring radiation contamination.

air pre-heater: (Mechanical) A system of tubes or passages heated by flue gas, or other waste heat, through which combustion air is passed for preheating before admission to the combustion chamber of a boiler or other apparatus. It returns usable heat to the combustion chamber that would otherwise be lost, therefore saving energy.

air pump: (Mechanical) A reciprocating or centrifugal pump used to move or remove air. Types of air pumps includes: air ejector, blower or fan, air compressor, and vacuum pump.

air receiver: (Mechanical) A pressure vessel in which compressed air is stored. Receiver in a pneumatic control system.

air shaft: An air passage, usually vertical or nearly vertical.

air separators: (Water Flow) Air separators free the air entrained in the water in a hydronic system. Two types of air separators centrifugal and dip tube. See Chapter 8.

air valve: (Airflow) A variable air volume device for controlling airflow volume. See Chapters 7 and 9.

air volume: (Airflow) Quantity of airflow expressed in cubic feet per minute.

Ak factor: (Airflow) The effective area of an air outlet or inlet. Also called K factor. Manufacturers of supply air outlets conduct various air-flow tests on their products. From these tests, they establish an effective area of the outlet. The effective area of an outlet is defined as the sum of the areas of all the vena contracta (the smallest area of an air stream leaving an orifice) existing at the outlet. The effective area is based on the number of orifices and the exact location of the vena contracta, along with the size and shape of the grille bars, diffuser rings, etc. Based on their findings, the manufacturers publish area correction factors or flow factors for their products. These "Ak" flow factors apply to a specific type and size of grille, register or diffuser, a specific air measuring instrument, and the correct positioning of that particular instrument.

alarm: (Control) A signal, either audible, visible or both, that warns of an abnormal and critical operating condition.

alert: (Control) A form of alarm that warns of an abnormal, but not critical, operating condition.

algorithm: (Computer) A set of rules, precisely defined, which specify a sequence of actions to be taken to solve a problem.

alignment: (Mechanical) Setting in a straight line, e. g., the driver (motor) sheave with the driven (fan) sheave.

alternating current: (Electrical) The type of electrical circuit in which the current constantly reverses (alternates) flow. The standard U. S. electrical service used in HVAC systems is single-phase or three-phase (polyphase) alternating current at 60 cycles per second.

alternating current motor: (Electrical) An electric motor that operates from single or polyphase alternating current supply. See alternating current, capacitor motor, induction motor, synchronous motor.

ambient air: (Airflow) The surrounding air.

ambient air temperature: (Thermodynamics) The temperature of the surrounding air.

ambient compensated: (Control) A control device designed to allow for ambient air temperatures.

ammeter: (Instrumentation) An instrument for measuring current Aka amperage.

amperage: (Electrical) Electron (current) flow of one coulomb per second past a given point. Current in amperes, especially the rated current of an electrical apparatus, such as a motor.

ampere, amp: (Electrical) A measure (unit) of electrical flow rate. One ampere is equal to one coulomb per second (6.3×10^{18} electrons per second).

amplifier: (Control) A device used to increase signal power or amplitude. *amplitude:* (Electrical) The maximum value of an alternating current. See peak value.

amplitude modulation: (Physics) Variation of amplitude of a transmitted wave in accordance with impressed modulation, the frequency remaining constant. Aka AM.

analog: (Control) An analog (aka analogue) control uses a continuous physical signal such as voltage or pressure to represent and manipulate the measurements it handles. Electrical, mechanical, pneumatic, hydraulic, and other systems use analog signals. See analog signal.

analog device: (Control) Analog devices include pneumatic controllers, transducers, relays and actuators.

analog input: (Control) An electrical input of variable value (temperature, humidity or pressure) provided by a sensing device.

analog point: (Control) A point that has a variable value, such as temperature, humidity or pressure, which will be measured by a sensing device to provide input to the control.

analog-to-digital: (Control) The conversion of a sampled analog signal to a digital code that represents the voltage level of the analog signal at the instant of sampling.

analog-to-digital converter: (Control) Device for converting analog signals into digital ones for subsequent computer processing or transmission. The part of a microprocessor based controller that changes analog input values to a digital number for use in a software program.

analog output: (Control) An electrical output of variable value such as temperature, humidity or pressure.

analog signal: (Control) A type of signal whose level varies smoothly and continuously in amplitude or frequency. Traditionally, HVAC control has been performed by analog devices such as pneumatic controllers, transducers, relays and actuators. An analog signal is any time continuous signal where some time varying feature of the signal is a representation of some other time varying quantity. HVAC analog signals typically measure response to changes in sound, light, temperature, humidity, pressure or position using a transducer.

anemometer: (Instrumentation) An instrument used to measure air velocities (speed) either by mechanical (e.g., swing-vane/deflecting-vane or rotating-vane anemometer) or electrical (e.g., hot-wire anemometer) methods.

annual load factor: (Electrical) The electrical load factor of a generating station or consumer, etc. taken over a whole year.

annular: (Definition) of, relating to, or forming a ring.

annular flow meter: (Instrumentation) A multi-ported flow sensor permanently installed in water pipes. The holes in the sensor are spaced to represent equal annular areas of the pipe in the same manner as the Pitot tube traverse is for round duct or water pipe. The flow meter senses the velocity of the water as it passes the sensor. The upstream ports sense high pressure and the downstream ports sense low pressure. The resulting difference or differential pressure is measured with an appropriate differential pressure gauge. Calibration data, which show flow rate in gallons per minute versus measured pressure drop, comes with the flow meter. See flow meters.

annunciator: (Instrumentation) Electrically controlled signal board with an arrangement of indicators which display details on operational conditions.

anticipating control: (Control) A control method which reduces the operating differential of the system by anticipating the temperature needed for the controlled variable to maintain space conditions. A boiler reset is an example of anticipating control. See Chapter 10.

anticipator: (Control) A method of reducing the operating differential of a space heating system by adding a small resistive heater inside the thermostat to raise the internal temperature of the thermostat faster than the surrounding room temperature. This causes the thermostat to shut off the heating equipment sooner than it would if affected only by the room temperature since some areas of the conditioned space reach temperature setpoint before the thermostat does.

apparatus dew point: (Psychrometrics) The temperature which would result if the psychrometric process line were carried to the saturation condition of the leaving air while maintaining the same sensible heat ratio.

apparatus dew point: (Refrigeration) The temperature of the cooling coil at saturation. Aka effective coil temperature.

apparent power: (Electrical) Apparent power is the product of voltage and current (amperage) expressed in volt-amperes or kilo volt-amperes. See power factor, real power.

artificial intelligence: (Computer) The concept that computers can be programmed to assume capabilities that are thought to be like human intelligence: learning, reasoning, adaptation, and self-correction.

as built drawing: (Definition) A drawing or set of drawings which depicts the actual as built conditions of the completed construction or changes found during an inspection of an installation.

as installed: (Fume Hood) After construction but before occupancy. An "as installed" fume hood test determines the hood performance rating when many of the variables affecting fume hood performance can be controlled. A well-designed and located fume hood in a properly air balanced laboratory should be able to achieve an acceptable performance rating.

as manufactured: (Fume Hood) A fume hood test to evaluate a fume hood in the relatively ideal conditions of the manufacturer's test lab where many of the factors that adversely affect hood performance can be eliminated.

aspect ratio: (Airflow) The aspect ratio in rectangular ducts and air inlets or outlets is the ratio of the adjacent sides (width to height). For example, a duct 24 inches x 18 inches has an aspect ratio of 1.33:1. To minimize friction losses for energy conservation considerations the aspect ratio should not exceed 3:1.

as used: (Fume Hood) A fume hood test intended to be conducted within the hood as it is typically used. The exposure to the user at the hood will be dependent upon various factors including: apparatus used in the hood, apparatus forces the user to conduct tests too far forward, hood is used as a storage cabinet, etc.

atmospheric pressure: (Physics)The pressure exerted by the atmosphere at the surface of the earth due to the weight of the air. Expressed as 14.696 (rounded to 14.7) pounds per square inch (psi) or 29.921 (rounded to 29.92 or 30) inches of mercury, or 33.9 (rounded to 34) feet of water. Variations in atmospheric pressure are measured with a barometer.

authority: (Control) The adjustment of a controller which determines the effect of the secondary signal as a percentage of the primary signal.

automatic air vent: (Water Flow) Removes entrained air from water systems. There are several types including hydroscopic and float. See Chapter 8.

automatic control: (Control) An HVAC control system that reacts to a change or unbalance in the controlled condition by adjusting the variables, such as temperature, pressure or humidity, to restore the system to the desired balance.

automatic drain: (Control) A drain installed on the refrigerated air dryer of a pneumatic air compressor which automatically removes moisture from the dryer.

automatic flow limiting valve: (Control) A self-powered water flow control valve that limits flow (gpm) to a preset value when the differential pressure across the valve is within a certain pressure (psi) range. Typically the valve is controlled by a spring. As the differential pressure across the valve depresses the spring it opens the valve's orifice and delivers a constant flow rate (gpm) within the limits of the spring. Aka self-limiting valve or automatic flow control valve. A valve selected for a given pipe size can be preset to the gpm flow to match the component's (coil, etc.) requirement. Therefore, care must be taken during installation to make sure the correct gpm valve is in the appropriate pipe.

automatic temperature control damper: (Control) An air damper controlled by temperature requirements of the system. Automatic temperature control dampers are opposed-blade or parallel-blade and can be either two-position or modulating. Two-position control means the damper is either open or closed. Example of two-position control: An automatic minimum outside air damper that is full open when the fan is on and shut when the fan is off. Modulating control provides for the gradual opening or closing of a damper. Example of modulating control: Automatic return and outside air dampers. The outside air damper closes as the return air damper opens and vice versa to allow more or less return or outside air to enter the mixed air plenum to satisfy the mixed air temperature requirements. For energy conservation automatic temperature control dampers should have a tight shutoff when closed.

automatic temperature control valve: (Control) Water valves controlled by temperature requirements of the system. Automatic temperature control valves are used to control flow rate or to mix or divert water streams. They are classified as two-way or three-way construction and either modulating or two-position. See two-way valve automatic temperature control valve and three-way valve automatic temperature control valve.

auxiliary air: (Fume Hood) Supply or supplemental air delivered to a laboratory fume hood to reduce room air consumption. Exhaust air requirements for laboratory fume hoods and other containment devices often exceed the supply air needed for the normal room air conditioning. To meet these needs, auxiliary air can be ducted directly into the fume hood or supplied to the room. The auxiliary air should be conditioned to meet the temperature and humidity requirements of the lab. Energy savings may be obtained when the heating, cooling and humidifying requirements are kept to a minimum. In some laboratory buildings, a corridor separates the offices from the labs. The offices are under positive pressure. Air leaves the offices and enters into the corridor. The corridor acts as a plenum. The air then flows into the negatively pressurized laboratory. The result is that it may be possible to reduce the auxiliary air or even eliminate it.

auxiliary authority: (Control) The amount of measured variable change at one sensing element compared to the change measured at the other sensing element, which causes the output signal to return to its original intermediate position, expressed as a percentage.

auxiliary contacts: (Control) A secondary set of electrical contacts mounted on a modulating motor or magnetic starter whose operation coincides with the operation of the motor or starter usually for pilot duty, low amp rating.

auxiliary potentiometer: (Control) A potentiometer ("pot") mounted on a modulating motor used to control other modulating devices in response to the motor's operation. Also known as "follow-up pot." See potentiometer.

auxiliary switch: (Control) A small switch operated from a main switch or circuit breaker also known as an "end switch."

averaging element: (Control) A sensing element that responds to the average temperature, pressure, etc., sensed at many different points in a fluid stream.

axial flow fan: (Airflow) Fan in which the airflow within the fan wheel is parallel to the fan shaft. Axial fans are further categorized as propeller, tube axial and vane axial.

B is for Backdraft Damper

backdraft damper: (Airflow) A damper that opens when there is a drop in pressure across the damper in the direction of airflow and closes under the action of gravitational force when there is no airflow.

backflow: (Water Flow) The undesirable reversal of flow (downstream pressure is greater than upstream or supply pressure) of non-potable water or other substances through a cross-connection and into the piping of a public water system or consumer's potable water system. Backflow is caused by an increase in downstream pressure, a reduction in the potable water supply pressure, or a combination of both. Increases in downstream pressure can be created by pumps, temperature increases in boilers, etc. Reductions in potable water supply pressure occur whenever the amount of water being used exceeds the amount of water being supplied, such as during water line flushing, fighting fires, or breaks in water mains. *backflow preventer:* (Water Flow) A mechanism to prevent backflow which provides a physical barrier to backflow. The principal types of mechanical backflow preventer are the reduced-pressure principle assembly, the pressure vacuum breaker assembly, and the double check valve assembly. A secondary type of mechanical backflow preventer is the residential dual check valve.

BACnet: A data communication protocol. Aka Building Automation and Control Networks.

back pressure: (Mechanical) The pressure exerted on a moving fluid (air, water, refrigerant) by obstructions or tight bends in the air vents or piping along which it is moving against its direction of flow. See capture hood, evaporator pressure.

baffle: (Definition) Any device to impede, direct, divide, deflect, check, or regulate flow or passage of a fluid, light, or sound. For example, baffles installed in a constant air volume terminal box to reflect sound back into the box where it can be absorbed by the box lining.

baffle: (Fume Hood) A panel located across the fume hood interior back which controls the pattern of air moving into and through the fume hood. The baffle is constructed so that it is impossible, by adjustment, to restrict airflow through the fume hood by more than 20%.

balanced current: (Electrical) All currents in a polyphase electric system are equal.

balanced load: (Electrical) A load connected to an electrical system in such a way that the currents taken from each phase, or from each side of the system are equal and at equal power factors.

balanced voltage: (Electrical) All voltages in a polyphase electric system are equal.

balancing station: (Fluid Flow) An assembly to measure and control airflow or water flow. It has a measuring device and a volume control device. For accurate measurements use the recommended lengths of straight ductwork or straight piping entering and leaving the station.

ballast: (Electrical) An auxiliary electrical device used to provide the starting voltage or to stabilize or limit the current in an electric circuit. Ballasts for fluorescent lamps or high intensity discharge lamps are examples.

ballast: (Lighting) Electric or electronic device that converts electrical current to the right quantity of voltage and amperage required to start the lamp safely and efficiently. The ballast then stabilizes the current, regulating the pace of electric flow after startup to ensure continued safe operation. Electronic lamp ballasts use solid state electronic circuitry to provide the proper starting and operating electrical condition to power one or more fluorescent lamps or high intensity discharge lamps. Compact fluorescent light technology requires ballasts that are either magnetic or electronic. Magnetic core and coil ballasts are less expensive but are considerably heavier than electronic ballasts and require longer to light and often produce a low hum. Electronic ballasts are lightweight and allow the lamp to start instantly, and are more energy efficient.

ball bearings: (Mechanical) A shaft bearing consisting of a number of hardened steel balls which roll in steel ring.

barometer: (Instrumentation) An instrument for measuring atmospheric pressure.

barometric pressure: (Definition) The pressure of the atmosphere as measured with a barometer. Expressed in inches of mercury.

bearings: (Mechanical) A support for rotating or reciprocating shafts. HVAC compressors, fans, motors and pumps are equipped with various types of bearings. Ball bearings are most common. On

double inlet, double wide fans the bearings are in the air stream
and they are large in relation to the inlet. They will reduce fan
performance if the fan is small. Single inlet, single wide fans have
one fan wheel and a single entry and the motor is opposite the inlet
therefore, the fan bearings are not in the air stream. This makes
single inlet, single wide fans more common in smaller sizes where
inlet duct needs to be attached and in applications where the air
temperature is relatively high or the air is dirty. On double suction
pumps the bearings are located on both sides of the pump casing
while on single suction pumps the bearings are located between
the pump and the motor.

belt drive: (Mechanical) The transmission of power from one shaft to
another by means of an endless belt running over pulleys (aka
sheaves).

belt driven: (Mechanical) A compressor, fan or pump equipped with a
pulley (sheave) on the compressor, fan or pump shaft and connected by belt to a pulley (sheave) on the motor shaft.

belt guard: (Mechanical) A safety device around drive belts to protect
workers from harm.

belt slip: (Mechanical) The slipping of a driving belt in a pulley or
sheave due to insufficient friction grip to overcome the resistance
to motion.

bench mounted fume hood: (Fume Hood) A fume hood that rests on a
counter top.

bimetallic strip: (Electrical) A bimetallic strip is comprised of two metals
having different temperature coefficients of expansion. The strip
deflects when one side of the strip expands more than the other
side. Bimetallic strip are used in thermal switches, thermometers
and thermostats.

bimetallic thermometer: (Instrumentation) Thermometer made with
two different metals which bend at different rates at the same temperature. See bimetallic strip. Bimetallic dial thermometers used
for commissioning or testing and balancing work are acceptable
when they are calibrated against a standard glass stem mercury
thermometer before each use.

bimetallic thermostat: (Control) Thermostat made with two different
metals which bend at different rates at the same temperature. See
bimetallic strip.

binary code: (Computer) A representation of information using a sequence of 0s and 1s. The figures 0 and 1 can be used in electric circuits to represent "on-off," or "pulse-no pulse." See character code.

biogas: (Energy) Gas, typically methane, produced by bacterial fermentation of organic matter.

biological safety cabinet: (Fume Hood) A special safety enclosure which uses air currents to protect the user. Biological safety cabinets (also known as safety cabinets, laminar flow cabinets and glove boxes) are used to handle pathogenic microorganisms.

biomass: (Energy) Organic matter (mostly from plants) harvested as a source of energy.

bit: (Computer) **B**inary dig**it**, a digit in binary code, i.e. 0 or 1. It is the smallest amount of storage. 8 bits (b) = 1 byte (B)

blueprint: (Image Technology) Process for reproducing plans and engineering drawings.

boiler: (Heating) A pressure vessel in which water is heated and then discharged either as hot water or low pressure steam for heating, or as high pressure steam for power generation, process function, etc.

boiler capacity: (Mechanical) Capacity stated in boiler horsepower or the weight of steam in pounds per hour which a boiler can generate at full load output.

boiler efficiency: (Energy Management) The ratio of the heat supplied by a boiler in the heating and evaporating of the feed water to the heat supplied to the boiler in the fuel. Boiler efficiency range is approximately 60% to 90%.

boiler feed water: (Mechanical) The water into the boiler. It is treated to remove air and impurities.

boiler horsepower: (Mechanical) Equal to 33,475 British thermal units per hour (9,810 watts) which is the energy rate needed to evaporate 34.5 pounds of water at 212 F in one hour. Also, a boiler's capacity to deliver stream.

boiler pressure: (Mechanical) The pressure in the boiler. It may range from atmospheric pressure for heating purposes to 1500 pounds per square inch and more for steam systems.

boiler scale: (Chemical) A hard coating (mainly calcium sulfate) deposited on the surfaces of plates and tubes in contact with water. If excessive, it leads to overheating of the metal and eventual metal failure. As the scale builds it reduces the heat transfer to the water. One-eight inch of scale reduces boiler efficiency by 10%.

boiler tubes: (Mechanical) Steel tubes forming part of the heat transfer surface in a boiler. See fire-tube and water-tube boilers.

boiler water treatment: (Chemical) Chemicals such as sodium compounds (soda ash), organics matter, and barium compounds introduced into boiler feed water to inhibit corrosion and the formation of scale.

boiling: (Physics) The very rapid conversion of a liquid into vapor. Boiling occurs when the liquid's temperature reaches such a value that the saturated vapor pressure of the liquid equals the pressure of the atmosphere. Boiling occurs only at the saturation temperature. Boiling is throughout the entire body of the liquid.

boiling point: (Physics) The temperature at which a liquid boils when exposed to the atmosphere. Since at the boiling point the saturated vapor pressure of a liquid equals the pressure of the atmosphere the boiling point varies with pressure.

booster fan: (Airflow) A fan used to increase the pressure of the airflow in some part of the duct system. A series (or in series) fan.

booster pump: (Water Flow) A pump used to increase the pressure of the water flow in some part of the pipe system. A series (or in series) pump.

bore: (Mechanical) The circular hole along the axis of a pipe or machine part (e. g., pulley or sheave), the internal wall of a cylinder, the diameter of such a hole.

bottom dead center: (Refrigeration) When the piston in a reciprocating compressor is at the bottom of its stroke.

Bourdon tube test gauge: (Instrumentation) A high quality test gauge with an accuracy of one-half of one percent of maximum scale reading. It has a mirrored back to help eliminate parallax error. Widely used for test and balance, commissioning, and energy management work to measure static pressures at pumps, terminal units, and primary heat exchange equipment such as chillers and condensers. Pressure test gauges are designed to measure pressures above atmospheric (pounds per square inch). Vacuum test gauges measure pressures below atmospheric (inches of mercury). Compound test gauges measure pressures above and below atmospheric and read in pounds per square inch above atmospheric and inches of mercury below atmospheric. Bourdon tube pressure test gauges, vacuum test gauges and compound test gauges are available in various ranges but all are calibrated to read zero at atmospheric pressure. Test gauges should be selected so the pressures measured fall in the upper one-half of the scale. A pulsation snubber, restrictor, or needle valve is used to stop system pressure pulsations and protect the test gauge from pressures above or below the limits of the scale. When using test gauges apply and remove pressure slowly by gradually opening the gauge cock. When using a Bourdon tube test gauge ensure that the gauge is at the same height for both the entering and leaving reading or that a correction is made for a difference in height. The need for correction and the possibility of error resulting from the correction can be eliminated by using only one test gauge with a manifold. With a single gauge connected in this manner the gauge is alternately opened to the high pressure side and then the low pressure side to determine the pressure differential eliminating any problem about gauge elevation.

brake: (Mechanical) A device for applying resistance to the motion of a body either to retard it (as with a vehicle brake) or to absorb and measure the power developed by an engine or motor. See brake horsepower.

brake horsepower: (Mechanical) The horsepower applied to the drive shaft of any piece of rotating equipment. The actual power required to drive a compressor, fan or pump under load. Air Horsepower divided by fan efficiency or water horsepower divided by pump efficiency.

branch duct: (Airflow) An air duct that comes off a main air duct. *branch pipe:* (Water Flow) A water pipe that comes off a main water pipe.

brightness: (Lighting) As a quantitative term brightness is depreciated. See luminance.

British thermal unit: (Heat) The heat required to raise the temperature of one pound of water one degree Fahrenheit. A unit of energy.

British thermal unit per hour: (Heat) Used to quantify heat losses and heat gains in the conditioned space and to identify the heating and cooling capacities of various types of equipment. A unit of power.

bug: (Computer) A defect in a computer program or equipment. *building commissioning:* See commissioning.

built-up system: (Definition) A built-up heating, ventilating and air conditioning system. It is custom designed for the building in which it is installed.

bump: (Mechanical) To momentarily start and stop a motor to visually determine the direction of the motor and attached fan or pump.

busbar: (Electrical) Length of constant voltage conductor in an electrical power distribution circuit consisting of thick strips of copper or aluminum that conduct electricity within a switchboard, distribution board, substation, or other electrical apparatus.

busduct: (Electrical) Metal enclosure for busbar. Aka busway.

bushing: (Mechanical) A removable cylindrical lining for an opening (e.g., the bore in a sheave) used to limit the size of the opening. For example, bushings for a sheave may range from 1 inch to 3 inches to accommodate various shaft sizes.

butterfly valve: (Water Flow) A manual valve used for regulating water flow. It has a heavy ring enclosing a disc that rotates on an axis. It has a low pressure drop and is used as a balancing valve but it does not have the good throttling characteristics of a ball valve or plug valve.

bypass: (Fume Hood) A compensating opening that maintains a relatively constant volume of exhaust through a fume hood regardless of sash position. The bypass functions to limit the maximum face velocity as the sash is closed.

bypass: (Mechanical) Any device for diverting fluid flow around instead of through; as in face and bypass dampers, hot-gas bypass, return air bypass dampers, three-way air valve, three-way water valve, variable air volume bypass.

byte: (Computer) Fixed number of bits, often corresponding to a single character. 1 B (byte) = 8 bits (b)

C is for Cone of Silence

calibrated balancing valve: (Instrumentation) A valve that is both a flow meter and a balancing valve. With the other types of flow meters such as orifice plate, Venturi and annular type, a balancing valve is also needed to set the flow. These valves are similar to ordinary balancing valves except the manufacturer has provided pressure taps in the inlet and outlet and has calibrated the device by measuring the resistance at various valve positions against known flow quantities. The valve has a graduated scale or dial to show the degree open. Calibration data which show flow rate in gallons per minute versus measured pressure drop are provided with the valve. Pressure drop is measured with any appropriate differential gauge. See flow meter, annular flow meter, orifice plate and Venturi.

calibration: (Instrumentation) Determining or correcting the error of an existing scale on an instrument.

California hood: (Fume Hood) A rectangular enclosure used to house distillation and other large research apparatus. It can provide visibility from all sides with horizontal sliding access doors along the length of the assembly. The California hood is not considered a laboratory fume hood.

calorific value: (Heat Energy) The number of heat units obtained by the combustion of a fuel in British thermal units per pound (solid and liquid fuels) or British thermal units per cubic foot (gaseous fuels). In fuels containing hydrogen, which burns to water vapor, there are two heating values: gross and net.

candlepower: (Lighting) Basic unit for measuring luminous intensity from a light source in a given direction. **Aka candela.**

canopy hood: (Fume Hood) A suspended ventilating device used to exhaust only heat, water vapor and odors. The canopy hood is not considered a laboratory fume hood.

capacitance: (Electrical) A measure of the amount of electric charge stored for a given electric potential.

capacitive load: (Electrical) A leading alternating current. Counteracts a lagging power factor. Aka leading power factor.

capacitor: (Electrical) An electric component having capacitance. It stores electrical energy between a pair of conductors (plates) and prevents rapid changes of voltage across its terminals. The process of storing energy in the capacitor is known as charging; when electric charges of equal magnitude, but opposite polarity, build up on each plate.

capacitor motor: (Electrical) A single phase induction motor arranged to start as a two-phase motor by connecting a capacitor in series with an auxiliary starting winding. The capacitor may be automatically disconnected when the motor is up to speed (capacitor start motor) or it may be left permanently in the circuit for power improvement (capacitor start run motor).

capacitor start: (Electrical) Starting unit for electric motor using series capacitance to advance phase of current.

capacity control: (Refrigeration) Various capacity control methods are used on compressors in refrigeration and air conditioning systems for reducing compressor cycling, decreasing the starting load, providing better dehumidification, and reducing power and energy consumption. Compressor capacity control is desirable for optimum system performance when loads vary widely due to changes in lighting, occupancy, product loading, ambient weather, etc. Capacity control methods include on-off cycling, variable speed, cylinder bypass (hot gas bypass) and cylinder unloading. Some applications will use two or more methods for smoother switching and better control such as unloading in conjunction with hot gas bypass. Capacity control also allows for a more continuous operation of the compressor which minimizes electrical problems and improves lubrication.

capillarity: (Physics) A phenomenon associated with surface tension which occurs in small bore tubes.

capillary action: (Physics) The effect that causes a liquid to form a concave meniscus in a small bore tube. Surface tension pulls the liquid (such as water or refrigerant) through a tube. See capillarity.

capillary pressure: (Physics) The pressure exerted by capillarity.

capillary tube: (Control) See temperature sensor.

capillary tube: (Refrigeration) See metering device.

capture hood: (Instrumentation) An instrument which captures the air of a supply, return or exhaust terminal and guides it over a flow-measuring device. It measures airflow directly in cubic feet per minute or liters per second. A small amount of back pressure is created when the hood is placed over the terminal. Some hoods have compensating devices for back pressure.

capture velocity: (Fume Hood) The air velocity at the fume hood face necessary to overcome opposing air currents and contain contaminated air within the fume hood or industrial hood (smoke hood, etc.).

carbon dioxide: (Chemical) A colorless gas produced by complete combustion of carbon.

carbon dioxide analyzer/recorder: (Instrumentation) An instrument which analyses automatically the flue gas leaving a furnace or boiler and records the percentage of carbon dioxide.

carbon monoxide: (Chemical) A colorless, odorless and poisonous gas. It is the product of incomplete combustion of carbon.

Carnot cycle: (Heat) An ideal heat engine cycle of maximum thermal efficiency. The Carnot cycle is the most efficient cycle possible for converting a given amount of thermal energy into work or, conversely, for using a given amount of work for refrigeration purposes. See heat engine.

cavitation: (Water Flow) The phenomena occurring in a flowing liquid

when the pressure falls below the vapor pressure of the liquid causing the liquid to vaporize and form bubbles. In an HVAC water
system the bubbles are entrained in the flowing water and are carried through the pump impeller inlet to a zone of higher pressure
where they suddenly collapse or implode with force. The following
are symptoms of a cavitating pump: snapping and crackling noises
at the pump inlet, severe vibration, a drop in pressure and brake
horsepower, and a reduction in flow, or the flow stops completely.

ceiling diffuser: (Airflow) A diffuser which typically provides a horizontal flow pattern which tends to flow along the ceiling producing a
high degree of surface effect. Typical square or rectangular ceiling
diffusers deliver air in a one, two, three or four-way pattern. Round
ceiling diffusers deliver air in all directions.

ceiling reflectance: (Lighting) See reflectance.

Celsius scale: (Thermodynamics) The International System of Units
name for Centigrade scale. The original Celsius scale was marked
zero at the boiling point and 100 at the freezing point but was inverted in 1750.

Centigrade scale: (Thermodynamics) The most widely used method
of graduating a thermometer. The fundamental interval of temperature between the freezing and boiling points of pure water is
divided into 100 equal parts each of which is a Centigrade degree.
Freezing is 0 and boiling is 100. To convert a temperature on the
Centigrade scale to the Fahrenheit scale multiply by 1.8 and add 32.
For the Kelvin scale add 273.

central processing unit: (Computer) The main or central control element
of a computer consisting of an arithmetic logic circuits unit and a
control unit which supervises the executions of instructions.

centrifugal fan: (Mechanical) A fan in which the airflow within the fan
wheel is radial or circular to the fan shaft.

centrifugal force: (Physics) Acting in a direction away from a center or
axis, the effects of inertia in connection with rotation, an outward
force away from the center of rotation.

centrifugal pump: (Mechanical) A pump in which the water flow within
the pump impeller is radial or circular to the pump shaft. Not positive displacement.

change of state: (Physics) A change from solid to liquid, solid to vapor,
liquid to vapor or vice versa.

chattering valve: (Water Flow) Water valves must be installed with the direction of flow opposing the closing action of the valve plug as the water pressure tends to push the valve plug open. If the valve is installed the opposite way it will cause chattering because as the valve plug is modulated by the controls to the closed position the velocity of the water around the plug becomes very high. Therefore, if the flow and pressure were with the closing of the plug then at some point near closing the velocity pressure would overcome the spring resistance and force the plug closed. Then, when flow is stopped the velocity pressure goes to zero and the control spring takes over and opens the plug. The cycle is repeated and chattering is the result.

check valve: (Water Flow) A check valve limits the direction of fluid flow. Used to control various fluids. The water check valve allows water to flow in one direction and stops its flow in the other direction. The swing check valve has a gate that opens when the system is turned on and pressure from the water flowing in the proper direction is applied. The gate closes due to its own weight and gravity when the system is off. The spring loaded check valve has a spring that keeps the valve closed. Water pressure from the proper direction against the spring opens the valve.

chiller: (Refrigeration) A chiller (aka water chiller or water cooler) is a machine that removes heat from a water using a vapor-compression (mechanical chiller) or absorption (absorption chiller) refrigeration cycle. The water may contain glycol and corrosion inhibitors. In air conditioning systems, chilled water is distributed to heat exchangers, or cooling coils, in air handling units. The cooling coils transfer sensible heat and latent heat from the air to the chilled water cooling and usually dehumidifying the air stream. The warmed water is returned to the chiller for cooling. Chilled water is used to cool and dehumidify air in mid- to large-size commercial, industrial, and institutional facilities. A typical chiller for air conditioning applications is rated between 15 to 1500 tons (180,000 to 18,000,000 British thermal units) in cooling capacity. See mechanical chiller.

chip: (Control) See integrated circuit.

chronometric tachometer: (Instrumentation) An analog mechanical contact tachometer. It's a combination of a precision stop watch and a revolution counter.

circuit: (Electrical) Arrangement of conductors and passive and active components forming a path or paths for electrical current. Arrangement of one or more complete paths for electron flow.

circuit breaker: (Electrical) Device for opening an electric circuit under abnormal operating conditions, e.g., excessive current or heat.

circuit diagram: (Electrical) Conventional representation of wiring system of electrical or electronic equipment.

circulating pump: (Mechanical) A centrifugal pump used to circulate water.

cleanroom: (Environmentally Controlled Area) A specially constructed, enclosed area environmentally controlled with respect to airborne particles, temperature, humidity, airflow patterns, air motion, sound, vibration, and lighting which contains one or more clean zones. A cleanroom is constructed and used in a manner to minimize the introduction, generation, and retention of particles inside the room and in which the concentration of airborne particles is controlled. High efficiency particulate air and ultra low penetration air filters are used to remove contaminants. A cleanroom has a controlled level of contamination that is specified by the number of particles per cubic meter at a specified particle size. See cleanroom classes and class 1 through class 100,000. Particle levels are usually tested using a particle counter. Other relevant parameters, e.g. temperature, humidity, and pressure, are controlled as necessary. Some cleanroom systems control humidity to prevent electrostatic discharges (e.g., static electricity) as well as maintain positive pressure so that air leaks out of the cleanroom keeping unfiltered air from coming in. Personnel enter and leave through airlocks and wear protective clothing (a cleanroom suit) such as hair covers, face masks, gloves, shoe covers and coveralls.

cleanroom class: United States cleanness classifications for cleanrooms and clean zones come from U. S. Federal Standard 209. International cleanroom cleanness classifications come from the ISO, which is the International Organization for Standardization, a worldwide federation of national standards members. ISO 14644 and Federal Standard 209 classes are based on the number of airborne micrometer-sized particles per unit volume (concentration) in the cleanroom or clean zone. Cleanness classifications, from most clean to least clean, are ISO Class 1 to ISO Class 9; Federal Standard 209, U. S. Class 1 to U. S. Class 100,000; and metric class M1 to M7.

cleanroom class 1: The particle count in this cleanroom cannot exceed 1 particle (particle size 0.5 micrometer) per cubic foot of air or (3 particles at 0.3 micrometer, 7.5 particles at 0.2 micrometer, 35 particles at 0.1 micrometer per cubic foot of air). Also, metric class M1.5, ISO class 3.

cleanroom class 10: The particle count in this cleanroom cannot exceed 10 particles (particle size 0.5 micrometer) per cubic foot of air or (30 particles at 0.3 micrometer, 75 particles at 0.2 micrometer, 350 particles at 0.1 micrometer per cubic foot of air). Also, metric class M2.5, ISO class 4.

cleanroom class 100: The particle count in this cleanroom cannot exceed 100 particles per cubic foot of air. Particle size 0.5 micrometer. Also 300 particles at 0.3 micrometer, 750 particles at 0.2 micrometer, 350 particles. Metric class M3.5, ISO class 5.

cleanroom class 1,000: The particle count in this cleanroom cannot exceed 1000 particles per cubic foot of air. Particle size 0.5 micrometer. Metric class M4.5, ISO class 6.

class 10,000 cleanroom: The particle count in this cleanroom cannot exceed 10,000 particles per cubic foot of air. Particles size 0.5 micrometer Also for class 10,000 clean rooms the particle count cannot exceed 70 particles per cubic foot if the particle size is 5.0 micrometer and larger. Metric class M5.5, ISO class 7.

cleanroom class 100,000: The particle count in this cleanroom cannot exceed 100,000 panicles per cubic foot of air. Particles size 0.5 micrometers Also for class 100,000 cleanrooms the particle count cannot exceed 700 particles per cubic foot if the particle size is 5.0 micrometers and larger. Metric class M6.5, ISO class 8.

cleanroom occupancy state, as-built: A cleanroom which is complete and operating with all services connected and functioning. It has no production equipment or personnel.

cleanroom occupancy state, at-rest: A cleanroom which is complete and operating with all services connected and functioning. It has production equipment but no personnel.

cleanroom occupancy state, operating: A cleanroom which is complete and operating with all services connected and functioning. It has production equipment and personnel.

cleanroom operational conditions: The environment that exists within a cleanroom.

clean zone: (Environmentally Controlled Area) A defined or dedicated space in which the concentration of airborne particles is controlled to meet a specified airborne particulate cleanliness class. A clean zone is constructed and used in a manner to minimize the introduction, generation, and retention of particles inside the zone. Other relevant parameters, e.g. temperature, humidity, and pressure, are controlled as necessary. A clean zone may be open or enclosed and may or may not be located within a cleanroom.

clearance: (Refrigeration) In a reciprocating compressor the space between the piston head and the end of the cylinder when the piston is at the top of its stroke.

clockwise rotation (Mechanical) Fan rotation as viewed from the drive side of a centrifugal fan.

closed circuit: (Electrical) The condition which exists when either deliberately or accidentally an electrical conductor or connection is closed (unbroken) with a switch or other device providing a flow of electricity. One in which there is zero impedance to the flow of any current, the voltage dropping to zero.

closed cycle: (Heat) Heat engine in which the working substance continuously circulates without replenishment.

closed cycle control system: (Control) One in which the controller is worked by a change in the quantity being controlled. See closed loop.

closed loop: (Control) Description of any system in which part of the output is fed back to the input to effect a control or regulatory action; also, the performance of such a system when the feedback path is connected. See feedback.

closed loop system: (Control) A control system in which information from the output is fed back and compared to the input to generate an error signal. This error signal is then used to generate the new output signal.

closed impeller: (Mechanical) A pump impeller having shrouds or side walls, designed primarily for handling clear liquids, such as water.

closed water system: (Water Flow) There is no break in the piping circuit and the water is closed to the atmosphere.

clutch: (Mechanical) A device by which two shafts or rotating members may be connected or disconnected, either while at rest or in relative motion.

coanda effect: (Fluid Flow) The tendency of a fluid jet (such as air) to attach itself to a downstream surface roughly parallel to the jet axis. If this surface curves away from the jet the attached flow will follow it, deflecting from the original direction. See surface effect.

coefficient of friction: (Physics) The force resisting the relative motion of solid surfaces, fluid layers, and material elements sliding against each other. Also, the ratio of the limiting friction to the normal reaction between the sliding surfaces.

coefficient of expansion: (Physics) The fractional expansion of the unit of length, area, or volume per degree rise in temperature.

coefficient of heat transfer: (Heat) The heat flow per hour through one square foot of building material when the temperature difference is 1degree between the two sides of walls, floors, etc. (Btu/sf-hr-F).

coefficient of performance: (Refrigeration) Ratio of work performed to energy used. It is the ratio of heat transfer to work input. It may be greater than unity.

coefficient of utilization: (Lighting) A term used in lighting calculations to denote the ratio of the useful light to the total output of the installation. The amount of light (lumens) delivered in a work plane as a percent of the rated lumens of the lamp.

cogeneration: (Definition) The use of heat engines or power stations (aka power plants) to simultaneously generate both electricity and useful heat. Also: The use of energy, e.g., natural gas, to produce both electricity and useful heat. Conventional power plants exhaust the heat created as a byproduct of electricity generation into the environment. Whereas, combined heat and power or bottoming cycle plants or facilities capture the heat byproduct for domestic and industrial heating purposes. Cogeneration is also called combined heat and power. Types of cogeneration include: topping cycle, polygeneration, trigeneration micro generation, distributed energy resource, and Rankine cycle.

coil: (Electrical) A length of insulated conductor wound around a core which can be iron or ferrite. Because current passing in the coil will create a magnetic field which couples with the winding the coil can be considered an inductor.

coil: (Mechanical) A heat transfer device (heat exchanger). They come in a variety of types and sizes and are designed for various fluid combinations. In hydronic applications, coils are used for heating, cooling or dehumidifying air. Hydronic coils are most often made of copper headers and tubes with aluminum or copper fins and galvanized steel frames. HVAC coils come in a variety of types, materials and sizes. The coil frame is designed to support the core and prevent strains caused by expansion and contraction of the core.

coil bypass factor: (Psychrometrics) The coil bypass factor represents the amount of air passing through the coil without being affected by the coil temperature. It is a ratio of the leaving air dry bulb temperature minus the coil temperature divided by the entering air dry bulb temperature minus the coil temperature. A more efficient coil, that is a coil with more rows per inch and/or more rows of tubes, has a lower coil bypass factor.

coil face area: (Airflow) The area (width times height) of a heating or cooling coil across which air flows.

cold: (Heat) A relative term to describe the temperature of an object or area compared to a known temperature. For instance, a 50F outdoor air temperature in the winter might be considered a warm temperature while in the summer it would be a cool temperature.

cold cathode lamp: (Lighting) An electric discharge lamp whose mode of operation is that of a glow discharge e. g., neon lights.

cold deck: (Airflow) In a multizone or dual duct system it is the air chamber after the cooling coil.

color: (Lighting) The color of light is determined by its wavelength. The range of wavelengths that comprise the visible light spectrum make up only a small portion of the entire electromagnetic spectrum. White light contains all the colors of the visible spectrum.

color rendition: (Lighting) Quantitative measure of the ability of a light source (lamp) to influence the color appearance of objects being illuminated. To reproduce the color of said objects faithfully in comparison with an ideal or natural light source (sunlight). It represents the ability of a lamp to render colors accurately and to show color shade variations more clearly. Color rendition is measured on the Color Rendering Index scale (CRI), ranging in value from 0 to 100. Light sources with a high CRI are desirable in color-critical applications.

color temperature: (Lighting) Describes how "warm" or "cool" a light source (lamp) is. Measured on the Kelvin scale it is based on the color of light emitted by an incandescent source. A lamp with a low color temperature 2700K to 3000K, will have a "warm" appearance (red, orange, or yellow) while a lamp with a high color temperature will have a "cool" appearance (blue or blue-white). Direct sunlight is about 5300 Kelvin. Daylight, (blue sky) is around 5000 Kelvin. Cool-colored light is considered better for visual tasks; warm-colored light is considered more flattering to skin tones and clothing.

combination valve: (Control) A valve that regulates flow and limits direction it is a combination of a check valve, calibrated balancing valve and shutoff valve. Made in a straight or angle pattern. The valve acts as a check valve preventing backflow when the pump is off and can be closed for tight shutoff or partly closed for water balancing. The valve has pressure taps for connecting flow gauges to read pressure drop across it. A calibration chart is supplied with the valve for conversion of pressure drop to gallons per minute. The valve also has a memory stop. A combination valve is also called multi-purpose valve or triple-duty valve.

combustion: (Heat) A chemical process, such as with the combining of oxygen with gaseous products from fuel, that produces heat and light. Combustion can be categorized as complete or incomplete. When a fossil fuel (hydrocarbon) burns in air, the products of combustion are carbon dioxide, water, carbon monoxide, pure carbon (soot or ash) and various other compounds including nitrogen. Incomplete combustion occurs when there is not enough oxygen to allow the fuel to react completely with the oxygen to give complete combustion. A deadly product of incomplete combustion is carbon monoxide, a poison. Complete combustion, on the other hand, does not produce carbon monoxide. For HVAC purposes combustion is a chemical reaction between a fossil fuel such as coal, natural gas, liquid petroleum gas or fuel oil and oxygen. Fossil fuels consist mainly of hydrogen and carbon molecules. These fuels also contain minute quantities of other substances (such as sulphur) which are considered impurities. When combustion takes place, the hydrogen and the carbon in the fuel combine with the oxygen in the air to form water vapor (H_2O) and carbon dioxide (CO_2). If the conditions are ideal, the fuel-to-air ratio is controlled at an optimum level, and the heat energy released is captured and utilized to the greatest practical extent. Complete combustion (a condition in which all the carbon and hydrogen in the fuel would be combined with all the oxygen in the air) is a theoretical concept and cannot be attained in HVAC equipment. Therefore, what is attainable is called incomplete combustion. The products of incomplete combustion may include unburned carbon in the form of smoke and soot, carbon monoxide (a poisonous gas) as well as carbon dioxide and water.

combustion air: (Heat) The air entering a furnace or boiler for combustion.

combustion chamber: (Heat) The space in which combustion takes place.

combustion control: (Control) The control, either by an attendant or by automatic devices, of the rate of combustion in a boiler in order to adjust it to the demand of the boiler.

comfort zone: (Indoor Condition) The range of effective temperature and humidity in which the majority of adults feel comfortable. Generally, between 68F to 79F and 40% to 60% relative humidity.

command language: (Computer) The language used to communicate with the operating system.

commissioning, building: Commissioning (aka building commissioning) ensures that a building's heating, ventilating, air conditioning systems, environmental control systems and life safety systems are designed, installed, and operating to maximum potential. The commissioning process uncovers potential problems with the building's equipment and systems and identifies solutions to avoid increased operating and maintenance cost, reduced comfort, loss of productivity and life and equipment safety associated with improperly functioning equipment and systems.

compact fluorescent lamp: (Lighting) An energy efficient replacement for incandescent lamp. See fluorescent lamp.

comparator: (Control) A device whose output is a digital 0 or 1 depending on whether the input signal is above or below a given reference point.

compensated induction motor: (Electrical) An induction motor with a commutator winding on the rotor, in addition to the ordinary, primary, and secondary winding; this winding is connected to the circuit in such a way that the motor operates at unity or at a leading power factor.

compensated series motor: (Electrical) Type of alternating current series motor in which a compensating winding is fitted to neutralize the effect of armature reaction and so give a good power factor.

compound gauge: (Instrumentation) A pressure gauge which measures above atmospheric pressures in pounds per square inch and below atmospheric pressures in inches of mercury. Compound gauges are calibrated to read zero at atmospheric pressure.

compressed air: (Control) The compressed air used to operate a pneumatic control system. This air must be kept clean, dry and oil-free in order for the pneumatic components to function correctly. A number of devices are installed in the system including filters and dryers to dry the air and remove oil, vapors, dirt and other contaminants.

compression ratio: (Mechanical) The ratio, in an internal combustion engine, of the total volume enclosed in the cylinder at the outer dead center to the volume at the end of compression; the ratio of the volume between the piston and cylinder head before and after a compression stroke.

compression ratio: (Refrigeration) The ratio of the discharge pressure to the suction pressure. By equation, it is the absolute discharge pressure (psia) divided by the absolute suction pressure (psia). Compression ratio is of primary importance when it approaches a high limit. A high compression ratio (a high head pressure and low suction pressure) results in a loss of efficiency and excessive superheating of the discharge vapor which can damage the compressor.

compression tank: (Water Flow) See expansion tank.

compressor: (Control) See air compressor.

compressor: (Refrigeration) The pump in the refrigeration system. It takes low temperature, low pressure refrigerant gas (vapor) from the evaporator and compresses it to a high temperature, high pressure gas. A centrifugal, reciprocating or rotary pump for raising the pressure of a gas.

compressor capacity control: (Refrigeration) See capacity control.

computer aided design: (Computer) Computer and program with high resolution graphics. Used in a wide range of design activities. Designs can be modified and evaluated rapidly and precisely. Aka CAD.

concentrated solution: (Refrigeration, Absorption System) A solution with a large concentration of absorbent as compared to the amount of dissolved refrigerant.

concentrator: (Refrigeration, Absorption System) A vessel containing a solution of absorbent and refrigerant to which heat is applied for the purpose of boiling away some of the refrigerant. Aka generator.

condensate: (Refrigeration) The liquid obtained as a result of removing from a vapor some portion of the latent heat of evaporation.

condensation: (Physic) Removing heat from a vapor so that it changes to the liquid state and/or increasing pressure on a vapor so that it changes to the liquid state.

condensation: (Psychrometrics) When moist air is cooled below its dew point, water vapor condenses.

condensation stage: (Refrigeration) The cooling of a refrigerant vapor to convert it to a liquid in the condenser.

condenser: (Refrigeration) A vessel with receives high temperature, high pressure refrigerant gas from the compressor and cools the gas to a high temperature, high pressure liquid. Condensers may be air-cooled or water-cooled.

condensing pressure: (Refrigeration) The saturation pressure corresponding to the temperature of the liquid-vapor mixture in the condenser.

condenser rise: (Water Flow) The difference between the temperature of the condenser water leaving the condenser and the temperature of the condenser water entering the condenser. The water leaving the condenser is the water entering the cooling tower. See cooling tower range.

condensing temperature: (Refrigeration) The temperature at which the refrigerant vapor condenses in the condenser and is the saturation temperature of the vapor corresponding to the pressure in the condenser.

conditioned space: (Building) An enclosed space provided with mechanical heating (capacity exceeding 10 British thermal units per hour per square foot) and/or mechanical cooling (capacity exceeding 5 British thermal units per hour per square foot) and typically operating within a 55 to 90 degree Fahrenheit temperature range.

conductance: (Electrical) The ratio of the current in the conductor to the potential difference between its ends (reciprocal of resistance). It is a measure of how easily electricity flows along a certain path through an electrical element, measured in Siemens, (SI unit). 1 amp per volt.

conductance: (Heat) Heat flow through non-homogeneous materials at the rate of Btu per hour per square foot per thickness for 1F difference in temperature. Symbol C. Btu/hr-sf-F

conduction: (Physics, Heat) Heat energy transmitted by direct contact, such as warm hand to cold pipe. The transfer of heat from one portion of a medium to another. Heat energy being passed from molecule to molecule. See thermal conductivity.

conduction current: (Electrical) Resulting from the flow of electrons in a conduction medium.

conductor: (Electrical) A material which offers a low resistance to the passage of an electric current. That part of an electric transmission, distribution or wiring system which actually carries the current.

A material which readily passes electrons. Good conductors are silver, copper and aluminum wire. Copper, the second best conductor, is the most often used in electrical wiring because of price and availability.

conductor: (Heat) A material which readily passes heat. Good conductors of heat are metals such as are iron, copper and aluminum.

conduit: (Electrical) A trough or pipe containing electric wires or cables in order to protect them against damage from external causes.

conduit: (Fluid Flow) A duct, pipe or passage way for the conveyance of a fluid.

cone of silence: (Definition) Cone-shaped space directly over the antenna of a radio-beacon transmitted in which signals are virtually undetectable. Ask Maxwell Smart.

connected load: (Electrical) The sum of the rated inputs of all apparatus connected to an electric power supply system.

conservation of energy: (Physics) Energy maybe converted from one form to another but it is not created or destroyed.

constant speed: (Physics) Unchanging rate of motion with regard for the direction of motion.

constant velocity: (Physics) Unchanging rate of motion and unchanging direction of motion.

constant volume air system: (Airflow) An air system (fans, ductwork, terminal boxes, dampers, etc.) which delivers a constant quantity of air (at varying temperature). See variable air volume system.

constant volume terminal box: (Airflow) A terminal box which delivers a constant quantity of air. Boxes may be single duct or dual (double) duct.

constant volume dual duct (aka double duct) terminal box: (Airflow) A terminal box supplied by separate hot and cold ducts through two inlets. The boxes mix warm or cool air as needed to properly condition the space and maintain a constant volume of discharge air. Dual duct boxes may use a mechanical constant volume regulator with a single damper motor to control the supply air. The mixing damper is positioned by the motor in response to the room thermostat. As the box inlet pressure increases, the regulator closes down to maintain a constant flow rate through the box. Another type of constant flow regulation uses two motors, two mixing dampers and a pressure sensor to control flow and temperature of the supply air. The motor connected to the hot duct inlet responds to the room thermostat and opens or closes to maintain room temperature. The motor on the cold duct inlet is also connected to the room thermostat but through a relay which senses the pressure difference across the sensor. This motor opens or closes the damper on the cold duct inlet to (1) maintain room temperature and (2) maintain a constant pressure across the sensor and therefore, a constant volume through the box.

constant volume single duct terminal box: (Airflow) A single inlet terminal box supplied with air at a constant volume and temperature (typically cool air). Air flowing through the box is controlled by a manually operated damper or a mechanical constant volume regulator. The mechanical volume regulator uses springs and perforated plates or damper blades which decrease or increase the available flow area as the pressure at the inlet to the box increases or decreases. A reheat coil may be installed. A room thermostat controls the coil.

contact: (Electrical) That part of either of two conductors which is made to touch the other when it is desired to pass current from one to the other as in a switch.

contact breaker: (Electrical) See circuit breaker.

contact tachometer: (Instrumentation) Chronometric and digital are two types of contact tachometers. Included with contract tachometers are rubber and metal tips for centering the instrument on the rotating shaft. *contactor:* (Electrical) Electro-mechanical devices that "open" or "close" contacts to control motors. Unlike manual starters, they can be remotely or automatically operated.

contactor controller: (Electrical) A controller in which the various circuits are made and broken by means of contactors. Required for operation of electrical circuits and apparatus such as motors for fans and pumps.

continuity: (Electrical) The existence of an uninterrupted path for current in a circuit.

continuous circuit: (Electrical) Early name for direct current.

control: (Control) General term for manual or automatic adjustment of power level.

control board: (Control) A switch board on which are mounted the operating handles, pushbuttons or other devices for operating switch gear situated remotely from the board. The board may have mounted on it indicating instruments, diagrams, and other accessory apparatus. See control panel.

control circuit: (Control, Electrical) One which performs the function of control. The operation of a piece of equipment or of an electrical circuit.

control hysteresis: (Control) Ambiguous control depending on previous conditions. Jump or snap action arising in electronic or magnetic amplifiers because of excessive positive feedback which occurs under certain conditions of load.

controlled device: (Control) A flow control device, such as a damper for air control or a valve for water or steam control.

controlled medium: (Control) The fluid—air, water or steam—whose controlled variable is being sensed and controlled.

controlled variable: (Control) The condition or quantity of the controlled medium—such as temperature, humidity, pressure or flow rate—being controlled.

controller: (Control) A proportioning device (electrical or pneumatic) designed to control dampers or valves to maintain temperature (thermostat), humidity (humidistat), or pressure (pressurestat), or to control other controllers (master/submaster controller).

controller: (Electrical) Controllers are also used for starting and stopping motors. These controllers are grouped into three categories: manual starters, contactors, and magnetic starters.

control current or voltage: (Electrical) Controls magnitude, direction or relative phase, or determines the operation of an apparatus or electrical circuit.

control limit switch: (Electrical) A limit switch connected to the control circuit of the motor whose operation is to be limited.

control loop: (Control) The basic control loop is comprised of a controlled variable, sensing element, controller, actuator and controlled device.

control panel: (Electrical) A panel containing a full set of indicating devices and remote control units.

control point: (Control) The actual temperature, pressure or humidity which the controller is sensing. The value of the controlled variable. Departure from this value causes a controller to operate to reduce the error and restore the intended steady state. See set point.

control relay: (Control) An "ON or OFF" control (switch) which uses one or more pairs of contacts to make or break a control circuit (i.e., permitting the next step in a control circuit).

convection: **(**Physics) Heat energy that moves from one place to another by means of currents that are set up within the fluid medium (vapor or liquid). The regions of higher temperatures, being less dense, rise while the regions of lower temperature move down to take their place. The convection current helps to keep the temperature more uniform than if the fluid was stagnant.

converter: (Electrical) A circuit for changing AC to DC or vice versa. Rating can be a few watts to megawatts.

cooling coil: (Mechanical) A chilled water coil or refrigerant coil for transferring heat from the conditioned space to the water or refrigerant.

cooling tower: (Water Flow) A tower of wood, concrete, etc., used to cool water after circulation through a condenser. The water is allowed to trickle down over wood slats thus exposing a large surface to atmospheric cooling. Cooling is accomplished by evaporation of a portion of the water thus removing heat from the remaining water. Cools heated water from a water-cooled condenser. As outdoor air passes through the cooling tower it removes heat from the condenser water. The water is cooled towards the wet bulb temperature of the entering air.

cooling tower range: (Water Flow) The difference between the temperature of the cooling tower water leaving the cooling tower and the temperature of the cooling tower water entering the tower. The water leaving the cooling tower is the water entering the condenser. See condenser rise.

Coriolis effect: (Physics) The effect of Coriolis force is to deviate a moving body perpendicularly to its velocity. So a body free falling towards the earth is slightly deviated from a straight line and will fall to a point east of the point directly below its initial position. Coriolis forces explain the direction of trade winds in equatorial regions.

coulomb: (Electrical) SI unit of electric charge, defined as that charge which is transported when a current of one ampere flows for one second.

counterclockwise rotation, fan: (Mechanical) Fan rotation as viewed from the drive side of a centrifugal fan.

counter flow coil: (Water Flow) For HVAC air and water systems counter flow means that the flow of air and water are in opposite directions to each other. Water coils are piped counter flow for the greatest heat transfer for a given set of conditions. The water enters on the same side that the air leaves. For cooling coils, this would mean that the coldest water is entering the coil at the point that the coldest air is leaving the coil.

critical surface: (Environmentally Controlled Area) The surface to be protected from particulate contamination.

cross draft: (Laboratory, Fume Hoods) A flow of air that blows across or into the face of the hood. Cross drafts, whether created by people moving about, the room ventilation system, or an open door (if located adjacent to the fume hood), can drastically disturb the flow of air into the hood face and even cause reverse flow of air out the front of the hood.

cubic feet per minute: (Airflow) A unit of measurement. The volume or rate of airflow. Found by multiplying the velocity of air in feet per minute times the area in square feet of the duct or face area of the coil (filter, etc.) through which air is flowing.

cupro-nickel: (Definition) An alloy of copper, nickel and other strengthening materials, such as iron and manganese. Cupro-nickel does not corrode in seawater or saltwater air. Aka cupronickel or coppernickel.

current: (*Electrical*) The transfer of electrical energy through a conductor. The flow of electrons.

cut-in: (Control) A control value to start a device, such as a compressor or boiler.

cut-out: (Control) A control value to stop a device. See differential.

cutoff plate: (Fan) The cutoff plate is part of centrifugal fan housing. It's found in the fan outlet. If the cutoff plate isn't properly positioned air will be drawn back into the fan wheel and the fan will lose efficiency.

cylinder bypass: (Refrigeration) A method of compressor capacity control used when other methods such as compressor on-off cycling or cylinder unloading may not be satisfactory. See hot gas bypass.

cylinder unloaders: (Refrigeration) On reciprocating compressors cylinder heads can be fitted with cylinder unloaders to remove a portion of the refrigeration load from the compressor to make the compressor more efficient and energy-saving on startup and when full cooling is not needed. Controls gas volumes due to fluctuations in cooling load requirements. Unloaders can be electrical or mechanical.

cylinder unloading: (Refrigeration) A method of compressor capacity control.

D is for Damper

dampener: (Mechanical) a device attached to both inlet and outlet sides of large reciprocating machinery to reduce pulsations which, if not reduced, would cause either mechanical damage or noise nuisance or both.

damper: (Airflow) A device used to regulate airflow. See volume damper.

> *damper*: (Boiler) A5n adjustable iron plate or shutter fitted across a boiler flue to regulate the draft (draught).

dead air space: (Airflow) Lack of air movement such as in a fume hood.

dead band: (Control) A control scheme or controlled device that won't allow the mixing of hot and cold air. The control set point when neither heating nor cooling occurs.

dead band pressure: (Control) The deadband pressure is the output pressure at which no heating or cooling occurs.

de-aerator: (Boiler) A vessel in which boiler feedwater is heated under reduced pressure in order to remove dissolved air.

deceleration: (Physics) Negative acceleration. Measured in feet per second squared. Incorrect use of acceleration. See acceleration.

decibel: (Sound) sound pressure level.

de-energize: (Electrical) to disconnect a circuit from its source of power.

deflector vane: (Laboratory, Fume Hoods) An airfoil-shaped vane along the bottom of the hood face which deflects incoming air across the work surface to the lower baffle opening. The opening between the work surface and the deflector vane is open, even with the sash fully closed.

deflecting vane anemometer: (Instrumentation, Air) The deflecting vane anemometer gives instantaneous direct readings in feet per minute. It's used most often for determining air velocity through supply, return and exhaust air grilles, registers or diffusers. It may also have attachments for measuring low velocities in an open space or at the face of a fume hood. With other attachments Pitot traverses and static pressures can be taken. To use this instrument refer to the manufacturer's recommendation for usage, proper attachment selection and sensor placement. A correction (Ak) factor is typically needed when measuring grilles, registers or diffusers. The old Alnor Velometer® is an example.

demand factor: (Electrical) Ratio of the maximum demand on a supply system to the total connected load.

demand limiter: (Electrical) A component which sets an upper limit to the current which can be passed also called current limiter.

demand meter: (Instrumentation) A meter reading or recording the load on an electrical system.

density: (Physics) The weight of a substance per unit volume. The reciprocal of specific volume. Density units are pounds per cubic foot. Mass per unit volume. Density equals Mass divided by Volume. Water density 62.4 lb/cf. Air density 0.075 lb/cf. (standard conditions).

depreciation: (Financial, Energy Management) Spreading the cost of an asset over the span of several years.

depreciation factor: (Lighting) The ratio of the light output when the lighting equipment is clean to that when it is dirty.

derivative control: (Control) A control "braking" action to eliminate overshoot by anticipating the convergence of actual and desired conditions. This, in effect, counteracts the control signal produced by the proportional and integral control actions.

design conditions: (Definition) The environmental conditions for which the conditioned room is designed.

desuperheater: (Boiler) A vessel in which superheated steam is brought into contact with a water spray in order to make saturated or less highly superheated steam.

desuperheating: (Refrigeration) Heat lost from the superheated vapor leaving the compressor. Heat is lost in the discharge pipe and condenser piping. Condenser pressure accommodates a temperature lower than the superheated temperature.

deviation: (Control) The difference between the set point and the value of the controlled variable at any instant. Aka offset or error.

dew point: (Psychrometrics) The temperature at which moisture will start to condense from the air.

differential: (Control) The difference between the cut-in value and the cut-out value. For example, if the cut-in temperature is 55F and the cut-out temperature is 40F the differential is 15F.

differential pressure gauge: (Instrumentation) An instrument that reads the difference between two pressures directly, therefore eliminating the need to take two separate pressures and then calculate the difference. A gauge which measures the difference between two fluid pressures applied to it.

diffuser: (Airflow) A supply air outlet generally found in the ceiling with various deflectors arranged to promote mixing of primary air with secondary air. Types of diffusers are round, square, rectangular, linear and troffers. Some diffusers have a fixed airflow pattern while others have field adjusted patterns.

diffuser: (Mechanical) A chamber surrounding the impeller of a centrifugal pump or compressor in which part of the kinetic energy of the fluid is converted to pressure energy by a gradual increase in the cross-sectional area of flow.

diffusion vane pump: (Water Flow) A pump built with a series of guide vanes or blades around the impeller. The diffusion vanes have small openings near the impeller and enlarge gradually to their outer diameter where the liquid flows into a chamber and around to the pump discharge.

digital: (Computer) Representation of a numerical quantity of a number of discrete signals or by the presence or absence of signals in particular positions. See bit.

digital control: (Control) A control application that uses digital computers and digital controllers in a feedback system. The rest of the system can either be digital or analog.

digital-to-analog: (Control) The conversion of a digital code into its equivalent analog signal level.

digital-to-analog converter: (Control) The part of a microprocessor-based controller that changes digital values from a software program to analog output signals for use in the control system.

digital point: (Control) A point that has an "either/or" value, such as on/off, which will be sensed to provide direct input to the energy management system.

digital signal: (Control) Representation of a numerical quantity by a number of discrete signals (not continuous) or by the presence or absence of signals in particular positions. Binary digital signals have one of two states (0 or 1) defined by voltage or current levels.

digital thermometer: (Instrumentation) Digital thermometers are acceptable for HVAC/R test measurements when they are calibrated against a standard glass stem mercury thermometer before each use or are capable of being field calibrated. They should read in tenths of degrees and be accurate to one-tenth of a degree. See thermometers.

dilute solution: (Refrigeration) A solution with a small concentration of absorbent as compared to a large amount of dissolved refrigerant. To dilute is to make a solution less concentrated.

diode: (Electronics) An electronic device that conducts current in one direction only.

dip stick: (Mechanical) A rod inserted in a tank or sump to measure the depth of liquid.

direct acting controller: (Control) Increases its branch output as the condition it is sensing increases.

direct control: (Control) See direct digital control.

direct current: (Electrical) The type of electrical circuit in which the current always flows in one direction.

direct current converter: (Electrical) a converter which changes direct current from one voltage to another.

direct digital control: (Control) The sensing and control processes directed by digital control electronics—A control loop in which a digital controller periodically updates a process as a function of a set of measured control variables and a given set of control algorithms.

direct drive fan or pump: (Mechanical) A fan or pump directly connected to its driver motor. Most small fans are direct drive. Larger fans are belt driven. Most water (hydronic) pumps no matter what the size are direct drive.

direct lighting: (Lighting) A system of lighting in which not less than ninety percent of the total light emitted is directed downwards.

directly conditioned space: (Building) An enclosed space provided with mechanical heating and/or mechanical cooling and typically operating within a 55F to 90F temperature range.

direct return: (Water Systems) In a direct return system, the return is routed to bring the water back to the pump by the shortest possible path. The terminals are piped "first in, first back; last in, last back." The direct return arrangement is popular because generally, less main pipe is needed. However, since water will follow the path of least resistance, the terminals closest to the pump will tend to receive too much water, while the terminals farthest from the pump will "starve." To compensate for this, balancing valves are required.

direct sound: (Sound) The sound intensity arising from a source to a listener as contrasted with the reverberant sound which has experienced reflection between the source and the listener.

dirt depreciation factor: (Lighting) A factor (used in illumination calculations) which relates the initial illumination provided by a clean new light fixture or lamp (luminaire) to the reduced illumination that it will provide as a result of the accumulation of dirt (on the luminaire at the time when it is next scheduled for cleaning).

discharge duct: (Airflow) Duct leaving a fan, plenum or terminal box.

discharge pipe: (Refrigeration) Pipe (line) leaving a compressor.

discharge pipe: (Water Flow) Pipe leaving a pump.

discharge pressure: (Airflow) Static, velocity or total pressure read (inches water gauge) at the outlet of a fan, terminal box or other component in the airflow system.

discharge pressure: (Refrigeration) Pressure read (psig) at the compressor outlet. Also called head pressure or high side pressure.

discharge pressure: (Water Flow) Static (head) pressure read (feet of water, psig) at the outlet of a pump, coil or other component in the water system.

discharge temperature: (Definition) The temperature at which air, water, refrigeration vapor or other fluid is discharged from some apparatus such as a fan, pump, compressor, coil, terminal box, etc.

discharge valve: (Mechanical) A valve for controlling the rate of discharge of fluid from a pipe, centrifugal pump or compressor, etc.

discount rate: (Financial, Energy Management) An interest rate used to discount future cash flows to their present values.

disk valve: (Mechanical) a form of suction and delivery valve used in pumps and compressors; it consists of a light steel or fabric disk resting on a ported flat seating; steel valve disks are usually spring loaded.

displacement: (Mechanical) The volume of fluid displaced by a pump, plunger, purse stroke or per unit time. The swept volume of a working cylinder.

diversity: (Airflow, Constant Air Volume Systems) The total output (cubic feet per minute) of the fan is greater than the maximum required volume through the cooling coil. For example, the cooling coil is sized for 8,000 cubic feet per minute and the fan has an output of 10,000 cubic feet per minute. Diversity is 0.80.

diversity: (Airflow, Fume Hoods, Laboratory) A diversity permits the exhaust system to have less capacity than that required for the full operation of all units.

diversity: (Airflow, Variable Air Volume Systems) The total volume (cubic feet per minute) of the Variable Air Volume boxes is greater than the maximum output of the fan. For example, the Variable Air Volume boxes total 40,000 cubic feet per minute and the fan output is 30,000 cubic feet per minute. Diversity is 0.75.

diverting airflow pattern: (Airflow) See parallel blade damper.

diverting tee: (Water Flow) A device to create the proper resistance in a one-pipe system to direct water to the terminal.

diverting valve: (Water Flow) A valve in a water pipe to divert water in different directions such as a diverting valve going to a water-cooled condenser that allows water to enter the condenser or diverts it to the cooling tower. See three-way automatic temperature control valve.

dioctyl phthalate test: (Environmentally Controlled Area) A test using dioctyl phthalate oil to determine the effectiveness of air filters. Aka DOP test.

double inlet-double wide fan: (Fan) Double Inlet-Double Wide fans have two single wide fan wheels mounted back-to-back on a common shaft in a single housing. Air enters both sides of the fan. On DIDW fans the bearings are in the air stream and they are large in relation to the inlet and will reduce fan performance if the fan is small. Therefore, DIDW fans are less common in smaller sizes. Generally, because of the double inlet, DIDW fans are more suited to open inlet plenums. They are used most often in high volume applications.

double suction pump: (Pump) A pump in which the liquid enters the impeller inlet from both sides. The impeller is similar to two single suction impellers, back to back. Double suction pumps have fixed suction and discharge openings. The suction connections are normally one or two pipe sizes larger than the discharge connection.

Doppler effect: (Physics) The apparent change of frequency (or wave length) because of the relative motion of the source of radiation and the observer. A change in the frequency with which waves (as of sound or light) from a given source reach an observer when the source and the observer are in motion with respect to each other so that the frequency increases or decreases according to the speed at which the distance is decreasing or increasing.

dosimeter: (Instrumentation) Instrument for measuring dose. It gives a measure of the radiation field and dosage experienced.

double pipe exchanger: (Energy Management) Heat exchanger formed from concentric pipes. The essential flow pattern is that two fluid streams are always parallel.

double pole: (Electrical) Switches, circuit breakers, etc., which can make or break a circuit on two poles simultaneously.

double throw switch: (Electrical) One which enables connections to be made with either of two sets of contacts.

downcomer: (Mechanical) A pipe or duct for conveying fluid in a downward direction. Aka downpipe or downtake.

draft: (Definition) Localized feeling of coolness caused by high air velocity, low ambient temperature, or direction of airflow.

drift: (Control) See drop (Control).

drift: (Cooling Towers) Water blown off the cooling tower.

drive: (Mechanical) Components to turn a fan, pump or compressor. Components include motors, sheaves (pulleys) and belts.

drive side: (Fan) On single inlet-single wide fans the drive side is the side opposite the inlet. On double inlet-double wide fans, the drive side is the side that has the drive.

droop: (Control) See drop (Control).

drop: (Airflow) The vertical distance that the lower edge of a horizontally projected air stream drops between the outlet and the end of its throw. *drop:* (Control) The difference between the high peak and low peak of a control pressure sine wave in any given cycle. A reduction in the fluid flow (controlled medium, air, water, steam) pressure, as in "pressure drop" across a control valve.

drop: (Electrical) A reduction in electrical pressure. Term commonly used to denote voltage drop.

drop: (Fluid Flow) A reduction in flow or pressure as in water, air or refrigerant flow or water pressure, etc.

dry bulb temperature: (Psychrometrics) The air temperature indicated on an ordinary thermometer. A measure of sensible temperature.

dry flue gas: (Boiler) Gaseous products of combustion from a boiler furnace excluding water vapor. See flue gas.

dry ice: (Definition) Solid (frozen) carbon dioxide. At ordinary atmospheric pressure it sublimes slowly. See sublimation.

dry saturated vapor: (Fluid) A saturated vapor completely free of liquid particles. See saturated vapor.

dual current motor: (Electrical) A motor that will operate safely at either of two nameplate amperages. The operating amperage depends on the voltage supplied.

dual duct terminal box: (Airflow) A dual duct or dual inlet terminal box supplied with any combination of heated, cooled, or dehumidified air. One duct is the "hot duct" and the other is the "cold duct." The hot duct supplies warm air which may be heated air, or return air from the conditioned space. The cold duct supplies cool air, which may be cooled and dehumidified when the refrigeration unit is operating, or cool outside air brought in by the economizer cycle. A room thermostat controls a mixing damper arrangement in the boxes that determines whether the discharge air will be cool air, warm air or a mixture of both. Dual duct boxes are pressure independent and may be constant or variable volume.

dual path system: (Airflow) An air system in which the air flows through heating and cooling coils in the air handling unit that are parallel to each other. The coils may be side-by-side or stacked. The heating coil is located in the hot deck and the cooling coil is located in the cold deck. Multizone duct systems and dual duct systems are dual path. Some systems may not have a heating coil but instead bypass return air or mixed air into the hot deck.

dual pressure system: (Control) A pneumatic system which requires two different main air pressures.

dual temperature water: (Water Flow) A heating and cooling water system such as a three-pipe system that requires two different water temperatures. A typical chilled water range is 45-55 degrees Fahrenheit. A typical heating water temperature range is 100-150 degrees Fahrenheit.

dual voltage motor: (Electrical, Motor) A motor that will operate safely from either of two nameplate voltages. Typical single-phase dual voltage motors are 110/220 volts or 115/230 volts. Typical three-phase motor dual voltage motors are 220/440 volts, 230/460 volts, or 240/480 volts.

duct (Airflow) A passageway made of sheet metal or other suitable material used for conveying air. See air duct.

duct pressure: (Airflow) For rating purposes, duct is designed, fabricated and installed as either low, medium, or high pressure duct to withstand designated air pressures. Low pressure ducts are fabricated for 2 inches, water gauge air pressure or less and up to 2,500 feet per minute air velocity. Medium pressure ducts are fabricated for 2 to 6 inches, water gauge air pressure and between 2,000 and 4,000 feet per minute air velocity. High pressure ducts are fabricated for air pressures above 6 inches, water gauge and air velocities above 2,000 feet per minute. For energy conservation, all ductwork should be sealed. Low and medium pressure ductwork may be sealed and pressure leak tested. High pressure ductwork is typically sealed and pressure leak tested.

duct system: (Airflow) A system or layout of supply and return air ductwork from the air handling unit to the supply air outlets and back to the unit. See dual path system and single path system.

ductwork: See air duct.

dumping: (Airflow) The rapidly falling action of cold air caused by a variable air volume box or other device reducing airflow velocity.

dust proof: (Definition) Electrical apparatus or other equipment which is constructed so as to exclude dust or particles.

dynamic discharge dead: (Mechanical, Pumps) Static discharge head plus friction head plus velocity head.

dynamic suction head: (Mechanical, Pumps) Static suction head minus friction head loss and velocity head.

dynamic suction lift: (Mechanical, Pumps) Static suction lift plus friction head loss plus velocity head.

E is for Earth

Earth: (Definition) The third planet from the sun (aka "blue planet"). It is the largest of the terrestrial planets in the Solar System in diameter, mass and density and the home to trillions of plants and animals. 4.5 billion years old.

earth potential: (Electrical) The electrical potential of the earth, usually regarded as zero so that all other potentials are referred to it.

ebullator: (Heat) A heated surface used to impart heat to a fluid.

ebullition: (Heat) The state or process of boiling. See boiling.

economizer: (Energy Management) A component or group of components used to reduce the energy to heat or cool fluids.

economizer: (Air Side) Natural cooling of the conditioned space using outside air (when appropriate) instead of refrigerated mechanical cooling. Opening outside air dampers while closing return air dampers to bring in up to 100% outside air. Relief air dampers or exhaust air dampers are used to maintain proper air pressure and air volumes coming into the mixed air plenum. The mechanical cooling is reduced or turned off.

economizer: (Boiler) A bank of tubes, placed across a boiler flue, through which feedwater is pumped. The feedwater is heated by the otherwise waste heat of the flue gases. The heated feedwater increases boiler efficiency.

economizer: (Refrigeration, Centrifugal Chiller) A pressure chamber off the condenser to flash (boil-off) refrigerant increasing capacity and efficiency of the system. Aka: intercooler. See intercooler.

economizer: (Water Side) Natural cooling of the conditioned space using cooling tower water instead of refrigerated mechanical cooling. The two methods are heat exchanger cycle and strainer cycle.

economizer: (Water Side, Heat Exchanger Cycle) "Cool" water from the cooling tower basin is pumped from the tower by the condenser water pump. This water bypasses the water-cooled condenser and goes through a plate heat exchanger and then back to the cooling tower (open loop system). The chilled water pump moves "warm" cooling coil water around the closed loop system into the heat exchanger. Heat is transferred from the cooling coil water to the cooling tower water. The now "cooled" cooling coil water is pumped back to the coil(s). The cooling tower water and the cooling coil water do not mix.

economizer: (Water Side, Strainer Cycle) "Cool" water from the cooling tower basin is pumped by the condenser water pump from the tower. It bypasses the water-cooled condenser and goes through a strainer and then directly through the cooling coil(s) and back to the cooling tower.

economizer control: (Control) A control system for the changeover between using natural cooling and refrigerated mechanical cooling.

eddy: (Fluid Flow) An interruption in the steady flow of a fluid, caused by an obstacle situated in the line of flow; the vortex or whirlpool formed.

eddy current: (Electrical) An electric current induced by an alternating magnetic field. The term comes from analogous currents seen in fluids, such as when dragging an oar through water. It is caused when a conductor is exposed to a changing magnetic field due to relative motion of the field source and conductor; or due to variations of the field with time. This can cause a circulating flow of electrons, or a current, within the conductor. These circulating eddies of current create electromagnets with magnetic fields that opposes the change of the magnetic field. The stronger the applied magnetic field, or the greater the electrical conductivity of the conductor, or the faster the field that the conductor is exposed to changes, then the greater the currents that are developed and the greater the opposing field. Eddy currents, like all electric currents, generate heat as well as electromagnetic forces. They are one of the main causes of heating in motors, transformers, etc. The heat can be harnessed for induction heating. The electromagnetic forces can be used for levitation, creating movement, or to give a strong braking effect. Eddy currents can be minimized by lamination of conductors with thin plates and thin electrically insulating layers between the laminations.

eddy flow: (Physics) See turbulent flow.

effective area: (Airflow) The effective area of an outlet or inlet grille or diffuser is the sum of the areas of all the vena contracta existing at the outlet and is affected by (1) the number of orifices and the exact location of the vena contracta and (2) the size and shape of the grille bars, diffuser rings, etc. Manufacturers have conducted airflow tests and based on their findings they have established flow factors or area correction factors for their products. Each flow factor, sometimes called "K-factor" or "Ak," applies to (1) a specific type and size of grille, register or diffuser, (2) a specific air measuring instrument and (3) the correct positioning of that instrument.

effective coil temperature: (Refrigeration) See apparatus dew point.

effective heating surface: (Heating) The total area of a boiler (or other heating) surface in contact with water (or other fluid) on one side and hot gases on the other side.

effective range: (Instrumentation) That part of the scale of an indicating instrument over which a reasonable precision may be expected.

efficacy: (Lighting) Lamp output per unit of energy input to the lamp. The ratio of lamp output to wattage consumption, expressed in lumens per watt.

efficiency: (Definition) A non-dimensional measure of the performance of a piece of apparatus, e.g. a fan or pump, obtained from the ratio of the output of a quantity, e.g., power, energy, to its input often expressed as a percentage. It must always be less than 100% (which would imply perpetual motion). Not to be confused with efficacy. Efficacy takes account only of the output of the apparatus and is not given an exact quantitative definition.

efficiency: (Energy Management) Useful energy output divided by the power input.

efficiency: (HVAC System Operation) Effective operation as measured by a comparison of positive output with cost in energy, time, and money.

efficiency ratio: (Mechanical) The ratio of a heat engine. The actual thermal efficiency to the ideal thermal efficiency corresponding to the cycle on which the engine is operating.

effluent: (Definition) Outflow, or flowing out; such as the fumes out of the exhaust stack of a laboratory hood.

ejector: (Definition) A device for exhausting a fluid by entraining it by a high velocity steam or air jet, e.g., an air ejector.

electric: (Electrical) Derived from the Latin word "electricus" meaning amber-like. Amber is a hard, translucent, yellow fossil resin found along the shores of the Baltic Sea. Amber, when rubbed with a cloth gives off sparks. The ancient Greeks used words meaning "electric force" in referring to the mysterious forces of attraction and repulsion exhibited by amber.

electrical: (Electrical) Descriptive of means related to, pertaining to, or associated with electricity, but not inherently functional.

electrical degrees: (Electrical) Angle, expressed in degrees of phase difference of vectors, representing currents or voltages arising in different parts of a circuit.

electrical engineering: (Definition) That branch of engineering chiefly concerned with the design and construction of all electrical machinery and devises, power transmission, etc.

electric circuit: (Electrical) Series of conductors forming a partial, branched or complete path around which either a direct or alternating current can flow.

electric conduction: (Electrical) Transmission of energy by flow of charge along a conductor.

electric generator: (Electrical) Electrostatic or electromagnetic device for conversion of mechanical energy to electrical energy. Aka generator.

electricity: (Electrical) The manifestation of a form of energy associated with static or dynamic electric charges. The effect created by the interaction of charged particles. Electricity is defined only by theory. However, these theories about the nature and behavior of electricity have been advanced, and have gained wide acceptance because of their apparent truth and demonstrated workability. The theory of electricity or electron flow states that negatively charged particles, called electrons, flow from point to point through a conductor.

electric potential: (Electrical) Capacity to do electrical work. It is equal to the work done by an electric field in carrying a unit positive charge from infinity to that point, expressed relative to zero potential (earth or ground). Voltage. See electromotive force.

electrode boiler: (Boiler) A boiler in which heat is produced by the passage of an electric current though the liquid to be heated.

electromagnetic spectrum: (Physics) All the different wave lengths, radio waves, microwaves to infrared, visible light, ultra-violet light, x-rays and gamma rays.:

electromotive force: (Electrical) A measure of electric force or potential, voltage.

electronic controller output signal: (Control) The output signal from electronic controllers is direct current voltage (volts direct current, vdc). Typically the range is 3 volts from 6 to 9 vdc with a 7.5 vdc midpoint. In the following example, a controller that has a 6 degree throttling range from 75F to 78F will have a setpoint temperature of 75F. See Chapter 10.

electronic instruments: (Instrumentation) Most of the mechanical analog instruments now have electronic digital counterparts. All instruments, analog or digital, should be maintained and calibrated to manufacturer recommendation before each use.

elevation view: (Definition) The view or representation of any given side of a building, duct, component or assembly drawn in projection on a vertical plane.

empirical formula: (Definition) A formula founded on experience or experimental data only, not deduced in form from purely theoretical considerations.

enclosed space: (Building) A space surrounded by solid surfaces.

energy: (Energy) A measure of power consumed. The ability to do work. The capacity of the body for doing work. Energy equals power times time. Energy can take many forms including electrical, heat, and mechanical. Both mechanical and electric energy convert into heat energy. Energy can be stored (potential energy) or in motion (kinetic energy). This book will use the United States system of units. That is, heat energy measurements and units are British thermal units. Electrical energy units and measurements are Watt-hours or kilowatt-hours. Mechanical energy measurements and units are foot-pounds.

energy management control system: See energy management system.

energy management system: (Energy Management, Control) A system based on a computer whose primary function is the controlling of energy-using equipment to reduce the amount of energy used.

enthalpy: (Heat) The measurement of heat content. Total heat content. Thermodynamic property of a working substance. Enthalpy equals

Internal Energy plus Pressure times Volume. Thermodynamic property of a working substance associated with the study of heat of reaction, heat capacity, and flow processes. Units of enthalpy, air, refrigeration, steam and water are in British thermal units per pound.

enthalpy control: (Control, Air Systems) Control devices that compare the enthalpy of the return air with the enthalpy of the outside air, and then position the economizer dampers to admit the air with the lowest enthalpy.

entrainment: (Fluid Flow) The capture and transport of small particles in a fluid, e.g., air in water, air in refrigerant, water in steam. See air entrainment, air separator, and manual air vent.

entropy: (Physics) Measure of a system's energy that is unavailable for work, or the degree of a system's disorder. When heat is added to a system held at constant temperature, the change in entropy is related to the change in energy, the pressure, the temperature, and the change in volume. Its magnitude varies from zero to the total amount of energy in a system. For a closed thermodynamic system, a quantitative measure of the amount of thermal energy not available to do work or the extent to which the energy of a system is available for conversion to work. Also: a measure of the disorder, chaos or randomness of a system. The tendency for all matter and energy in the universe to evolve toward a state of inert uniformity. The inevitable and steady deterioration of a system or society—The degradation of the matter and energy in the universe to an ultimate state of inert uniformity. A process of degradation or running down or a trend to disorder.

entropy of fusion: (Physics) The measure of the increased randomness that accompanies the transition from solid to liquid or liquid to gas equal to the latent heat divided by the absolute temperature, vaporization. *envelope*: (Energy Management) The exterior of a building including walls, window, doors, roof, etc.

environment: (Definition) the physical and chemical surroundings of an object, e.g., the temperature and humidity, the physical structures, the gases.

environmental control: (Control) the control of temperature, humidity, atmosphere, and contamination.

equalized spring coupling: (Pump) The equalized spring type coupling is used where quiet, smooth operation is required. The motor drives the pump shaft through four springs. An equalizing bar balances the tension on the springs. This coupling needs no maintenance when the alignment is proper. Check alignment if the coupling breaks or noisy operation is observed.

equivalent duct diameter: (Math) The equivalent round duct diameter for a rectangular duct.

eutectic: (Refrigeration) Relating to a mixture of two or more substances having a minimum melting point. See eutectic mixture.

eutectic change: (Refrigeration) The transformation from the liquid state to the solid state in a eutectic mixture.

eutectic mixture: (Refrigeration) A mixture of chemical compounds or elements that has a single chemical composition that solidifies at a lower temperature than any other composition. For example, sodium chloride (common salt) and water or sodium chloride and ice form a eutectic mixture. Salt water freezes at 0 degrees F. Common uses of the eutectic mixture of salt and water include spreading salt (typically mixed with sand or gravel) on roads to aid snow removal; or when salt is mixed with ice (to produce low temperatures) when making "homemade" ice cream.

evaporation: (Physics) Evaporation is the heating of a liquid (water or refrigerant) to convert it to a vapor (gas). Evaporation occurs only at the surface of the liquid at any temperature below the saturation temperature. That is, the conversion of a liquid into a vapor at temperatures below the boiling point. The rate of evaporation increases with rise of temperature. Evaporation depends on the saturated vapor pressure of the liquid which rises until it is equal to the atmospheric pressure at the boiling point.

evaporation stage: (Refrigeration) The heating of a liquid refrigerant to convert it to a vapor in the evaporator.

evaporative cooling: (Psychrometrics) The process of evaporating part of a liquid (water) by supplying the necessary latent heat from the main bulk of liquid which is thus cooled. The adiabatic exchange of heat between air and water spray or wetted surface. As a fan blows air across a wet pad some of the water is evaporated and the air is cooled. The wet bulb temperature of the air remains constant but the dry bulb temperature is decreased. A human body is cooled as air moves across and evaporates some of the sweat on the skin.

evaporator: (Refrigeration) The part of the refrigeration system in which the refrigerant is vaporized and absorbs heat. The evaporator receives the liquid from the condenser by way of the metering device. The metering device reduces the pressure on the liquid. Therefore, the temperature is lowered. This is done by evaporating (aka boiling off or flashing) some of the liquid refrigerant. In the evaporator, the low temperature, low pressure liquid is heated and changes state to a low temperature, low pressure gas. If the evaporator is in an air conditioning unit, it is also called an evaporator coil, an evap. coil, or a DX coil. Water chillers can have a DX evaporator or flooded evaporator. DX evaporators have refrigerant in the tubes and water around the tubes. Flooded evaporators have water in the tubes and refrigerant around the tubes.

evaporator coil: (Refrigeration) A coil containing a refrigerant other than water which is used for cooling air, aka refrigerant coil, DX coil, evaporator, evap. coil, chiller, or water cooler. See evaporator.

evaporator pressure: (Refrigeration) The pressure of the refrigerant vapor in the suction line. Aka suction pressure, back pressure or low-side pressure.

excess air: (Boiler) The proportion of air that has to be supplied in excess of that theoretically required for complete combustion of a fuel. Because of the imperfect conditions under which combustion takes place in practice a few percent of excess air is needed.

exhaust air inlet: (Air Systems) An exhaust air grille, register or other opening to allow air from the conditioned space into the exhaust air duct.

exhaust collar: (Laboratory, Fume Hoods) The connection between the duct and the fume hood through which all exhaust air passes.

exhaust fan: (Airflow) A fan used to extract foul air, fumes, suspended paint particles, etc. from a working area. A fan placed in a draft systems. For example, the smoke uptake of a boiler to draw air through the firing chamber and exhaust the flue gases.

expansion tank: (Water Flow) A tank which compensates for the normal expansion and contraction of water in a hydronic system. Water expands when heated in direct proportion to its change in temperature. In hydronic systems, an allowance must be made for this expansion, otherwise when the system is completely filled, the water has nowhere to go and there is the possibility of a pipe or piece of equipment breaking. Expansion tanks are required in a closed loop heating or chilled water HVAC system to absorb the expanding fluid and limit the pressure within a heating or cooling system. Closed expansion tanks are also known as compression tanks. Pressure in the system will vary from the minimum pressure required to the maximum allowable working pressure. The minimum pressure requirement, anywhere in the system, must always be (1) greater than the vapor pressure of the water to avoid cavitation problems and (2) greater than atmospheric so air doesn't enter the system. Maximum allowable working pressure must not exceed the construction limits. On a low temperature boiler, for example, the pressure relief valve is set for 30 pounds per square gauge (15 pounds per square absolute) which is therefore, the maximum allowable working pressure. When the system is filled with water the air in the compression tank (about 2/3 of the tank is water, 1/3 is air) will act as a spring or cushion to keep the proper pressure on the system to accommodate the fluctuations in water volume and control pressure change in the system. A properly sized expansion, or compression tank, will accommodate the expansion of the system fluid during the heating or cooling cycle without allowing the system to exceed the critical pressure limits of the system. The expansion tank uses compressed air to maintain system pressures by accepting and expelling the changing volume of water as it heats and cools. Some tank designs incorporate a diaphragm or bladder to isolate the expanded water from the pressure controlling air cushion. As water is expanded, it is contained in the bladder preventing tank corrosion and water logging potentials. The pressure controlling air cushion is pre-charged at the factory and can be adjusted in the field to meet critical system requirements. This design and operation of this style of expansion tank allows for the reduction of tank sizes. The compression tank is the oldest type used. It works well when the air is controlled and kept in the tank, not in the system. Bladder and diaphragm are other styles of tanks. The bladder style tank (aka replaceable diaphragm tank) has a membrane between the air and water. This allows the tank to be pre-charged with air to the minimum operating pressure and thereby reduces the tank size and the installed weight. In a diaphragm style tank (aka fixed diaphragm tank) the diaphragm cannot be replaced if the tank fails. The advantage is the initial cost because with smaller tanks the cost of replacing a bladder may exceed the cost of replacing the tank. An open expansion tank may be used in an open water system, i.e., open to the atmosphere. As the water temperature increases (and therefore, the total volume of water in the system increases) the water level just rises in the tank. Open expansion tanks are typically limited to installations having operating temperatures of 180F or less because of water boiling and evaporation problems. Being open to the atmosphere also has the drawback of continual exposure to the air where there may be low (possibly freezing) air temperatures, and pollution or contamination of the water, etc.

expanded metal: (Definition) A metal network formed by suitable stamping or cutting sheet metal and stretching it to form open meshes.

expansion joint: (Mechanical) a joint arranged between two parts (pipe, duct, buildings, etc.) to allow them to expand with temperature rise, without distorting laterally.

expansion valve: (Refrigeration) A type of metering device.

extraction fan: (Airflow) An exhaust fan.

extractor: (Airflow) A device used in low pressure systems to divert air into branch ducts or outlets.

F is for Freudian Slip

face and bypass dampers: (Airflow) Dampers placed before a coil or other apparatus to allow air to go through the coil (apparatus) or to bypass some or all the air around the coil for temperature control.

face velocity: (Airflow, Fume Hoods) The average velocity of the air entering or leaving (as appropriate) various devices including: coils, filters, supply air outlets, return air inlets, exhaust air inlets, duct openings, fume hood opening, etc. For cooling coils, face velocity over 600 feet per minute may result in condensate blown off the coil. Recommended face velocities for outlets and inlets are given in the equipment manufacturer published data (expressed in feet per minute).

Fahrenheit scale: (Thermodynamics) The method of graduating a thermometer in which freezing point of water is marked at 32 degrees and boiling point at 212 degrees, the fundamental interval being 180 degrees. Most countries use the Celsius (aka Centigrade) and Kelvin scales. To convert degrees Fahrenheit to degrees Celsius subtract 32 and multiply by 5/9 (0.556). For the Rankine equivalent add 459.67 (typically rounded to 460) to degrees Fahrenheit; this total multiplied by 5/9 gives the Kelvin equivalent.

fail-safe: (Control) A design in which power supply or control is able to return to a safe condition in the event of failure or malfunction by automatic operation of protective devices. In HVAC, a fail- safe system is a damper or valve which will go to its normal position to minimize damage to the HVAC equipment and components in case of a control failure. This could include placing heating valves open and humidifier valves closed upon loss of control power. In energy management control systems (EMCS) terminology a fail-safe system means returning all controlled devices to conventional control in case of load management control failure.

fan: (Mechanical) A power-driven, constant volume machine that moves a continuous flow of air by converting rotational mechanical energy to increase the total pressure of the moving air.

fan air volume: (Airflow) The rate of airflow expressed at the fan inlet in cubic feet per minute of air produced, independent of air density.

fan blast area: (Fan) The fan outlet area less the area of the cutoff.

fan characteristic: (Fan) A graph showing the relation between fan pressure and delivery, used as a basis for fan selection. The characteristic is determined by the shape of the fan blades.

fan coil unit: (Mechanical) See air conditioning unit.

fan curve: (Fan) See fan performance curve.

fan efficiency: (Fan) The output of useful energy divided by the power input; air horsepower divided by brake horsepower.

fan guard: (Fan) A sheet metal or expanded metal guard surrounding a fan's drives (sheaves and belts) to protect personnel from the moving parts.

fan outlet area: (Fan) The gross inside area of the fan outlet expressed in square feet.

fan outlet velocity: (Fan) The theoretical velocity of the air as it leaves the fan outlet. It is calculated by dividing the air volume by the fan outlet area. Since all fans have a non-uniform outlet velocity, the "fan outlet velocity" doesn't express the velocity conditions that exist at the fan outlet but is a theoretical value of the velocity that would exist in the fan outlet if the velocity were uniform.

fan performance curve: (Fan) A fan performance curve is a graphic representation of the performance of a fan from free delivery to no delivery.

fan sheave: (Fan) The driven pulley on the fan shaft.

fan static efficiency: (Fan) Static air horsepower divided by brake horsepower.

fan static pressure: (Fan) The fan total pressure less the fan velocity pressure. The fan inlet velocity head is assumed equal to zero for fan rating purposes.

fan total efficiency: (Fan) Total air horsepower divided by brake horsepower.

fan total pressure: (Fan) The rise in total pressure from the fan inlet to the fan outlet; the measure of total mechanical energy added to the air by the fan.

fan total static pressure: (Fan) The static pressure rise across the fan calculated from static pressure measurements at the fan inlet and outlet.

fan velocity pressure: (Fan) The pressure corresponding to the average air velocity at the fan outlet.

fan pulley: (Fan) A pulley affixed to a shaft by a key or a set bolt (screw), as distinct from a loose pulley which can revolve freely on the shaft.

feedback: (Control) The transmission of information about the results of the control action. Aka closed loop. A closed transmission path is maintained between the input and output signals of the loop.

feedback signal: (Control) That which is responsive in an automatic controller to the value of the controlled variable.

feedwater: (Boiler) the water, previously treated to remove air and impurities, which is supplied to a boiler for evaporation.

feedwater heater: (Boiler) An arrangement for heating boiler feedwater by means of steam or other heat supply which has done work in an engine or turbine or other apparatus.

feet per minute: (Airflow) A unit of measurement. The velocity of air.

feet per second: (Water Flow) A unit of measurement. The velocity of water.

fenestration: (Building) The arrangement of windows or other openings in the outer walls of a building. The controlling of light emission into a room or building.

filament: (Electrical) A fine wire of high resistance, which is heated to incandescence by the passage of an electric current. In an electric filament lamp (incandescent) it acts a s the source of light.

filter-dryer: (Refrigeration) A combination device used as a strainer and moisture remover in a refrigeration system—Removes moisture and solid particles from the refrigerant before entering the metering device. Usually found in the liquid line. Aka strainer-dryer.

filter: (Indoor Air Quality) A device typically composed of fibrous materials which remove solid particulates such as dust, pollen, mold, and bacteria from the air. Air filters are used in applications where air quality is important, notably in building ventilation systems.

fins: (Mechanical) Fins on a coil increase the area of heat transfer surface to improve the efficiency and rate of transfer and are generally spaced up to 14 fins per inch. As with coil tubes, the more fins, the more heat transfer, but also the more resistance to airflow. Aluminum is usually picked over copper for fin material for reasons of economy. However, when cooling coils are sprayed with water copper fins are needed to prevent electrolysis between the dissimilar copper tubes and aluminum fins. Coils wetted only by condensation are seldom affected by electrolysis and are usually copper tube, aluminum fin.

firebox: (Boiler) That part of a locomotive-type boiler containing the fire.

fire door: (Boiler) The door of a boiler furnace.

fire door: (Building) A fire-resisting door of various materials.

firestat: (Control) A device in the air handling unit or ductwork to shut down air conditioning or ventilating fans when air temperature goes above a preset limit.

fire tube boiler:: (Boiler) A fire tube boiler has hot flue gases from the combustion chamber (the chamber or space where combustion takes place) flowing through tubes (fire tubes) and out the boiler stack. The fire tubes are surrounded by water. Heat from the hot gases transfers through the walls of the tubes and heats the water. Fire tube boilers may be further classified as externally fired, meaning that the fire is entirely external to the boiler, or they may be classified as internally fired, in which case, the fire is enclosed entirely within the steel shell of the boiler. Two other classifications of fire tube boilers are wet-back or dry-back. This refers to the compartment at the end of the combustion chamber. This compartment is used as an insulating chamber so that heat from the combustion chamber, which can be several thousand degrees, does not reach the boiler's steel jacket. If the compartment is filled with water it is known as a wet-back boiler, if the chamber contains only air it is called a dry-back.

firewall: (Building) A fire-resisting wall of various materials.

firing: (Boiler) The process of adding fuel to a boiler furnace.

first air: ((Airflow, Environmentally Controlled Area) The air coming directly from the high efficiency particulate air filter before it passes over any work location.

first work location: (Airflow, Environmentally Controlled Area) The work location in the path of the first air stream.

fixed pulley: (Mechanical) A non-adjustable pulley keyed to its shaft. Pulleys are used on smaller drive systems. Aka fixed sheave.

fixed sheave: (Mechanical) A pulley that has fixed belt grooves. Fixed sheaves are normally used on the fan. Size-for-size, fixed sheaves are less expensive than variable pitch sheaves and there's less wear on the belts. Sheaves are used on bigger drive systems.

fixed air system: (Mechanical) A fixed air system is one in which there are no changes in the system resistance resulting from closing or opening of dampers, or changes in the conditions of filters or coils, etc. For a fixed system, an increase or decrease in system resistance results only from an increase or decrease in cubic feet per minute and this change in resistance will fall along the system curve.

fixed water system: (Mechanical) A fixed water system is one in which there are no changes in the system resistance resulting from closing or opening of valves, or changes in the condition of the coils, etc. For a fixed system, an increase or decrease in system resistance results only from an increase or decrease in gallons per minute and this change in resistance will fall along the system curve.

flash gas: (Refrigeration) Instantaneous evaporation of some liquid refrigerant in the metering device which cools the remaining liquid refrigerant to a desired evaporator temperature. The liquid refrigerant, which is boiled off or flashed, changes state to a vapor. Unwanted flashing may occur in the liquid line or in the metering device (excess flash gas).

flexible connectors: (Water Flow) Flexible connectors are used between piping and pumps and other pieces of equipment to reduce noise and vibration. Flexible connectors can be made of rubber, plastic or braided metal.

flexible disc coupling: (Pump) The flexible disc is used in heavy duty applications where extremely quiet conditions are not needed. The motor drives the pump shaft through the flexible disc. If rough operation or noise is observed, check the coupling for wear. The flexible disc should never be tightly bound between the two coupler halves. Check with the manufacturer for clearance specifications.

floating control: (Control) The controller moves the controlled device at a constant speed toward either its open or closed position. A neutral zone between the two positions allows the controlled device to stop at any position whenever the controlled variable is within the controller's differential. When the controlled variable is outside the differential, the controller moves the controlled device in the proper direction.

flow characteristics: (Airflow) The relationship between the position of a damper and its percent of airflow.

flow meter: (Instrumentation) Instrument for measuring or giving an output signal proportional to the rate of flow of a fluid in a pipe or duct. To determine flow a pressure drop is measured across the flow meter. The pressure drop is usually measured with a differential pressure gauge. The pressure drop may be applied to a flow capacity curve. If a flow capacity curve (or slide rule) is used to determine flow care must be taken when reading capacity curves as they are logarithmic and can be misleading if close attention is not paid to the numbering on the curves. Water flow meters such as orifice plates, Venturi, annular type and calibrated balancing valves are permanently installed devices used for flow measurements of pumps, primary heat exchange equipment, distribution pipes and terminals. For flow meters to give accurate, reliable readings they should be installed far enough away from any source of flow disturbance to allow the turbulence to subside and the water flow to regain uniformity. The manufacturers of water flow meters usually specify the lengths of straight pipe upstream and downstream of the meter needed to get good readings. Straight pipe lengths vary with the type and size of flow meter but typical specifications might be between 5 to 25 pipe diameters upstream and 2 to 5 pipe diameters downstream of the flow meter. With most types of water flow meters (Venturi, orifices plates and annular flow meters) a balancing valve is also required to set the flow. Some other types of water flow sensors and flow meters are: Doppler Effect meters, Pitot tubes and manometers, magnetic flow meters, vortex shedding meters and turbine meters.

flue: (Boiler) A passage through which flows the products of combustion of a boiler.

flue gas: (Boiler) The gaseous exhaust products of a boiler containing in general carbon monoxide, carbon dioxide, oxygen, nitrogen, and water vapor whose analysis is used as a check of boiler efficiency.

flue gas analyzer: (Instrumentation) Measures and records furnace and boiler flue gas products.

flue gas temperature: (Heat) The temperature of the flue gas at the point in the flue where it leaves the boiler furnace.

fluid: (Fluid Flow) A vapor (gas) or liquid. Vapors, air for example, are considered compressible fluids. Liquids, water or liquid refrigerant for example, are non-compressible fluids. Refrigerants are fluids which change from a vapor to a liquid and back to a vapor. The fluids in HVAC work are air, water, steam, and refrigerants.

fluid dynamics: (Physics) The condition of a fluid in motion. The velocity of a fluid is based on the cross-sectional area of the conduit and the volume of fluid passing through the conduit.

fluidity: (Physics) The quality or state of being fluid; the physical property of a substance that enables it to flow. The less viscous the fluid is, the greater its ease of movement (fluidity). See viscosity.

fluorescent lamp: (Lighting) A gas-discharge (mercury vapor) electric lamp having the inside of the bulb or tube coated with fluorescent material so that the ultraviolet radiation from the discharge is converted to light of an acceptable color. It uses electricity to excite mercury vapor in argon or neon gas, resulting in a plasma that produces short-wave ultraviolet light. This light then causes a phosphor to fluoresce, producing visible light. Unlike incandescent lamps, fluorescent lamps always require a ballast to regulate the flow of power through the lamp. In common tube fixtures the ballast is enclosed in the fixture. Compact fluorescent lamps may have a conventional ballast located in the fixture or they may have ballasts integrated in the lamps, allowing them to be used in lamp holders normally used for incandescent lamps. See ballast.

foot-pound: (Energy) A unit of mechanical energy. The movement of one pound over a distance of one foot.

force: (Physics) A push or pull. Anything that has a tendency to set a body in motion, cause a moving body to rest, or to change the direct motion. Force equals Mass times Acceleration. A force is that which can cause an object with mass to change its velocity. Force has both magnitude and direction, making it a vector quantity.

fossil fuel: (Energy Management) Fossil source fuel. Hydrocarbons found within the top layers of the earth's crust. Coal, petroleum, and natural gas are types of fossil fuels.

four-pipe system: (Water Flow) A four-pipe arrangement is used where independent heating and cooling is required. The four-pipe system is two separate two-pipe arrangements—one, two-pipe arrangement for chilled water, and one for heating water. No mixing occurs. The return connections from the terminals can be made either direct or reverse return. The air handling unit is usually provided with two separate water coils, one for heating and one for chilled water. Each coil has its own control valve. However, some older air handling units have only one coil. When this is the case, modulating three-way valves, like the ones used in the three-pipe system, are installed in the supply branches. A three-way, two-position valve on the return branch line diverts the leaving hot or cold water to the proper main. Not energy efficient.

fourth wire: (Electrical) A name given to the neutral wire in a 3-phase, 4-wire distribution system.

four-wire system: (Electrical) A system of distribution of electric power requiring 4 wires. In a 3-phase system, the 4 wires are connected to the 3 line terminals of the supply transformer and the neutral point.

fractional horsepower v-belts: (Mechanical) A light-duty, flexible belt in sizes 2L through 5L. Fractional horsepower belts are generally used on smaller diameter sheaves because they are more flexible than the industrial belt for the same equivalent cross-sectional size.

freezestat: (Control) A device in the unit or ductwork to protect against coil freeze up when the air temperature drops below a preset limit.

freezing point: (Physics) The temperature at which a liquid solidifies, which is the same at which the solid melts (melting point). The freezing point of water is used as the lower fixed point in graduating a thermometer. Its temperature is defined at zero degrees Celsius, 32 degrees Fahrenheit and 273 Kelvin.

frequency: (Electrical) The number of complete cycles per second of alternating current. See Hertz.

frequency modulation: (Physics) Variation of frequency of a transmitted wave in accordance with impressed modulation, the amplitude remaining constant.

Freudian slip: (Definition) An error in speech which Freud believed revealed unconscious ideas or wishes.

friction: (Physics) The resistance to motion which is called into play when it is attempted to slide one surface over another with which it is in contact. The frictional force opposing the motion is equal to the moving force up to a value of the limiting friction. Any increase in the moving force will cause slipping. Static friction is the value of the limited friction just before slipping occurs. Kinetic friction is the value of the limiting friction after the slipping has occurred. This is slightly less than the static friction.

frictional electricity: (Electrical) Static electricity produced by rubbing materials together.

friction head: (Pump) The pressure required to overcome the resistance to flow, expressed in pounds per square inch or feet of head.

friction horsepower: (Mechanical) The difference between the indicated horsepower and the brake horsepower.

full load amperage: (Motors) The full load operating current at rated voltage and horsepower. Aka full load amps.

fume hood exhaust system: (Laboratory, Fume Hoods) An arrangement consisting of a laboratory fume hood, its adjacent room environment and the equipment (such as the ductwork and the blower) required to make the hood and system operable.

furnace: (Heating) A device used for heating. An enclosure in which energy in a non-thermal form is converted to heat, especially such an enclosure in which heat is generated by the combustion of a suitable fuel.

fuse: (Electrical) A device used for protecting electrical apparatus against the effect of excess current; it consists of a piece of fusible metal, which is connected in the circuit which melts and interrupts the circuit when an excess current flows.

fusible metals: (Definition) Alloys which melt in the 47 to 248 degree Celsius (117 to 478 degree Fahrenheit) temperature range; used as solders and for safety devices such as fire dampers.

future value: (Energy Management) The future worth of a present amount of money.

G is for Guillotine Damper

gas: (Physics) A state of matter in which the molecules move freely, thereby causing the matter to expand indefinitely occupying the total volume of any vessel in which it is contained.

gas turbine: (Mechanical) A simple, high speed machine used for converting heat energy into mechanical work in which stationary nozzles discharge jets of expanded gas (usually products of combustion) against the blades of a turbine wheel.

gate: (Computer) A circuit which controls the flow of binary signals.

gate: (Control) An electronic decision making circuit. An electronic circuit that performs a logical function (such as AND or OR) and passes signals when permitted by another independent source.

gate valve: (Control) A manual valve used for tight shutoff to service or remove equipment. It has a straight through flow passage that results in a low pressure drop. It regulates flow only to the extent that it is either fully open or fully closed. Not used for throttling purposes because the internal construction is such that when it's partly closed the resulting high velocity water stream will cause erosion of the valve seat. This erosion, called "wiredrawing" leads to eventual leakage when the valve is fully closed.

gauge: (Instrumentation) An instrument for measuring pressure.

gauge glass: (Instrumentation) Tube fitted vertically between a pair of gauge fittings and used to indicate the liquid level in a tank or boiler.

gauge number: (Definition) An arbitrary number denoting the gauge or thickness of sheet metal or diameter of wire, rods or drill bits, etc. in one of many gauge number systems.

gauge pressure: (Instrumentation) The pressure indicated on a gauge. The pressure of a fluid as shown on a pressure gauge. The amount by which the pressure exceeds atmospheric pressure, the sum of the two giving the absolute pressure (gauge pressure plus atmospheric pressure).

generator: (Electrical) Electrostatic or electromagnetic device for conversion of mechanical energy to electrical energy. Aka electric generator.

generator: (Refrigeration, Absorption System) A vessel containing a solution of absorbent and refrigerant to which heat is applied for the purpose of boiling away some of the refrigerant. Aka concentrator.

glass stem thermometer: (Instrumentation) Glass stem thermometers are generally limited to hand-held immersion readings in applications where precise readings are required. These thermometers can also be used as a standard for calibrating and verifying other types of thermometers if the error of the glass stem thermometer doesn't exceed plus or minus one scale division.

globe valve: (Control) A manual valve used in water makeup lines. It can be used in partially open positions and therefore, can be used for throttling flow. However, the globe valve has a high pressure drop even when fully open which unnecessarily increases the pump head and therefore, should not be used for water balancing.

glove box: (Laboratory, Fume Hood) An enclosure used to confine and contain hazardous materials with user access by means of gloved portals or other limited openings. Glove boxes require far less exhaust air than laboratory fume hoods or other biological safety cabinets. Glove boxes provide greater protection but are more restrictive than laboratory fume hoods or other biological safety cabinets.

governor: (Control) Speed regulator on rotating machine.

greenhouse effect: (Energy Management) Phenomenon by which thermal radiation from the sun is trapped within the atmospheric environment of earth causing a higher surface temperature. The heating of the atmosphere by the absorption of the infrared energy reemitted by the earth from incoming solar energy. The trapping by water vapor and carbon dioxide of long wave radiation emitted by the earth which leads to the temperature at the earth's surface being considerably higher than would otherwise be the case.

Greenwich Mean Time: (Definition) Greenwich Mean Time is the basis for scientific and navigational purposes and refers to the zero meridian of longitude which passes through Greenwich, England.

Gregorian calendar: (Definition) Name commonly given to the civil calendar now used in most countries.

grille: (Airflow) A wall-, ceiling- or floor-mounted louvered or perforated covering for an air opening. To control airflow pattern some grilles have a removable louver. Reversing or rotating the louver changes the air direction. Grilles are also available with adjustable horizontal or vertical bars so the direction, throw, and spread of the supply air stream can be controlled. Also, a plain or ornamental openwork of wood or metal used as a protecting screen or grating.

groove: (Mechanical) The "V" shaped or "U"shaped channel formed by flanges on an adjustable or fixed sheave or pulley.

grooving: (Boiler) Cracking of the plates of steam boilers at points where stresses are set up by the differential expansion of hot or cold parts.

ground: (Electrical) The lowest potential, or voltage, of an electrical system. Some use the term "earth" instead of ground.

grounded circuit: (Electrical) Circuit which is deliberately connected to the earth for safety or testing purposes.

ground water: (Definition) Water contained in and saturating the subsoil.

guide pulley: (Mechanical) A loose pulley used to guide a driving belt past an obstruction or to divert its direction. Aka idler pulley.

guide vane: (Mechanical) A general term for a foil to guide airflow in a duct or fan, or refrigerant gas in a compressor.

guillotine damper: (Airflow) A manual single-blade volume damper installed either vertically or horizontally in the duct and moved in or out to control airflow volume.

guy wire: (Definition) A tensioned cable designed to add stability to structures, as on an exhaust stack. One end of the cable is attached to the structure, and the other is anchored to the ground at a distance from the structure's base. Aka guy-rope or guy.

H is for HVAC

hair line: (Instrumentation) A fine, straight line in the optical system; used for the positive location of an image or correlation on to a measuring scale.

hard wiring: (Electrical) Any permanent wiring.

harmonic: (Physics) Sinusoidal component of repetitive complex wavelength with frequency which is an exact multiple of basic repetition frequency.

harmonic absorber: (Physics) Arrangement for removing harmonics in current or voltage wave forms using tuned circuits or a wave filter.

harmonic analysis: (Physics) The process of measuring or calculating the relative amplitudes of all the significant harmonic components present in a given complex wave form.

harmonic distortion: (Physics) Production of harmonic components from a pure sine wave signal as a result of nonlinear response of a transducer or amplifier.

harmonic filter: (Physics) One which separates harmonics from a pure sine curve. Aka harmonic suppressor.

harmonic wave: (Physics) A wave whose profile is a pure sine curve. Aka sinusoidal wave.

harmonic: (Instrumentation) For a strobe light tachometer, harmonics are frequencies of light flashes that are a multiple or sub-multiple of the actual rotating speed. For example, if the light frequency is either exactly two times or exactly one-half the actual speed of the rotating equipment, the part will appear stationary but the image will not be as sharp as when the rpm is correct.

hazardous chemical: (Laboratory, Fume Hood) A chemical for which there is significant evidence, based on at least one study conducted in accordance with established scientific principles, that acute or chronic effects may occur in exposed employees. Hazardous chemicals include carcinogens, toxic agents, irritants, corrosives and agents that damage the lungs, skin, eyes or mucous membrane.

head: (Fluid Flow) Pressure.

head pressure: (Refrigeration) Pressure of the refrigerant vapor on the discharge of the compressor. Aka discharge pressure or high-side pressure.

header: (Fluid Flow) A manifold supplying fluid to a number of tubes, pipes or passages or connecting them in parallel.

heat: (Energy, Physics)) A form of energy transferred by a difference in temperature. Heat always flows from a higher (warmer) temperature to a lower (cooler) temperature. In HVAC systems, fluids such as air, water, and refrigerants are used to carry or transfer heat from one place to another.

heat anticipator: (Control) Example: A control that shuts the space heater off a little early to give the room-heat time to reach the thermostat. See anticipator.

heat balance: (Energy Management) Evaluation of operating efficiency of a furnace, boiler, refrigeration system or other system or appliance. Total heat input being apportioned as to heat in the work, heat stored in refractory materials, lost by conduction, radiation, unburned gases and waste products, sensible heat, and latent heat of water vapor, thus determining the quantity and percentage of heat usefully applied in the sources of heat losses.

heat detector: (Control) An indirect acting thermostat for operation in conjunction with a gas flow control valve for controlling working temperatures in furnaces and heating appliances up to about 1000 degrees centigrade.

heat exchange: (Heat) The process of using two streams of fluid for heating or cooling one or the other either for conservation of heat or for the purpose of adjusting process streams to correct processing temperatures.

heat exchanger: (Equipment) A device specifically designed to transfer heat between two physically separated fluids. The term heat exchanger can describe any heat transfer device such as a coil or a particular category of devices. Heat exchangers are made in various sizes and types (shell and tube, U-tube, helical and plate) and are designed for several fluid combinations such as steam to water (converter), water to steam (generator), refrigerant to water (condenser), water to refrigerant (chiller), water to water (heat exchanger), air to refrigerant (coil), and air to water or water to air (coil).

heat pipe: (Mechanical) Means of cooling where heat is transferred along a tube from a heat source to a heat sink of small temperature difference. Heat transfer occurs as a liquid vaporizes at a desired temperature.

heat pump: (Mechanical) Machine operating on a reverse engine cycle to produce a heating effect. Energy from a low temperature source e.g., outside air is absorbed by the working fluid which is mechanically compressed resulting in a temperature increase. The high temperature energy is transferred in the heat exchanger. Used for HVAC heating and cooling.

heat transfer coefficient: (Heat Flow) The heat per hour through one square foot of material when the temperature difference between the two sides is 1F.

heat of absorption: (Refrigeration, Absorption) The heat released when two liquids are mixed. See absorption chiller.

heat of compression: (Fluid Flow, Refrigerant, Air) Mechanical energy of pressure transformed into heat energy. The energy equivalent of the work done on the refrigerant vapor to compress it. Heat of compression occurs in the compressor. It is added to the heat absorbed in the evaporator plus any superheat to calculate the total amount of heat that needs to removed from the refrigeration system. Also heat added to air as it passes through a centrifugal fan—approximately 0.5 degree Fahrenheit for every inch of static pressure.

heat of dilution: (Refrigeration) The heat released when two liquids are mixed. See absorption chiller, heat of absorption.

heaters: (Electrical) See overload protection.

heating coil: (Definition) A hot water, steam or electric coil. Heating coils may be termed "preheat" for a heating coil after the filters and before the cooling coil or "reheat" for coils after the cooling coil in the AHU or in the duct system downstream of the AHU.

heating element: (Electrical) A heating device in which heat is produced by the passage of electric current through a resistance such as a coil. Aka heating coil.

heating load: (Definition) The quantity of heat per unit time that must be provided to maintain the temperature in a building at a given level—The heat to add to, or remove from, a building to maintain temperature—Heat quantity from equipment or structure, e.g., heat load from the lights, windows, computers, etc.

heating resistor: (Electrical) The wire or other suitable material used as the source of heat in an electric heater.

hermetic compressor: (Refrigeration) In the hermetic, and most semi-hermetic compressors (aka accessible hermetic compressor) the compressor and motor driving the compressor are integrated and operate in the refrigerant system. The motor is cooled by the refrigerant. Generally, hermetic units range from fractional horsepower sizes to about 10 tons in a single unit. If more tonnage is needed, several compressors may be installed in the same air conditioning unit. Above 10 tons, the construction is often the open type or the semi-hermetic type. Semi-hermetic compressors are almost completely accessible. Semi-hermetic compressors range in capacity from 2 tons to about 150 tons and offer the advantage of direct drive and hermetic sealing plus serviceability. The hermetic compressor is a welded hermetic and the semi-hermetic compressor is a bolted hermetic. One advantage of the hermetic compressor is that it's taller, more compact and has less vibration than the open compressor. However, the main advantage of the hermetic compressor that it has no external shaft. This eliminates mechanical problems with shafts, belts, and sheaves, and concern for refrigerant leakage. In a hermetic unit, the motor is located within the refrigerant atmosphere. Therefore, another advantage is that the motor is continuously cooled by the refrigerant vapor flowing to the compressor suction valves. Lubrication is also simplified since the motor and the compressor operate in the same closed space with the oil. Disadvantages of hermetic compressors are that they are limited on capacity, and limited on speed because the compressor has to run at the motor speed (direct drive). Another major disadvantage is that it is not field-serviceable. If the motor fails and the motor drive cannot be serviced then the entire compressor must be removed. Also burned out windings can contaminate the refrigeration system requiring the system to be entirely pumped down and the refrigerant replaced. Therefore, if a motor burns out in a hermetic compressor, or any other internal problem occurs, the maintenance trend is for a complete replacement of the compressor. Otherwise, the entire unit must be returned to the shop or factory to be dismantled and reconditioned.

Hertz: (Electrical) Hertz is the number of complete cycles per second of alternating current. Aka frequency or cycles per second (CPS). The USA uses 60 Hertz, while many other countries operate at 50 Hertz.

heuristic: (Computer) Describes an approach based on common sense rules and trial and error rather than on comprehensive theory.

heuristic program: (Computer) One which attempts to improve its own performance as the result of learning by previous actions within the program

high efficiency particulate air filter: (Airflow) A filter with efficiency in excess of 99.97% for 0.30 micrometer and larger particles as determined by a Dioctyl Phthalate oil test or other tests. Known as HEPA filters they utilize spun-hooked glass fiber rolled into a paper-like material. This material is pleated to increase the fiber surface area and bonded, or potted, into a frame. Hot melt is used to hold the pleats far enough apart to allow air to flow between them. A HEPA filter can last three to five years or more in a standard cleanroom environment depending on ambient conditions and maintenance of prefilters. Since airflow capacity decreases and static pressure increases over time, HEPA filters actually become more efficient as the filter loads. The filter should be changed once it has reached its capacity. For example, if a HEPA unit is set for 90 feet per minute on but is only achieving 70 feet per minute then the HEPA filter should be changed.

high intensity discharge: (Lighting) Electrical lamps producing light by means of an electric arc between tungsten electrodes housed inside a translucent or transparent fused quartz or fused alumina tube. This tube is filled with both gas and metal salts. The gas facilitates the arc's initial strike. Once the arc is started, it heats and evaporates the metal salts forming a plasma, which greatly increases the intensity of light produced by the arc and reduce its power consumption. Compared with fluorescent and incandescent lamps, high-intensity discharge lamps have much higher efficacy; they give a greater amount of light output per watt of electricity input. Types of high intensity discharge lamps include: mercury vapor, metal halide, and high-pressure and low-pressure sodium vapor. Mercury vapor lamps were the first commercially available high intensity discharge lamps. Originally they produced a bluish-green light, but more recent versions can produce light with a less pronounced color tint. However, mercury vapor lamps are falling out of favor and being replaced by sodium vapor and metal halide lamps. Metal halide and ceramic metal halide lamps can be made to give off neutral white light useful for applications where normal color appearance is critical. Low-pressure sodium vapor lamps are extremely efficient. They produce a deep yellow-orange light and have an effective color rendition index of nearly zero; items viewed under their light appear monochromatic. High-pressure sodium lamps tend to produce a much whiter light, but still with a characteristic orange-pink cast. New color-corrected versions producing a whiter light are now available, but some efficiency is sacrificed for the improved color.

high limit control: (Control) A limit control that monitors the condition of the controlled medium and prevents the operation of equipment when it would cause dangerous or undesirably high temperatures, pressures or relative humidity.

high pressure air systems: (Control) Static pressures above 6 inches water gauge with velocities above 2000 feet per minute.

high pressure gauge: (Control, Boiler, Refrigeration) A pressure gauge installed on equipment for safety to equipment, property and people. Indicates pressure and typically shuts off equipment when pressures reaches or exceeds set point.

high pressure gauge: (Control, Pneumatic) A pressure gauge installed in the main supply line before the pressure reducing valve which indicates the pressure of the air stored in the receiver tank.

high temperature water: (Water Flow) Water temperature range of 350 to 450 degrees Fahrenheit.

hit and miss ventilators: (Airflow) A ventilating device consisting of a slotted plate over which may be moved another slotted plate, so that the opening for access of air may be restricted as required.

hood face: (Laboratory, Fume Hoods) The plane of minimum area at the front portion of the hood through which air enters when the sash is fully open.

hopec laboratory fume hood: (Laboratory, Fume Hoods) Hand Operated Positive Energy Control laboratory fume hood. A hood with a combination vertical-horizontal sash.

horsepower: (Mechanical) A unit of power.

hot deck (Airflow) In a multizone or dual duct unit it is the air chamber after the heating coil.

hot gas bypass: (Refrigeration) A method of compressor capacity control.

In this method of capacity control the compressor discharge gas (hot gas) bypasses the condenser to prevent the compressor suction pressure from failing below a desired setting. Hot gas bypass provides an artificial load on the evaporator by introducing a portion of high pressure, high temperature gas to the evaporator (suction side) of the system. The use of hot gas bypass can be accomplished in several ways. The standard or external method of hot gas bypass consists of a valve located in the condensing section. The inlet of the hot gas bypass valve is piped from a tee in the discharge line between the outlet of the compressor and inlet of the condenser. They open in response to a decrease in downstream pressure and modulate from full open to full close over a given range. Introduction of the hot, high pressure gas into the low pressure side of the system at a metered rate prevents the compressor from lowering the suction pressure further.

hot leg: (Electrical) Any wire in an electrical circuit which is at a higher voltage than neutral or ground. Aka hot lead or hot wire.

hot wire: (Instrumentation) Said of an electrical indicating instrument (e.g. hot wire anemometer) whose operation depends on the thermal expansion of, or change in resistance of a wire or strip when it carries a current. The current heats the wire which expands and moves the indicating needle.

hot wire anemometer: (Instrumentation) Measures instantaneous air velocity in feet per minute using an electrically heated wire. As air passes over the wire, the wire's resistance is changed and this change is shown as velocity on the instrument's scale. This instrument is very position sensitive when used to measure air velocities. Therefore, it is important to ensure that the probe is held at the correct angle to the airflow. The hot wire anemometer is most often used to measure low velocities such as found at the face of fume hoods; however, some instruments can also measure temperatures and static pressures.

humidifier: (Mechanical) An apparatus for maintaining desired humidity conditions in the air supplied to a conditioned space.

humidity: (Psychrometrics) See relative humidity and specific humidity.

humidity ratio: (Psychrometrics) The ratio of the mass of water vapor in a sample of moist air to the mass of dry air with which it is associated. See specific humidity.

humidistat: (Control) A humidistat (hygrometer) operates on the principle that hygroscopic materials such as nylon, silk, wood, leather and human hair will expand when exposed to moisture. The change in the material is detected mechanically and converted to a control signal.

humidity sensors: (Control) Humidity sensors are used to measure relative humidity or dew point of the air. See humidistat.

hunting: (Control) The condition which happens when a controller changes or cycles continuously resulting in fluctuation and loss of control. The desired set point condition can't be maintained. An undesirable condition where a controller is unable to stabilize the state of the controlled medium and cycles rapidly.

hvac: (Definition) Heating, Ventilating and Air Conditioning (cooling) an occupied or non-occupied space using the liquid and vapor fluids: water, air, steam, and refrigerants. Aka "h-vac," (sounding out the letter h and the word vac).

hydronics: (Water Flow) The science of heating and cooling with liquids.

hygristor: (Electrical) Resistant element sensitive to ambient humidity.

hygrometer: (Instrumentation) Instrument for measuring or giving output signal proportionate to atmospheric humidity. Electrical hygrometers make use of hygristors.

hygrometric, hygrometry: (Instrumentation) Measurement of the hygrometric state, or relative humidity of the atmosphere.

hygroscopic: (Definition) Water absorbing.

I is for Ice

ice: (Definition) Ice is formed when water is cooled below its freezing point. It is a transparent, crystalline solid with a specific heat capacity of approximately 0.5 Btu/lbF. Because water attains its maximum density at 4 degrees Celsius (39.2 degrees Fahrenheit) ice is formed on the surface of ponds and lakes during frosts and thickens downwards.

impedance: (Electrical) The total opposition to the flow of alternating current by any combination of resistance, inductance and capacitance. Impedance is measured in Ohms.

impeller: (Pump) The rotating component of a centrifugal pump which imparts kinetic energy to the fluid.

incandescence: (Physics) the emission of light by substance because of its high temperature, e.g. a glowing electrical lamp filament.

incandescent lamp: (Lighting) A lamp in which light is produced by heating a filament to a white or red heat.

inches of water column or inches of water gauge: (Airflow) A unit of air pressure measurement equal to the pressure exerted by a column of water one inch high.

inclined manometer, inclined-vertical manometer: (Instrumentation) The inclined manometer has an inclined scale which reads in inches of water gauge in various ranges such as 0 to 0.25 inches water gauge, 0 to 0.50 inches water gauge, or 0 to 1.0 inches water gauge. The inclined-vertical manometer has both an inclined scale that reads 0 to 1.0 inches water gauge and a vertical scale for reading greater pressures such as 1.0 to 10 inches water gauge. The inclined and inclined-vertical manometer has a left and right tube connection for attaching tubing from a Pitot tube or other sensing device to the manometer. The left tubing connector is called the "high" side of the manometer and the right connector is the "low" side. The manometers are filled with colored oil which is lighter than water therefore, the oil will move a greater distance for a given pressure allowing more precise readings.

indicated horsepower: (Mechanical) The horsepower developed by the pressure-volume changes of the working agent within the cylinder of a reciprocating engine. It exceeds the useful or brake horsepower at the crank shaft by the power lost in friction or pumping.

indirect heating: (Heating) A system of heating by convection, as opposed to direct heating by radiation.

indirect lighting: A system of lighting in which 90% of the total light from the lamps is emitted away from the work surface.

inductance: (Electrical) Property in an electrical circuit where a change in the current flowing through that circuit induces an electromagnetic field (magnetic flux) that opposes the change in current. This magnetic flux tends to act to oppose changes in the flux by generating a voltage (aka "back EMF") that counters or tends to reduce the rate of change in the current. The ratio of the magnetic flux to the current.

induction: (Electrical) The process of producing electron flow by the relative motion of a magnetic field across a conductor. In a transformer, current flowing through the primary coil sets up a magnetic field. This magnetic field produces current flow in the secondary coil.

induction motor: (Electrical) Alternating current motor in which currents in the primary winding (connected to the supply) set up a flux which causes currents to be induced in the secondary winding (the rotor); these currents interact with the flux to resist rotation. A non-synchronous motor.

induction terminal box: (Airflow) Induction boxes may be constant or variable volume, pressure dependent or pressure independent. One type of box uses supply air aka primary air. It is forced through a discharge device such as a nozzle or Venturi to induce room or return air, "secondary air," into the box where it mixes with the supply air. The high velocity of the primary air creates a low pressure region which draws in or induces the higher pressure secondary air. The mixed air (primary air plus secondary air) is then supplied to the conditioned space. Some induction boxes have heating or cooling coils through which the secondary air is induced. See VAV Fan Powered Terminal Box

inductive load: (Electrical) A lagging current associated with inductive (ac) motors and transformers. Aka lagging load.

industrial v-belts: (Mechanical, Fan) HVAC fan belt sizes. Typically, sizes "A" through "E." See Chapter 14, Tables.

inertia: (Physics) The property of a body proportional to its mass, which opposed a change in the motion of the body.

infiltrated air, building: (Airflow) Air coming into a building by way of doors, windows, etc., or any other means in the building envelope. Typically, unwanted air in commercial and industrial temperature and pressure controlled buildings.

infiltrated air, laboratory: (Airflow) Auxiliary air induced from the corridor or other spaces into the lab.

insulation: (Definition) To separate from conducting bodies by means of nonconductors so as to prevent transfer of electricity, heat, or sound.

insulator: (Electrical) Material that does not readily pass electrons. Some of the best insulators: rubber, glass and plastic.

insulator: (Heat) A material that resists heat flow. Some of the best insulators: fiberglass, vapor barrier, foam and still air. *integral control:* (Control) Control action designed to eliminate the offset in proportional control.

integrated circuit: (Electronic) Device in which all the components of an electrical circuit are fabricated on a single piece of semiconductor (silicon) material called a chip.

intercooler: (Refrigeration, Water Chiller) A water chiller component.

Hot liquid refrigerant drains from the condenser into the high pressure chamber of the intercooler. It then passes through a high side float valve into the intermediate chamber of the intercooler. As the refrigerant goes through the valve the pressure on the refrigerant is reduced. This reduction in pressure reduces the boiling point of the refrigerant. However, the temperature of the liquid refrigerant is still above the new boiling point. Because the liquid refrigerant is hotter than its boiling point, a part of the liquid refrigerant begins to boil off. This boiling off of the liquid refrigerant is called "flashing." One of the purposes of the intercooler is to preflash liquid refrigerant in its intermediate pressure chamber to reduce the temperature of the liquid refrigerant to the lower temperature corresponding to the pressure in the intermediate chamber. Another purpose of the intercooler is to take the preflashed vapor from the intermediate chamber and send it to the suction side of the second stage compressor to be compressed. The pressure of this vapor is above the evaporator pressure. Therefore, the power required to compress it to the condensing pressure is less. Also, the temperature of this vapor is cooler than the temperature of discharged vapor from the first stage compressor. When the two vapors are mixed, the temperature of the vapor going into the second stage compressor is reduced, thereby increasing the capacity and efficiency of the system. From the intermediate chamber, the lower temperature liquid refrigerant passes through a float valve into the evaporator. The refrigerant is now at an intermediate pressure. Its temperature is somewhere between the higher pressure and temperature of the compressor and the lower pressure and temperature of the evaporator. As the liquid refrigerant passes through the intermediate float the pressure is reduced to the evaporator pressure. This causes some of the liquid to flash, cooling the remainder of the liquid to the evaporator temperature. The liquid-vapor refrigerant goes to the evaporator through the liquid line. The intercooler reduced the total volume of flash gas required to cool the refrigerant to its evaporator temperature. This reduction in the volume of flash gas in the intercooler means that more of the liquid refrigerant is available for use in the evaporator. This makes the chiller system more efficient by increasing net refrigerant effect. It also means that there is less load on the first stage of the compressor, thereby reducing the horsepower requirements. Aka economizer.

interest rate: (Energy Management) Fee paid on borrowed capital.

internal rate of return: (Energy Management) Interest rate where the present value (worth) of the investment equals the present value (worth) of the (annual) savings. The project break-even interest rate.

International System of Units: (Definition) The modern metric system. It is the world's most widely used system of units. Aka SI from "Système International d'Unités." The SI is dynamic as units are created and definitions are modified through international agreement among nations as the technology of measurement progresses.

intrinsically safe circuit: (Electrical) A circuit in which any spark or thermal effect, produced either normally or in specified fault conditions, is incapable, under the test conditions of causing ignition of a mixture of flammable or combustible material in the air in the mixture's most easily ignitable concentration.

intrinsically safe device: (Electrical) An apparatus in which all the circuits are intrinsically safe.

isenthalpic: A process carried out at a constant enthalpy.

isentropic: A process carried out at a constant entropy. Any process in which no change in energy occurs is said to be isentropic.

isothermal: Occurring at a constant temperature.

J is for Joule

joint of duct: (Airflow) Ductwork of some length, typically 48".

Joule: (Energy) Unit of energy in the International System of Units. The work done by a force of one Newton traveling through a distance of one meter. The work required to move an electric charge of one coulomb through an electrical potential difference of one volt; or one coulomb volt. The work done to produce power of one watt continuously for one second; or one watt second. The kinetic energy of a 2 kg mass moving at a velocity of 1 meter per second. The energy is linear in the mass but quadratic in the velocity.

K is for Kilo

Kelvin: (Thermodynamics) The Kelvin scale is an absolute temperature scale where absolute zero is 0 Kelvin or minus 273 degrees Celsius.

kinetic energy: (Physics) The energy in a body due to its motion.

kilo: (Mathematics) Greek for 1000

kilovolt-ampere: (Math Conversion) 1000 volt-amperes or 1000 volt-amps *kilovolt-ampere reactance:* For capacitors.

kilowatt: (Math Conversion) 1000 Watts

kilowatt-hour: (Math Conversion) 1000 Watt-hours

L is for Lexicon

laboratory: (Definition) A facility where relatively small quantities of hazardous chemicals are used on a nonproduction basis.

laboratory fume hood: (Laboratory, Fume Hood) A ventilated box-like structure enclosing a work space intended to capture, contain, and exhaust fumes, vapors, particulate matter, and other contaminants generated inside the enclosure. The fume hood consists of side, back and top enclosure panels and one side—called the face—which is open or partially open. The fume hood also has a sash and an exhaust plenum equipped with a baffle system for regulation of airflow. The fume hood is generally mounted on a bench or table. Air is brought into the fume hood to contain and exhaust contaminants from the hood. The conventional constant air volume fume hood has a sash which moves vertically up and down. At the sash's full open position the free area at the hood face is generally about 10 to 13 square feet with the minimum face velocity being about 100 feet per minute. The volume of air through the hood is therefore about 1000 to 1300 cubic feet per minute. As the sash is lowered the volume of air is reduced but the velocity of the air through the opening is increased (sometimes as high as 400 feet per minute). These high face velocities can cause unwanted turbulence which can induce contaminants out of the hood into the laboratory space. In order to try to maintain constant face velocities around 100 feet per minute, a bypass hood is used. The bypass hood also has a vertically moveable sash. The construction of the hood is similar to the conventional hood described above with the addition of the bypass to allow for a constant volume of airflow through the fume hood as the sash is closed. In other words, as the sash is pulled down the air volume through the hood face is reduced. At the same time the sash is being closed the bypass is being opened and more air is then drawn through the bypass. This keeps the volume of air through the hood and the hood face velocities relatively constant. Another variation of the conventional hood uses a combination horizontal/vertical sash. The auxiliary air fume hood also has a vertically movable sash. With the sash open, auxiliary air is ducted to the hood and distributed across the face area prior to its passage into the hood. With the sash closed the auxiliary air is introduced directly into the fume hood interior. Auxiliary air fume hoods are designed to reduce the amount of conditioned laboratory air required for exhaust air. For instance, a hood might use 70% auxiliary air and only 30% room air as the total fume hood exhaust. In theory, this makes this type of hood more energy efficient because the auxiliary air is only nominally heated or cooled as compared to the room air. However, tests have shown that the best safety performance for this type of hood occurs when the auxiliary air is slightly warmer than the laboratory room air. Therefore, a concern about this type of fume hood is that the auxiliary air may enter the hood only nominally treated. Another concern is that the auxiliary air enters the hood in such a way that it may create turbulence, which can cause the air to reverse flow back out the hood face. Variable air volume fume hoods may use either a vertical sash or a combination sash. This type of hood is equipped with special controls to allow the volume of exhaust air to vary while still maintaining a constant face velocity.

laboratory fume hood system: (Laboratory, Fume Hood) Laboratory fume hood systems may be constant air volume, variable air volume or a combination of the two. They may be auxiliary air systems, partial return air systems or 100% room air systems. With the 100% room air system the exhausted air is replaced entirely by the conditioned air from the laboratory space. Since there is no return air there is an added energy burden placed on the heating and cooling components to condition the outside air. To try to reduce energy costs variable air volume systems are used.

laboratory use: (Laboratory, Fume Hood) The handling or use of chemicals in which the chemical manipulations are carried out on a laboratory scale using multiple chemical procedures and/or chemicals. The procedures involved are not of a production process. They do not, in any way, simulate a production process. Adequate protective laboratory equipment is available and in common use to minimize the potential for employee exposure to hazardous chemicals.

ladder logic diagram: (Electrical) The diagram that is used to describe the logical interconnection of the electrical wiring of control systems.

lag: (Control) A time delay between an initiating action and a desired effect. The delay in action of the sensing element of a control due to the time required for the sensing element to reach equilibrium with the variable being controlled.

lagging power factor: (Electrical) When the load is inductive the inductance tends to oppose the flow of current, storing energy then releasing it later in the cycle. The current wave lags behind the voltage wave (sine wave).

laminar airflow: Airflow in parallel flow lines with uniform velocity and minimum eddies.

laminar flow cabinet: (Environmentally Controlled Area) A clean bench or biological safety cabinet that uses smooth directional airflow to capture and carry away airborne particles. The laminar flow cabinet is not considered a laboratory fume hood.

lamp: (Definition) Various devices for producing light such as a vessel with a wick for burning an inflammable liquid (oil), or a decorative appliance housing a lamp that is usually covered by a shade.

lamp: (Heat) Various devices for producing heat, e.g., a heat lamp for heating food.

lamp: (Lighting) A glass bulb or tube that emits light produced by electricity, for example, incandescent or fluorescent lamps. See light bulb.

lamp depreciation factor: (Lighting) The fractional loss of lumens radiated by a lamp at rated operating conditions. This loss increases progressively during the lifetime of the lamp.

latent heat: (Psychrometrics) Hidden heat. The heat energy which when supplied to or removed from a substance causes a change of state without any change in temperature. Air: The units of latent heat are Btu per pound of dry air. Water: The units of latent heat are Btu per pound.

latent heat of condensation: (Heat, Refrigeration) Latent heat released when a vapor changes to a liquid. The amount of heat released by a pound of refrigerant to change its state from a vapor to a liquid. See latent heat of vaporization.

latent heat of fusion: (Heat) Latent heat released when a solid changes to a liquid or vice versa.

latent heat of vaporization: (Heat, Refrigeration) Latent heat released when a liquid changes to a vapor. The amount of heat required by a pound of refrigerant to change its state from a liquid to a vapor. The heat that a liquid will absorb when going from the liquid state to the vapor state. The heat that a vapor will give up when going from the vapor state to the liquid state (aka latent heat of condensation).

law of conservation of energy: (Physics) Energy is neither created nor destroyed. Energy can be converted from one form to another.

leading power factor: (Electrical) When the load is capacitive (installing capacitors) the current sine wave leads the voltage waveform. See capacitive load.

Leadership in Energy and Environmental Design: (Energy Management) Inception in 1998. A Green Building Rating System, developed by the United States Green Building Council (USGBC) a non-profit organization. A set of standards for environmentally sustainable construction. An open process where the technical criteria proposed by the Leadership in Energy and Environmental Design committees are publicly reviewed for approval by membership organizations.

lexicon: (Definition). A book, chapter, or an appendix in a book containing an alphabetical arrangement of words or vocabulary of a language or subject along with their definitions. For example, "Chapter 12 in this book is a lexicon."

life-cycle costing: (Energy Management) Method of calculating the cost of a system over its entire life span. Some considerations for determining life-cycle costing are amount of energy consumed, maintenance required, cost of equipment, installation costs, and federal regulations and requirements.

light bulb: (Lighting) A glass envelope enclosing the light source of an electric lamp or such an envelope together with the light source it encloses. A bulb-shaped light.

light emitting diode: (Lighting) A low current, low voltage light. It is a semiconductor diode that emits light across the visible, ultraviolet, and infrared wavelengths when voltage is applied. Becoming more popular as technology and uses increase and cost decrease. Aka LED.

light fixture (Lighting) An electrical device (luminaire) used to create artificial light with an electric lamp. Light fixtures have (1) a body and a light socket to hold the lamp(s) and allow for lamp replacement, (2) various devices to control the light, and (3) require an electrical connection to a power source. Permanent lighting fixtures are typically wired directly while moveable lamps generally have a cord and a plug. Light fixtures may also have reflectors for directing light.

light troffer: (Airflow) A type of ceiling diffuser which fits over a fluorescent lamp fixture and delivers air through a slot along the edge of the fixture. They are available in several types. One type delivers air on both sides of the lamp fixture and another type provides air to only one side of the fixture.

limit: (Control) Control applied in the line or low voltage control circuit to break the circuit if conditions move outside a preset range. A switch which cuts off power to a motor's windings when the motor reaches its full open position. A temperature, pressure, humidity, dew point, or other control that is used as an override to prevent undesirable or unsafe conditions in a controlled system.

limit shutdown: (Control) A condition in which the system has been stopped because a pre-established limit value has been exceeded.

linear slot diffuser: (Airflow) This type of diffuser is manufactured in various lengths and numbers of slots and may be set for different throw patterns.

liner: (Laboratory, Fume Hoods) Interior lining of a laboratory fume hood used for the side, back and top enclosure panels. Also for exhaust plenum and baffles.

liner: (Heat) An insulation attached to the inside of galvanized duct work.

line voltage: (Electrical) In the U. S. control industry the normal electric supply voltages are usually 110-120 or 208 volts for lighting and small appliances and 220 to 480 or higher volts to motors and large appliances.

lintel: (Laboratory, Fume Hood) The portion of a laboratory fume hood located directly above the access opening.

liters per minute: (Measurement) A unit of volume (liter in US and litre in UK). One liter/min is equal to 0.0353 cubic feet per minute (cfm)

liters per sec: (Measurement) A unit of volume (liter in US and litre in UK). One liter per second (l/s) is equal to cubic feet per minute 2.12 (cfm)

lithium bromide: (Refrigeration) A salt used as an absorbent in an absorption refrigeration system.

liquid slugging: (Refrigeration) When liquid refrigerant enters the compressor. Liquid slugging causes noisy operation, loss of capacity, an increase in power required and possible damage to the compressor.

load: (Electrical, Heat) Electric load: An electric or electronic component or device that uses power. See heat load.

load factor: (Energy Management) Load, as a percentage of work, on a piece of equipment such as a motor, chiller, boiler, etc.

locked rotor amperage: (Motors, Electrical) Locked rotor amperage occurs between zero and full motor speed when the starting current is drawn from the line with the rotor locked and with rated voltage supplied to the motor. During this short time, a fraction of a second for small motors to a second or longer for large motors, the locked rotor amperage far exceeds the full load operating current. Locked rotor amperage will generally be 5 to 6 times the full load amperage. This inrush of current will continue to decrease until the motor reaches full operating speed.

logic: (Control) The decision making circuitry on an integrated circuit.

log mean temperature difference: (Mathematics) The logarithmic average of the temperature differences when two fluids are used in a heat transfer process. The temperature difference after the process will be less than at the beginning and the exchange of heat will follow a logarithmic curve. It is this logarithmic average of the temperature differences which establishes which heat exchanger is best. The higher the log mean temperature difference number, the greater the heat transfer. For example, a coil piped parallel flow has a lower log mean temperature difference and would need more surface area to do the same heat transfer as a coil piped counter flow which has a higher log mean temperature difference.

low limit control: (Control) A limit control that monitors the condition of the controlled medium and interrupts system operation if the temperature, pressure, humidity drops below the desired minimum value.

low pressure gauge: (Control) Pneumatics: A pressure gauge installed in the main supply line after the pressure reducing valve which indicates the pressure of the main air.

low pressure air systems: (Airflow) Static pressures to 2 inches water gauge with velocities to 2500 feet per minute.

low temperature water: (Water Flow) Temperature range to 250 degrees Fahrenheit.

low voltage: (Electrical, Control) In the control industry, a power supply of 25 volts or less.

lumen: (Lighting) The unit of light quantity (luminous flux in the International System, SI) equal to the amount of light having a strength of one candle radiating equally in all directions on a unit surface.

luminaire: (Lighting) A complete light fixture.

luminance: (Lighting) An indicator of how bright a surface will appear. Luminance indicates how much luminous power will be perceived by an eye looking at the surface from a particular angle of view. The photometric measure of the density of luminous intensity in a given direction it is the amount of light that passes through or is emitted from a particular area.

M is for Manometer

magnetic relay: (Control) Solenoid operated relay or contact; a switching relay that utilizes an electromagnet (solenoid) and an armature to provide the switching force.

magnetic starters: (Control) Contactors with overload protection relays. Aka "mags" or "mag" starters.

magnitude: (Definition) Size, extent, dimensions.

make-up air: (Airflow) Air introduced into the secondary air system for ventilation, pressurization, and replacement or "make-up" of exhausted air.

make-up water: (Water Flow) The water that replaces the water lost through leakage and evaporation. To prevent air problems the make-up water to a closed system should be introduced into the system at some point in the air line to the compression tank or at the bottom of the compression tank.

manometer: (Instrumentation) An instrument for measuring air or water pressures. Essentially, a U-tube partly filled with a liquid, usually water, mercury or a light oil. The pressure exerted on the liquid is indicated by the liquid displaced. A manometer can be used as a differential pressure gauge. The manometer is used in the HVAC industry for measuring air pressure. Other pressure measuring instruments may be checked for calibration against a properly set up and accurately read manometer. Both analog and digital manometers are used. Analog types include inclined manometer, inclined vertical manometer, and U-tube manometer. The digital type is an electronic manometer and micro-manometer combination. Analog manometers contain no mechanical linkage and don't need calibration. The digital manometer, on the other hand, needs periodic calibration as recommended by the manufacturer or certified calibration agency. The electronic manometers are battery-powered which need to be checked before each use.

manual air vent: (Mechanical, Water Systems) A manual valve that is opened to allow air entrained in a water system to escape.

manual drain: (Mechanical, Pneumatic System) A manual device on a compressed air (pneumatic) receiver tank to remove moisture and other contaminants from the air.

manual starters: (Control, Motors) Motor-rated switches that have provisions for overload protection. Generally, manual starters are limited to motors of 10 horsepower or less and are normally located close to the motor they control.

manual valves: (Control, Water Flow) Manual valves regulate flow rate or limit the direction of flow. Gate, globe, plug, ball and butterfly valves regulate flow rate while a check valve limits the direction of flow. A combination valve does both.

manual volume dampers: (Control, Airflow) See volume dampers.

mass: (Physics) Weight of a body in pounds. The quantity of matter in a body measured by the ratio of the force required to produce given acceleration. Mass equals Force divided by Acceleration. Mass equals Volume divided by Specific Volume. Mass equals Volume times Density.

mass flow rate: (Units) Pounds per unit of time.

matched belts: (Mechanical) A set of belts whose exact lengths and tensions are measured and matched by the supplier in order for each belt to carry its proportionate share of the load.

mechanical energy equivalent: (Energy) 778 foot-pounds equal 1 British thermal unit.

mechanical chiller: (Refrigeration) One of two types of water chillers commonly used in HVAC refrigeration systems. The mechanical chiller is the most often used chiller with either a mechanical reciprocating, screw or centrifugal compressor. The other type of chiller is the absorption chiller. The absorption chiller does not have a mechanical compressor, but uses a generator and an absorber instead. Major components in a mechanical water chiller package are: compressor, condenser, evaporator, accessories, intercooler, purge system and controls.

mechanical seal: (Pump) A water seal. Mechanical seals have a stationary ring, usually made of hard ceramic material, and a rotating graphite ring. The stationary ring fits into a recess in the pump body and has a rubber gasket behind it which forms a watertight seal. Behind the molded graphite ring is a rubber bellows and a seal spring. This spring keeps the rotating graphite ring tight against the face of the ceramic ring making the water seal. No maintenance or adjustments are needed as the spring continually pushes the graphite ring forward to make up for wear. It is important that a water film be between the two surfaces to provide lubrication and cooling. Running or "bumping" a pump, without water in the system will damage mechanical seals.

mechanical work: (Physics) When a force acting on a body moves the body some distance. Work equals Force times Distance. Expressed in foot-pounds.

medium: (Control) Term used to for the fluid being controlled for temperature, pressure, etc. The controlled medium in an air duct is the air.

medium pressure air systems: (Airflow) Static pressures between 2 and 6 inches water gauge with velocities between 2500 and 4000 feet per minute.

medium temperature water: (Water Flow) Temperature range of 250 to 350 degrees Fahrenheit.

memory: (Computer) A computer subsystem used to store instructions and data.

memory stop: (Control, Water Flow) Some plug valves, calibrated balancing valves and combination valves have an adjustable memory stop. The memory stop is set during the final water balance so that, if the valve is closed for any reason, it can later be reopened to the "water balanced" setting.

meniscus: (Instrumentation) The curved surface of the liquid column in a liquid-filled manometer. In manometers that measure air pressures the liquid is either water or a light oil. In manometers that measure water pressures the liquid is mercury.

metering device: (Refrigeration) A metering device may be a capillary tube, thermostatic expansion valve, automatic expansion valve, manual valve or other device that reduces the liquid refrigerant's pressure and corresponding temperature and controls the flow of refrigerant into the evaporator coil. The metering device controls the flow of refrigerant and changes the high temperature, high pressure liquid refrigerant from the condenser to a low temperature, low pressure liquid.

micro-manometer: (Instrumentation) Typically, the micro-manometers used in HVAC test and balance and commissioning today are electronic digital manometers. They are used to measure very low pressures accurately to one ten-thousandth inch water gauge. See manometer.

micrometer: (Instrumentation) A mechanical or digital instrument for measuring very small distances, objects, or angles. Aka "mike."

micrometer: (Measurement) A unit of length equal to one millionth (10^{-6}) of a meter or one thousandth (10^{-3}) of a millimeter. Aka micron.

microprocessor: (Computer) A multifunctional integrated circuit that contains logic functions. A small computer used in load management to analyze energy demand and consumption so that loads are turned on and off according to a predetermined program.

microprocessor-based controller: (Computer) A device consisting of a microprocessor unit, digital inputs and outputs, A/D and D/A converters, a power supply and software to perform direct digital, programmable logic control.

minimum attractive rate of return: (Energy Management) An interest rate.

mixed air: (Airflow) Primary (supply) air plus secondary (room) air. Also, outside air plus return air.

mixed flow impeller: (Pump) The impeller in a mixed flow centrifugal pump. Some mixed flow impellers use double curvature vanes which are wider than the radial- or plain-flow impeller. See radial flow impeller in this chapter and Chapter 6.

mixing valve: (Control, Water Flow) A valve in a water pipe that mixes two different water steams such as return water and bypassed water. See three-way automatic temperature control valve.

modulating control: (Control) Tending to adjust by increments and decrements. A mode of automatic control in which the action of the controlled device is proportional to the deviation from set point of the controlled variable. See proportional control.

modulating motor: (Control) A reversible electric motor used to drive a damper or valve. It can position the damper or valve anywhere between fully open or fully closed in proportion to a deviation of the controlled variable.

moisture content: (Refrigeration) Percent by weight of the liquid in any mixture of liquid and vapor.

morning warm up: (Control) A control system that keeps outside air dampers closed until a desired space temperature is achieved.

motor controller: (Motors, Control) Motor controllers are used for starting and stopping motors. They can be grouped into three categories: manual starters, contactors and magnetic starters.

motor overload protection: (Electrical, Motors) Thermal overload protection devices (aka "heaters" or "thermals") prevent motors from overheating. If a motor becomes overloaded or one phase of a three-phase circuit fails (single-phasing) there will be an increase in current through the motor. If this increased current drawn through the motor is well above the full load current rating and lasts for any appreciable time the windings will overheat and damage may occur to the insulation, resulting in a burned-out motor. Because most motors experience various load conditions from no load to partial load to full load to short periods of being overloaded the overload protection devices must be flexible enough to handle the various conditions under which the motor and its driven equipment operate. Single-phase motors often have internal thermal overload protection. This device senses the increased heat load and breaks the circuit, stopping the motor. After the thermal overload relays have cooled down a manual or automatic reset is used to restart the motor. Some single-phase and three-phase motors require external overload protection.

motor sheave: (Motors, Mechanical)) The driver pulley on the motor shaft. The motor sheave may be either a fixed or adjustable groove sheave. Generally, after fans have been adjusted for the proper airflow variable pitch motor sheaves are replaced with fixed sheaves.

motor speed: (Motors) The rated motor speed. Motors operate at different speeds according to their type, construction, and the number of magnetic poles in the motor. Some single-phase motors are designed for multiple speeds by switching the winding connections. Typically, two to four different speeds are available. Wiring diagrams are usually provided with or on the motor.

motor voltage: (Motors, Electrical) The rated operating voltage.

motor windings: (Motors, Electrical) Motors have two windings. One winding is the stator. It is a stationary outside ring made of laminated steel. The other winding is the rotor which is a rotating cylindrical core inside, but separated from, the stator by a uniform air gap.

multi-meter: (Instrumentation) An instrument that measures more than one electrical component. A multi-meter may measure alternating current and voltage, direct current and voltage, resistance in Ohms and power factor.

multi-stage pump: (Pump) A pump having two or more impellers on a common shaft acting in series in a single casing. The liquid is conducted from the discharge of the preceding impeller through fixed guide vanes to the suction of the following impeller causing a head (pressure) increase at a given flow rate as it passes through each stage. Head can also be increased by connecting separate single stage pumps in series.

multi-stage thermostat: (Control) A temperature control that sequences two or more switches in response to the amount of heating or cooling demand.

multizone air system: Dual path air systems usually having a cooling coil and heating coil. The air passes through the coils into mixing dampers and then into zone ducts to the various conditioned spaces. The traditional multizone systems were designed as constant volume systems but the actual volume may vary up to 10% during normal operation because of the changes in resistance between the smaller heating coil and the larger cooling coil. Currently, multizone systems are designed for variable air volume. Generally, multizone systems have between 4 and 12 zones. In most cases the zones will be similar in flow quantities.

N is for Nameplate

nameplate amperage: (Motors, Electrical) The rated motor amperage. See full load amperage.

nameplate horsepower: (Motors, Power) The rated horsepower. Because of efficiency and other factors nameplate horsepower cannot be achieved. Operating horsepower is brake horsepower.

nameplate rpm: (Motors, Speed) The rated motor speed. Motors operate at different speeds according to their type, construction, and the number of magnetic poles in the motor. For example, a four-pole synchronous motor would be rated at 1800 rpm, but because of slip, about 2 to 5%, this example four-pole induction motor is rated at 1725 rpm. Some single-phase motors are designed for multiple rpm by switching the winding connections (two to four different speeds are available). Wiring diagrams are usually provided on the motor. See motor speed.

nameplate voltage: (Motors, Electrical) The rated operating motor voltage.

net positive suction head: (Pump) The minimum suction pressure at the pump to overcome all the factors limiting the suction side of the pump—internal losses, elevation of the suction supply, friction losses, vapor pressure and altitude of the installation. It is unlikely there will be a problem with net positive suction head in air conditioning chilled water closed circuit systems. It's also not ordinarily a factor in open systems or hot water systems unless there is considerable friction loss in the pipe or the water source is well below the pump and the suction lift is excessive. If there's insufficient net positive suction head for either system check the suction line for undersized pipe, too many fittings, throttled valves, or clogged strainers.

net positive suction head available: (Pump) A characteristic of the system in which the pump operates. It's dependent upon such conditions as elevation of the suction supply in relation to the pump, the friction loss in the suction pipe, the altitude of the installation, the pressure on the suction supply, and vapor pressure. In determining the net positive suction head available these considerations must be evaluated and a pump selected for the worst conditions likely to be encountered in the installation. In addition, as a safety factor, the net positive suction head available should always exceed the net positive suction head required by at least two feet.

net positive suction head required: (Pump) The actual absolute pressure needed to overcome the pump's internal losses and allow the pump to operate satisfactorily. Net positive suction head required is determined by the pump manufacturer through laboratory tests. It's a fixed value for a given capacity and doesn't vary with altitude or temperature. It does vary with each pump capacity and speed change. Net positive suction head required for a specific pump is available from the manufacturer either on submittal data, from a pump curve or from a catalog. A pump curve will give the full range of net positive suction head required values for each impeller size and capacity.

net present value: (Energy Management) The total present value of a time series of cash flows. It is a standard method for using the time value of money to appraise long-term projects. It measures the excess or short-fall of cash flows, in present value terms, once financing charges are met.

neutral: (Electrical) That part of an electrical system which is at zero voltage difference with respect to the earth or "ground."

night setback: (Control) The control scheme which incorporates various control functions such as lowering space temperature, closing outside air dampers, and intermittently operating blowers (fans) to reduce heating expense during unoccupied hours.

non-contact tachometer: (Instrumentation) Non-contact tachometers are used for measuring rotational speeds when the shaft isn't accessible. Two types non-contact tachometers used in HVAC work are the strobe light tachometer and the photo tachometer.

non-diverting airflow pattern: (Airflow) See opposed blade damper.

non-overloading fan or pump: (Fluid Flow) A term used with fans and pumps to describe the motor operation in relation to fluid flow. Fans: The horsepower curve increases with an increase in air quantity but only to a point to the right of maximum efficiency and then gradually decreases. If a motor is selected to handle the maximum brake horsepower shown on the performance curve it will not overload in any condition of fan operation. Backward curved and backward inclined fans are "non-overloading" fans. Pumps: The horsepower curve increases with an increase in water quantity but only to a point near maximum capacity, and then gradually decreases. If a motor is selected to handle the maximum brake horsepower shown on the performance curve it will not overload in any condition of pump operation. See non-overloading motor.

non-overloading motor: (Motors) A motor selected for a fan or pump where the horsepower curve increases with an increase in fluid flow quantity, but only to a point to the right of maximum efficiency, and then gradually decreases. If a motor is selected to handle the maximum brake horsepower shown on the fan or pump performance curve the motor will not overload in any condition of fan or pump operation.

non-uniform flow: (Fluid Flow) Fluid flow varying in velocity across the plane perpendicular to flow.

normally closed: (Control) The position of a controlled device when the power source is removed. A controlled device that moves toward the closed position as the controller output pressure decreases is normally closed. The position of the damper or valve when the actuator is de-energized.

normally closed: (Electrical) There is flow of electricity in a closed circuit. For relays, normally closed contacts are closed when the relay is de-energized.

normally open: (Control) The position of a controlled device when the power source is removed. A controlled device that moves toward the open position as the controller output pressure decreases is normally open. The position of the damper or valve when the actuator is de-energized. *normally open:* (Electrical) In an open circuit there is no flow of electricity. For relays, normally open contacts are open when the relay is de-energized.

O is for Ohm

occupied zone: (Definition)The conditioned space from the floor to about 6 feet above the floor.

offset: (Control) A sustained deviation between the actual control point and the set point under stable operating conditions.

ohm: (Electrical) A measure of resistance in an electrical circuit.

oil filter: (Control) An oil filter is installed upstream of the air dryer (for the air compressor) on a pneumatic air control system to collect any oil vapor or particles of dirt.

one-pipe main: (Water Flow) The one-pipe main is used for individual space control in residential and small commercial and industrial heating applications. This piping arrangement uses a single loop main but differs from the series loop arrangement since each terminal is connected by a supply and return branch pipe to the main. Because the terminal has a higher pressure drop than the main the water circulating in the main will tend to flow through the straight run of the tee fittings. This starves the terminal. To overcome this problem a diverting tee is installed in either the supply branch, return branch or sometimes, both branches. Diverting tees create the proper resistance in the main to direct water to the terminal. The advantage of the one-pipe main arrangement over the series loop is that each terminal can be separately controlled and serviced by installing the proper valves in the branches. However, as with the series loop arrangement, if there are too many terminals the water temperature at the terminals farthest from the boiler may not be adequate. See two-pipe system.

on-off control: (Control) A simple control system consisting of a switch in which the device being controlled is either on or off and no intermediate positions are available.

open circuit: (Control) The condition that exists when either deliberately or accidentally an electrical conductor or connection is broken or opened with a switch, safety, or interlock. An open circuit stops the flow of electricity.

open compressor: (Refrigeration) An open compressor has a motor which is outside the refrigeration system. The motor drives the compressor by means of an input shaft through seals. Advantages: Open compressor motors are air cooled using ambient air. Easier to exchange or repair without removing the refrigerant. In the event of a motor burnout open compressors are easier to service than a hermetic compressor. If the open compressor is belt driven, the V-belt drive allows the speed ratio between the motor and the compressor to be easily changed. This means that a single compressor can used on several different units by merely changing the size of the motor sheave and compressor valve clearance. Simpler in design and more reliable, especially in high pressure applications where compressed gas temperatures can be very high. Disadvantages: Failure of the shaft seals leading to loss of refrigerant. Crankshaft must emerge from the crankcase for installation of the driven sheaves. This means that the shaft must be sealed to separate the refrigerant from the atmosphere to prevent loss of refrigerant vapor. Mechanical seals, similar to the ones in water pumps, are generally used. One part of the mechanical seal is attached to the compressor housing and the other part is attached to the rotating crankshaft. A spring exerts pressure on the assembly to hold parts together. The assembly is flooded with oil from the lubrication system which forms a vapor-tight seal between the refrigerant and the atmosphere. Leaking shaft seals are a common source of compressor problems.

open impeller: (Pump) An impeller without side walls, consisting essentially of a series of vanes attached to a central hub. Used for handling liquids containing abrasives or solids.

open loop system: (Control) A control system whose output is a function of only the input to the system. No feedback.

open water system: (Water Flow) An open system is when there is a break in the piping circuit and the water is open to the atmosphere.

operating load point: (Measured or Tested Condition) Actual system operating capacity when an instrument reading is taken.

operating point: (Airflow, Water Flow) The point of fan (pump) operation. It is found on the fan rpm curve (pump impeller curve) where the operating static pressure and cfm (total dynamic pressure and gpm) intersect. Air: The intersection of the system curve with the fan performance curve. Water: The intersection of the system curve with the pump performance curve.

operator: (Control) See actuator.

opposed blade damper: (Airflow, Control) A multiple blade damper with a linkage which rotates the adjacent blades in opposite directions resulting in a series of openings that become increasingly narrow as the damper closes. This type of blade action results in a straight, uniform flow pattern sometimes called "non-diverting." Generally, opposed blade dampers are used in volume control and mixing applications.

optimize: (Definition) To make the most efficient use of.

optimum start/stop: (Control) A refined form of HVAC control that automatically adjusts the programmed start/stop schedule depending on inside and outside air temperature and humidity, resulting in the latest possible start and earliest possible stop of the HVAC equipment.

orifice plate: (Instrumentation) An orifice plate is essentially a fixed circular opening installed in a conduit (duct or pipe). A measurable "permanent" pressure loss is created as the fluid (air or water) passes from the larger diameter conduit through the smaller opening in the orifice. This results in an abrupt change in velocity which creates turbulence and a measurable amount of friction resulting in a pressure drop across the orifice. Calibration data, furnished with the orifice plate, show flow rate in cubic feet per minute (air) or in gallons per minute (water) versus measured pressure drop. A differential pressure gauge (such as a manometer or water differential pressure gauge) is connected to the pressure taps and flow is measured.

outlet velocity: (Airflow) The average velocity of air emerging from a fan, outlet or opening.

overload protection: (Electrical, Motors) See motor overload protection.

override: (Control) A manual or automatic action taken to bypass the normal operation of a device or system.

overshoot: (Control) The tendency to overcompensate for an error condition causing a new error in the opposite direction. Undershoot.

P is for Parallax

paradox: (Definition) A statement that contains conflicting ideas or apparently contradicts itself but actually is or may be true. For example, "When I shut down this computer I first click start." Alternate definition, "Two doctors."

parallax: (Instrumentation) A false reading that happens when the eye of the reader is not exactly perpendicular to the lines on the instrument scale (analog instrument).

parallax view: (Movie) The Parallax View. A journalist inadvertently discovers a security organization which is in the business of hiring socially unacceptable people to assassinate political figures.

parallel blade damper: (Airflow, Control) Generally, parallel blade dampers are used in a mixing application. Because the blades rotate parallel to each other, a parallel blade damper produces a "diverting" type of air pattern and when in a partially closed position the damper blades throw the air to the side, top or bottom of the duct. This flow pattern may adversely affect coil or fan performance or the airflow into branch ducts if the damper is located too close upstream.

parallel flow coils: (Water Flow) Parallel flow means that the flow of air and water are in the same direction to each other. The water and air enter on the same side. For parallel flow cooling coils, the coldest water enters the coil at the point where the warmest air enters the coil, therefore, less heat transfer. See log mean temperature difference. For parallel flow heating coils, the warmest water or steam enters the coil at the point where the coldest air enters the coil. Parallel heating coils are effectively used in HVAC units where freezing is possible due to outside air temperature.

particle size: (Environmentally Controlled Area) The maximum linear dimension of a particle.

particulate: (Environmentally Controlled Area) Of, pertaining to, or formed of separate particles.

parts per million: (Laboratory, Fume Hoods) Parts of tracer gas per million parts of air by volume.

peak load: (Electrical) The maximum instantaneous rate of consumption in the load circuit.

peak value: (Physics) The maximum positive or negative value of an alternating quantity. See amplitude.

perchloric acid hood: (Laboratory, Fume Hoods) A special purpose hood designed primarily to be used with perchloric acid. Perchloric acid fume hoods are mandatory for research in which perchloric acid is used because of the explosion hazard associated with this chemical.

perforated ceiling: (Airflow) This is an air distribution device where perforated ceiling panels or filter pads are used to distribute the air uniformly throughout the ceiling or a portion of the ceiling.

perforated duct: (Airflow) Exposed (no ceiling) perforated spiral round duct used to distribute air uniformly throughout a laboratory space or similar.

perforated face diffuser: (Airflow) Perforated face diffusers are used with lay-in ceilings and are similar in construction to the standard square ceiling diffuser with an added perforated face plate. They are generally equipped with adjustable vanes to change the flow pattern to a one-, two-, three-, or four-way throw.

performance curve: (Fans, Pumps) A pump performance curve is a graphic representation of the performance of a pump from free delivery to no delivery. Pumps are generally selected so their design operating point falls about midway, plus or minus one-quarter of the published curve. This allows changes in installation conditions.

performance rating: (Laboratory, Fume Hoods) The performance rating of a fume hood is a series of numbers consisting of a two digit number, which indicates the release rate of the tracer gas in liters per minute. The letters AM indicate "as manufactured" and a two or three digit number indicates the level of control of the tracer gas in parts per million of air by volume as established by the test. For example, a rating of 1.0 AM 10 would indicate that the hood controls to 10 parts per million at a release rate of 1 litre per minute. Aka Test Rating.

phase: (Electrical) The number of separate highest voltages alternating at different intervals in the circuit. Most HVAC motors are 1 or 3 phase.

photo tachometer: (Instrumentation) This instrument measures revolutions per minute by flashing a light at the moving part and counting the reflections.

pilot duty relay: (Control) A relay used for switching loads, such as another relay or solenoid valve coils. The pilot duty relay contacts are located in a second control circuit. Pilot duty relays are rated in volt-amperes.

pipe: (Water Flow) A passageway made of steel, iron or other suitable material that is used for conveying water.

pitch diameter: (Drives) Approximately where the middle of a V-belt rides in the sheave groove.

Pitot tube: (Instrumentation) A sensing device used to measure total pressures in a fluid stream. It was invented by a French physicist, Henri Pitot, in the 1700s. The standard Pitot tube has a double tube construction with a 90-degree radius bend near the tip and measures both total and static pressures. It is 5/16" in diameter and is available in various lengths from 12" to 60". The standard tube is recommended for use in duct 10" and larger diameter while a smaller (1/8" diameter) "pocket" Pitot tube is used in ducts smaller than 10". To help in taking Pitot tube traverses, the outer tube is marked with a stamped number at the even-inch points and a 1/8" line at the odd-inch intervals. The inner tube, or impact tube, senses total pressure and runs the length of the Pitot tube to the total pressure connection at the bottom. Static pressure is sensed by the outside tube through eight equally spaced holes around the circumference of the tube. These small openings must be kept clean and open to have accurate readings. They're located near the tip of the Pitot tube. The air space between the inner and outer tube serves to transmit the static pressure from the sensing holes to the static pressure connection at the bottom side of the Pitot tube. The static pressure connection is parallel with the tip of the Pitot tube as an aid to aligning the tip properly. To ensure the accurate sensing of pressures, the Pilot tube tip must be pointed so it faces directly into and parallel with the air stream.

plenum: (Airflow) An air chamber or air compartment.

plug fan: (Airflow) A fan without a housing (casing or shroud).

plug valve: (Water Flow) Plug valves are manual valves used for balancing water flow. Plug valves have a low pressure drop and good throttling characteristics and therefore, add little to the pumping head. Plug valves can also be used for tight shutoff.

pneumatic: (Definition). A technology to study and apply the use of pressurized gas to effect mechanical motion. HVAC pneumatic systems use centrally-located, electrically-powered compressor to compress ambient air to operate thermostats and other controllers, relays, switches, operators (actuators and motors), and controlled devices (dampers and valves) to maintain the proper temperature, pressure and humidity in conditioned spaces. Alternate definition: having a well-proportioned feminine figure; *especially* having a full bust.

pneumatic controller output pressure: (Control) The output pressure from pneumatic controllers is in pounds per square inch (psi). Typically, the range is 12 psi from 3 to 15 psi with a 9 psi midpoint. See Chapter 10.

point of no pressure change: (Water Flow) The point where the compression tank connects to the water system.

positive positioner: (Control) A positive positioner is used in pneumatic systems where accurate positioning of the controlled device is required. Positive positioners provide up to full main control air to the actuator for any change in position required by the controller.

potential energy: (Energy) Stored energy.

potentiometer: (Control) An electro-mechanical variable resistance device having a terminal connected to each end of the resistive element, and a third terminal connected to the wiper contact. The electrical input is divided as the contact moves over the element making it possible to mechanically change the resistance.

power: (Definition) The rate of flow of energy past a given point or the rate of doing work. Power equals work divided by time. Electrical power is measured in watts or kilowatts. Mechanical units of power are horsepower and foot-pounds. Heat power is measured in British thermal units per hour.

power: (Electrical) The unit for all components of electrical power is the watt. A watt is the product of voltage and amperage However, watt is used with real or actual power and volt-ampere is used to express units of apparent power. See apparent power, power factor and real power.

power factor: (Electrical) The ratio of real power to apparent power or watts divided by volt-amperes, expressed as a decimal or percentage. In an electric power system, a load with a low power factor draws more current than a load with a high power factor for the same amount of useful power transferred. The higher currents increase the energy lost in the distribution system and require larger wires and other equipment. Because of the cost of larger equipment and wasted energy, electrical utilities will usually charge a higher cost to industrial or commercial customers where there is a low power factor. Power factors are usually stated as "leading" or "lagging" For energy management purposes the power factor on motors is generally read at the motor control center or at the disconnect box using a portable clamp-on digital or analog power factor meter. The power factor is measured on only one phase of a single-phase system and on each phase of a three-phase induction system. The power factor should be 0.85 or greater. See lagging power factor and leading power factor. Another definition of power factor is the ratio of the average (or active) power to the apparent power (root-mean-square voltage times root-mean-square current) of an alternating-current circuit. Aka phase factor.

power factor correction: (Electrical) Correcting power factor by adding capacitors which act to cancel the inductive effects of the load. Power factor correction brings the power factor of an AC power circuit closer to 1.0. Power factor correction may be done by the electrical utility or by the individual electrical customer. Utilities make corrections to improve the stability and efficiency of the transmission network reducing transmission losses and improving voltage regulation at the load. Customers make corrections to reduce electrical costs. An automatic power factor correction device has of a number of capacitors that are used to improve power factor. The capacitors are switched on by contactors controlled by a regulator that measures power factor in the electrical network.

power factor meter: (Instrumentation) An instrument for measuring power factor. For a three-phase system read all phases from left to right. Also, read the corresponding amperage of each phase and the voltages between each phase. A separate wattmeter and volt ammeter can also be used. Divide the power in watts by the product of measured voltage and current.

power line subcarrier: (Electrical) A device to allow the use of a building's existing electrical power system to carry the signals of the energy management system.

present value: (Energy Management, Financial) The present value of an amount that will be received in the future.

present worth: (Energy Management, Financial) The attractiveness of single investment opportunity.

pressure: (Definition) The force exerted per unit of area. Pressure equals force divided by area.

pressure dependent terminal box: (Airflow) The quantity of air passing through the box is dependent on the inlet static pressure.

pressure drop: (Electrical, Fluid Flow) A reduction in fluid flow (vapor or liquid) pressure, as in "pressure drop" across a control valve. Electrical *pressure gauge:* (Instrumentation) Measures pressures above atmospheric and read in pounds per square inch, gauge. Pressure gauges are calibrated to read zero at atmospheric pressure. Permanently installed gauges deteriorate because of vibration and pulsation and are not reliable for test and balance measurements. Use only recently calibrated test gauges. See Bourdon tube test gauge.

pressure independent terminal box: (Airflow) The quantity of air passing through the box is independent (within design limits) of the inlet static pressure.

pressure range: (Control) The change in the pressure necessary for the controller output to change over the throttling range. For example, in a 3 to 15 psi system, the pressure range is 12 psi.

pressure reducing valve: (Control, Pneumatic Air) The pressure reducing valve reduces the pressure of the air in the receiver tank to the main air pressure to be used in the controlling devices. Main air pressure is normally set for 18 to 20 psig.

pressure reducing valve: (Water Systems). Pressure reducing valves are installed in the piping that supplies fill water or make-up water to the system. Example: A valve which reduces the pressure from the city water main to the proper pressure needed to (1) completely fill the system and (2) maintain this pressure. Pressure reducing valves generally come set at 12 pounds per square inch (about 28 feet). This is adequate for one- and two-story buildings. However, for three-story or higher buildings, the pressure reducing valve should be adjusted so there's a minimum of 5 psi additional pressure at the highest terminal. For example, if a coil was at an elevation of 30 feet, the pressure reducing valve should be set for 18 pounds per square inch (30 feet is equal to 13 pounds per square inch plus 5 pounds per square inch safety).

pressure relief valve: (Boilers, Control) Pressure relief valves are safety devices to protect the system and human life. The pressure relief valve opens on a preset value so the system pressure cannot exceed this amount.

pressure sensor: (Control) A control device that senses changes in absolute, gauge or differential pressures. The change is detected and converted to a control signal.

pressurestat: (Control) A pressure sensor or sensor and controller

pressure switch: (Control, Pneumatics) A switch that starts and stops the air compressor at predetermined set points. For example, the switch may be set to start the compressor when the pressure in the receiver tank falls to 65 pounds per square inch gauge (psig) and stop the compressor when the pressure in the tank reaches 85 psig.

primary air: (Airflow) Supply air.

primary fan: (Airflow) Main fan.

primary-secondary pumps: (Pump) The function of the primary pump in a primary-secondary circuit is to circulate water around the primary circuit. The function of the secondary pump is to supply the terminals.

primary-secondary circuits: (Water Flow) Primary-secondary circuits reduce pumping horsepower requirements while increasing system control. The primary pump and the secondary pump have no effect on each other when the two circuits are properly interconnected. Flow in one circuit will not cause flow in the other if the pressure drop in the pipe common to both circuits is eliminated. The secondary flow may be less than, equal to, or greater than the primary flow.

prime mover: (Definition) The initial source.

process: (Definition) A systematic series of actions directed to some end. A work process.

process function: (Definition) A systematic series of actions directed to the special purpose or activity for which a thing exists or is used. The manufacturing of some product.

processor: (Computer) See central processing unit.

process plant: (Control) The controlled variable conditioning apparatus such as a coil, boiler, chiller, fan or humidifier being controlled.

programmable read only memory: (Computer) A type of memory whose locations can be accessed directly and read. The computer operator or IT person can change parameters such as set points, limits and minimum off times within the control routines, but the program logic cannot be changed without replacement of the memory chips.

propeller fan: (Airflow) An axial fan that produces large volumes of airflow at low pressures, typically, three-quarters inch water gauge or less. Airflow is parallel to the shaft. The housing for a typical propeller fan is normally a simple ring enclosure and the fan will usually have two or more single thickness blades. Propeller fans are generally not very efficient. A characteristic of HVAC propeller fans is that the operating brake horsepower is lowest at maximum airflow and highest at minimum airflow. Used for general air circulation or exhaust without any attached ductwork. A typical commercial application of propeller fans would be general room air circulation or exhaust ventilation. Very large propeller fans are used for air circulation in cooling towers.

proportional band: (Control) The range of values of a proportional controller through which the controlled variable must pass to move the final control element through its full operating range. The ratio of the controller throttling range to the sensor span. Aka modulating range. See throttling range.

proportional control: (Control) Proportional control is when controlled device is positioned proportionally in response to changes in a controlled variable. The control signal is based on the difference between an actual condition and a desired condition. The difference in the conditions is called the "error" or the "offset." Some small amount of offset is always present. The controller creates an output signal related directly to the error's magnitude.

proportional-integral-derivative: (Control) The type of action used to control modulating equipment such as valves, dampers and variable speed devices. See individually proportional control, integral control and derivative control.

psychrometer: (Instrumentation) Digital, sling and aspiration psychrometers are used to measure wet bulb and dry-bulb temperatures. The wick of the wet-bulb thermometer must be kept clean and wetted with distilled water only.

psychrometric chart: (Psychrometrics) A chart showing various conditions, including: dry bulb, wet bulb and dew point temperatures, relative and specific humidity, amount of moisture in grains and pounds, and air specific volume.

pulsation suppressor: (Water Flow) A restrictor placed in the water line. See snubber.

pulley: (Mechanical) A sheave or small wheel with a grooved rim used to transmit power by means of a band, belt, cord, rope, or chain passing over its rim. See sheave.

pump: (Pump) A machine for imparting energy to a fluid. The addition of energy to a fluid makes it do work such as rising to a higher level or causing it to flow. In a hydronic system the pump is the component which provides the energy to overcome system resistance and produce the required flow. All-bronze pumps are used mainly for pumping high temperature water, caustics, sea water, brines, etc. These pumps have the volute, impeller and all parts of the pump coming in direct contact with the liquid made of bronze. The shaft has a bronze sleeve. All-iron pumps are used for pumping caustics, petroleum products, etc. As the name implies, all parts of the pump coming in direct contact with the liquid pumped are made of iron or ferrous material. Bronze-fitted pump have a cast iron casing and are equipped with a brass impeller. The metal parts of the seal assembly are made of brass or some other non-ferrous materiel. The shaft is steel. Bronze-fitted pumps are used for most hydronic applications.

pump efficiency: (Pump) The output of useful energy divided by the power input. Water horsepower divided by brake horsepower.

pyrometer: (Instrumentation) A pyrometer is thermometer used to measure the temperature of a surface. Pyrometers are generally used to measure pipe surface temperatures.

Q is for Quadrillion

quadrant handle regulator: (Mechanical) A duct damper handle used to adjust a manual volume damper at any position from full open to full close and lock it in position.

quadrillion: (Math) Unit sometimes used to express energy usage (in Btu) in the USA and other countries. Aka "quads." Scientific notation, 1.0×10^{15} The USA uses in excess of 100 quadrillion Btu (100 quads or 100×10^{15} Btu). See Chapter 14, Tables.

quality of vapor: (Refrigeration) Percent by weight of the vapor in any mixture of liquid and vapor.

quartermaster: (Bond, James Bond) An individual or a unit (department) specializing in distributing supplies and equipment. In the James Bond novels and films "Q" is the head of Q Branch—Research and Development. "M" and "Q" are job titles rather than personal names.

R is for Range

radial flow impeller: (Pump) The radial or plain flow impeller is the most frequently used for HVAC work. This impeller design has single curvature vanes which curve backwards and are used in pumps with speed ranges below 2,500 rpm. See mixed flow impeller.

radiation: (Heat) Heat energy transmitted from one body to another without the need of intervening matter.

radioisotope hood: (Laboratory, Fume Hoods) A special purpose hood design primarily used with radiochemicals or radioactive isotopes. Special filtering and shielding are required. Radioisotope hoods have exhaust ducts with flanged, neoprene gaskets with quick disconnect fasteners that can be quickly dismantled for decontamination.

radius of diffusion: (Airflow) The horizontal distance an airstream travels after leaving the outlet before it is reduced to its specified terminal velocity.

random access memory: (Computer) A type of memory that can be read from or written into. It can be accessed directly from any location as fast as any other location.

range: (Control) The range on a control is the difference between the cut-in value and the cut-out value but it is not the same as the "differential." For example, the cut-in temperature is 55F and the cut-out temperature is 40F. The differential is 15F. However, the range is between 40F and 55F.

Rankine (Thermodynamics) An absolute temperature scale. Zero on the Rankine scale is absolute zero. The Rankine degree is defined as equal to one degree Fahrenheit. A temperature of minus 459.67F is precisely equal to zero degrees Rankin (0R).

rate: (Energy Management, Financial) Quantity, amount, or degree of something measured per unit of something else.

rate of return: (Energy Management, Financial) An interest rate where present value (worth) of the investment equals the present value (worth) of the (annual) savings. The ratio of money gained or lost (realized or unrealized) on an investment relative to the amount of money invested. (The gain or loss on an investment over a specified period, expressed as a percentage increase over the initial investment cost.

read only memory: (Computer) A type of memory whose locations can be accessed directly and read, but cannot be written into.

reactive power: (Electrical) The energy flowing backwards and forwards in an alternating circuit expressed as volt-amperes reactive or kilovolts-amperes reactive.

real power: (Electrical) The measured watts or the product of power factor and apparent power, expressed in watts or kilowatts.

receiver, receiver tank: (Control) A tank which receives and stores the compressed air from the compressor for use throughout a pneumatic control system.

receiver, receiver tank: (Refrigeration) A tank which receives and stores liquid refrigerant from the condenser for use throughout the system. It is a temporary storage tank for liquid refrigerant. Refrigeration systems that have large load variations are not always called upon to remove heat at a constant rate. Such variations in load conditions may cause the refrigeration to accumulate in the condenser. The receiver stores the refrigerant not required in the system.

reflectance: (Lighting) Expresses the percentage of light that is reflected from a surface such as ceiling, floor or wall.

refrigerant: (Refrigeration) Any substance that acts as a cooling agent by absorbing heat from another body or substance. Fluids that change from a vapor to a liquid and back to a vapor. Fluids used in refrigeration systems to absorb heat by evaporation (vaporization) and release heat by condensation. The refrigerant for absorption chillers will typically be water with lithium bromide as the absorbent, or ammonia as the refrigerant with water as the absorbent.

refrigerating effect: (Refrigeration) The quantity of heat that each unit mass of refrigerant absorbs from the conditioned space. Refrigerating effect per unit mass of liquid refrigerant is potentially equal to its latent heat of vaporization. Due to the flash gas process the refrigerating effect per unit mass of liquid refrigerant circulated is always less than the total latent heat of vaporization.

refrigeration: (Heat, Refrigeration) The transfer of heat from one place where it is not wanted to another place where it is unobjectionable. This transfer of heat is through a change in state of a fluid. The branch of science that deals with the process of reducing and maintaining the temperature of a space or material below the temperature of the surroundings.

refrigeration cycle: (Refrigeration) For a mechanical vapor-compression refrigeration system the cycle is: evaporation, compression, condensation and expansion.

refrigeration load: (Refrigeration) The rate at which heat must be removed from the refrigerated space or material in order to maintain the desired temperature conditions.

register: (Airflow) A grille with a built-in or attached damper assembly.

relative humidity: (Psychrometrics) The ratio of the moisture present in the air to the total moisture that the air can hold at a given temperature. Relative humidity is expressed as a percentage.

relay: (Control, Electrical) A device with a coil and an isolated set of contacts that opens or closes contacts in response to some controlled action. Relay contacts are normally open or normally closed.

relay: (Control, Pneumatic) A switching device.

release rate: (Laboratory, Fume Hoods) The rate of release in liters per minute of tracer gas during a fume hood test.

relief valve: (Boiler) The function of a boiler relief valve is to prevent a boiler from exploding. Should the boiler's internal pressure rise to the maximum working pressure, as established by the manufacturer and tested and confirmed by the American Society of Mechanical Engineers, the relief valve will open and excess water or steam is released into a relief pipe.

relief valve: (Pneumatic Control) Generally, there are two safety relief valves. One high pressure relief valve is installed on the receiver tank to protect the tank from excessive pressure. A second relief valve is installed in the supply line to protect the control devices from excessive line pressure.

reset: (Control) A process of automatically adjusting the setpoint of a given controller to compensate for changes in another variable. For example, in HVAC the hot deck control point is normally reset upward as the outdoor temperature decreases, while the cold deck control point is normally reset downward as the outdoor temperature increases. *reset ratio:* (Control) The ratio of change in outdoor temperature to the change in control point temperature. For example, a 1:2 reset ratio means that the control point for a boiler hot water system will increase 2 degrees for every 1 degree change in outdoor temperature.

residual velocity: (Airflow) Room air velocity.

resistance: (Definition, Electrical) The opposition to the passage of an electric current through a conductor. Ohm (Ω) is the SI unit of resistance. The inverse quantity is electrical conductance, the ease with which an electric current passes. Siemens (S) is the unit of conductance.

resistance: (Definition, Heat) The resistance to heat flow is defined as the reciprocal of a heat transfer coefficient ($R = 1/U$).

resistance temperature detector: (Instrumentation) A sensing element whose resistance varies significantly and predictably with temperature.

resistor: (Electricity) An electronic component that slows current. The current is proportional to the voltage applied and to the electrical conductivity of the base material.

return air inlet: (Airflow) A return air grille, register, or other inlet. Typically, a perforated face opening, linear slot, light troffer, plastic grid (egg crate) or other opening to allow air from the conditioned space into a ceiling return air plenum or return air duct. Inlets are generally chosen and located to suit architectural design requirements for appearance and compatibility with supply outlets. In most cases, the location of ceiling inlets doesn't significantly affect air motion and temperature except when the inlet is positioned directly in the primary air stream from the outlet. This "short circuits" the supply air back into the return system without properly mixing with the room air. That being said, the flow of air is from the supply to the return. So in some spaces, a cleanroom is an example, placement of the returns gains importance.

return on investment: (Energy Management, Financial) An interest rate (I) where present value (worth) of the investment (P, or PV) equals the present value (worth) of the (annual) savings (A). A performance measure used to evaluate the efficiency of an investment or to compare the efficiency of a number of different investments. To calculate ROI, the benefit (return) of an investment is divided by the cost of the investment. The result is expressed as a percentage or a ratio.

reverse-acting controller: (Control) A reverse-acting controller decreases its branch output as the condition it is sensing increases.

reverse air flow: (Laboratory, Fume Hoods) Air movement toward the front of the hood.

reverse return: (Water Flow) In a reverse return system, the return is routed so the length of the circuit to each terminal and back to the pump is essentially equal. The terminals are piped "first in, last back; last in, first back." Because all the circuits are essentially the same length, reverse return systems need more piping than direct return systems, but are considered more easily balanced.

room cavity ratio (Lighting) Quantifies how effective a room area is at using the light from the fixtures or how light will interact with room surfaces. RCR is used in lighting calculations. The room cavity is from the light fixture to the work plane between the ceiling cavity and the floor cavity.

room velocity: (Airflow) The air velocity in the room's occupied zone. See residual velocity.

rotating vane anemometer: (Instrumentation) An instrument used in air balancing and commissioning to measure air velocity. The RVA is generally used for determining air velocity through supply, return and exhaust air grilles, registers or openings. It is also sometimes used to measure airflow through coils to get an approximation of total airflow. The analog rotating vane anemometer measures the linear feet of the air passing through it so a stopwatch or other timing device must be used to find velocity in feet per minute. The useful velocity range of the RVA is between 200 to 2000 fpm and the accuracy of the instrument depends on the precision of use, the type of application, and its calibration. Digital anemometers read in feet per minute or cubic feet per minute.

S is for Sublimation

salvage value: (Energy Management, Financial) The estimated value of an asset at the end of its useful life. The value of the equipment at the end of the project life—can be positive or negative.

sash: (Laboratory, Fume Hoods) The moveable, normally transparent panel set in the fume hood entrance. Sashes may be vertical or a combination of horizontal and vertical. The combination horizontal/vertical sash has horizontally sliding sashes set in a vertical rising sash. With the combination sash, the vertical sash allows for easier setup or removal of hood equipment or apparatus, while the horizontal sash facilitates user operations and also reduces total exhaust air volume.

saturated fluid: (Thermodynamics) The fluid is at its boiling point.

saturated liquid: (Thermodynamics, Refrigeration, Water) A liquid at the saturation temperature. A liquid cannot exist as a liquid at any temperature above its saturation temperature corresponding to the pressure. A liquid cannot exist as a liquid at any pressure below its saturation pressure. Also, fully in the liquid state, about to vaporize.

saturated liquid-vapor mixture: (Thermodynamics) The state in which liquid and vapor coexist.

saturated vapor: (Thermodynamics, Refrigeration) A vapor at the saturation temperature. A vapor cannot exist as a vapor at any temperature below its saturation temperature corresponding to the pressure. A vapor cannot exist as a vapor at any pressure above its saturation pressure. A saturated vapor is vapor whose temperature and pressure are such that any compression of its volume at constant temperature causes it to condense to liquid at a rate sufficient to maintain a constant pressure. Fully in the vapor state, about to condense.

saturation temperature: (Thermodynamics, Refrigeration) The temperature at which a fluid will change from a liquid to a vapor or vice versa. Increasing the pressure on the fluid raises the saturation temperature. The saturation temperature depends on the pressure on the fluid. Saturation temperature increases with an increase in pressure and decreases with a decrease in pressure.

scalar: (Definition) Magnitude only. Compare to vector.

scientific notation: (Definition) Standard Scientific Notation is a number from 1 to 9 followed by a decimal point, the remaining significant numbers, and an exponent of 10 to hold place value. See Chapter 14, Tables.

secondary air: (Airflow) Room air.

semi-hermetic compressor: (Refrigeration) See hermetic compressor.

semi-open impeller: (Pump) An impeller having a shroud or side wall on one side only. Used for handling liquids containing solids.

sensible heat: (Psychrometrics) Heat that causes a temperature change in a substance. The units of sensible heat are Btu per pound.

sensible heat factor: (Psychrometrics) See sensible heat ratio.

sensible heat ratio: (Psychrometrics) The ratio of the sensible heat to the total heat in the air.

sensing element: (Control) A sensing element, either internal or remote of the controller, measures the controlled variable (temperature, humidity or pressure) and sends a signal back to the controller. See temperature, humidity and pressure sensor.

sensitivity: (Control) The change in the output of a controller per unit change in the controlled variable. Example #1: A pneumatic controller with a range of 12 pounds per square inch and a 4 degree Fahrenheit throttling range has a sensitivity of 3 pounds per square inch per degree Fahrenheit. Example #2: The change in pressure in a remote sensor or transmitter per unit change in the sensed medium. A transmitter with a pressure range of 12 pounds per square inch and a span of 100 degrees Fahrenheit will have a sensitivity of 0.12 pounds per square inch per degree Fahrenheit.

sensitivity: (Instrumentation) A measure of the smallest incremental change to which an instrument can respond.

sensor: (Control) See sensing element.

setback: (Control) Reduction of heating or cooling at night or during hours when a building is unoccupied. See night setback.

sequencer: (Definition) An electronic device that may be programmed or set to initiate a series of events and make the events follow in sequence.

sequencing control: (Control) A control that energizes successive stages of heating or cooling equipment as its sensor detects the need for increased heating or cooling capacity. May be electronic or electro-mechanical.

series loop: (Water Systems) The series loop piping arrangement is generally limited to residential and small commercial heating applications. In a series loop, supply water is pumped through each terminal in series and then returned back to the boiler. The advantages to this type of piping arrangement are it is simple and inexpensive. The disadvantages are: (1) if repairs are needed on any terminal the whole system must be shut down and (2) it is not possible to provide a separate capacity control to any individual terminal since closing the valve to one terminal reduces flow to the terminals down the line. However, space heating can be controlled through dampering airflow. These disadvantages can be partly remedied by designing the piping with two or more circuits and installing balancing valves in each circuit. This type of arrangement is called a split series loop. The series loop circuit length and pipe size are also important because they directly influence the water flow rate, temperature and pressure drop. For instance, as the heating supply water flows through the terminals, its temperature drops continuously as it releases heat in each terminal. If there are too many terminals in series, the water temperature in the last terminal may be too cool.

service factor: (Electrical) The number by which the horsepower or amperage rating is multiplied to determine the maximum safe load that a motor may be expected to carry continuously at its rated voltage and frequency. Typical service factors are 1.0, 1.10, 1.15 for large motors, and 1.20, 1.25, 1.30 and 1.40 for small motors. For example, a 100 horsepower motor with a service of 1.10 would have a safe load of 110 horsepower.

setpoint: (Control) The point at which the controller is set to maintain the desired temperature, pressure or relative humidity.

shaft couplings: (Mechanical, Water Systems) Shaft couplings compensate for small deviations in alignment between the pump and motor shafts within the tolerances established by the manufacturer. Couplings are made in "halves" so the pump and motor may be disconnected from each other. It's important that shafts be aligned as closely as possible for quiet operation and the least coupling and bearing wear. The coupling can accommodate small variations in alignment, but its function is coupling, not compensating for misalignment. Severe misalignment between the shafts will lead to noisy operation, early coupling failure, and possible pump or motor bearing failures.

shaft sizes: Motor shaft sizes are in 1/8" increments and fan shafts are in 1/16" increments.

sheave: (Mechanical) A grooved pulley on a shaft.

shielded cable: Special cable used with equipment that generates a low voltage output. Used to minimize the effects of frequency "noise" to the output signal.

short circuit: (Electrical) The condition which occurs when a hot wire comes in contact with neutral or ground.

shut-off head: (Pump) The pressure developed by the pump when its discharge valve is shut. On the pump curve it is the intersection of the head-capacity curve with the zero capacity line.

side view: (Definition) The view or representation of any given side of a building, component, duct or assembly drawn in projection on a horizontal plane.

silicon controlled rectifier: (Electrical) A three terminal electronic semi-conductor switching device.

simple payback period: (Energy Management, Financial) The number of years before the savings on an energy retrofit project is paid back but does not consider the time value of money.

single duct terminal box: (Airflow) A terminal box usually supplied with cool air through a single inlet duct. The box may be constant or variable volume, pressure dependent or pressure independent. It may also contain a water coil, steam coil, or electric reheat in the box or in the downstream duct.

single inlet-single wide fan: (Fan) A single inlet-single wide (SISW) fan has one fan wheel and a single entry. The bearings are out of the air stream. They are more suited to having inlet duct attached to it than a double inlet- double wide (DIDW) fan.

single path system: (Airflow) An air system in which the air flows through coils essentially in series to each other. Single zone heating and cooling units and terminal reheat units are examples.

single-phase motor: (Motors) A motor supplied with single-phase current.

single phasing: (Electrical, Motors) The condition which results when one phase of a three-phase motor circuit is broken or opened. Motors won't start under this condition, but if already running when it goes into single-phase condition the motor will continue to run with a lower power output and possible overheating.

single pressure system: (Control) A pneumatic control system which requires only one main air pressure.

single stage pump: (Pump) A pump with one impeller.

single suction pump: (Pump) A pump in which the liquid enters the impeller inlet from one side. Single suction pumps are usually built with the inlet at the end of the impeller shaft. The casing is made so the discharge may be rotated to various positions. The suction connection is normally one or two pipe sizes larger than the discharge connection.

slot velocity: (Laboratory, Fume Hoods) The speed of the air moving through the fume hood baffle openings.

slugging: (Refrigeration) When liquid refrigerant enters the compressor. Slugging causes noisy operation, loss of capacity, an increase in power required and possible damage to the compressor.

smoke candle: (Airflow) One of various types of smoke producing devices used to allow visual observation of airflow through fume hoods, ductwork, and other spaces.

smoke detector: (Control) A device in the air handling unit or ductwork to shut down air conditioning or ventilating fans when smoke is sensed.

smudging: (Airflow) The black markings on ceilings and outlets usually made by suspended dirt particles in the room air which is then entrained in the mixed air stream and deposited on the ceilings and outlets. Anti-smudge rings are available which lower the outlet away from the ceiling and cover the ceiling area a few inches beyond the diffuser.

solenoid: (Electric) A coil wrapped around a metallic core which produces a magnetic field when an electric current is passed through it (an electromagnet).

solenoid valve: (Control) An electromechanical valve for use with liquid or gas. The valve is controlled by an electric current through a solenoid. With a two-port valve the flow is switched on or off. Example #1: A two-way valve in the liquid line of a refrigeration system. With a three-port valve the outflow is switched between the two outlet ports. Example #2: A three-way air valve used in pneumatic control systems for control of outside air dampers. See three-way air valve.

snubber: (Mechanical) A restrictor placed in the water line to a permanently installed gauge to suppress pulsating or fluctuating pressures.

span: (Control) The difference between the lowest possible set point and the highest possible set point.

specified rating: (Laboratory, Fume Hoods) A fume hood's performance rating as specified, proposed or guaranteed.

specific enthalpy: (Heat) The enthalpy of one pound of a substance. See enthalpy.

specific gravity: (Physics) Specific gravity is unit-less. It is the ratio of the density (weight or mass) of the substance to some standard substance. HVAC: specific gravity equals density of substance divided by density of water. The ratio of the mass of a substance to the mass of an equal volume of water at 4C (approx. 40F). Water has a density of 62.4 lb/cf at 4C. Therefore, the specific gravity of water at 40 degrees Fahrenheit is 1.0, aka standard conditions. Typically, for temperatures between freezing (0C) and boiling (100C), a specific gravity of 1.0 is used.

specific heat: (Heat, Psychrometrics) The ratio of heat required to raise the temperature of one pound of substance 1 degree Fahrenheit as compared to the heat required to raise one pound of water 1F. Therefore, specific heat of water is 1 Btu/lbF and the specific heat of air is 0.24 Btu/lbF

specific humidity: (Psychrometrics) The weight of water vapor associated with one pound of dry air. Specific humidity is measured in grains of moisture per pound of dry air or pounds of moisture per pound of dry air. 7000 grains equal one pound.

specific volume: (Physics) The volume of a substance per unit weight (or mass). Specific volume units are cubic feet per pound. Specific volume is the reciprocal of density.

split-system: (Definition) A split system air conditioning unit has an indoor section (fan, cooling coil, heating unit, and filter) and an outdoor section (compressor and condenser) connected by refrigerant tubing.

splitter damper: (Airflow) A device used in low pressure systems to divert airflow. Splitter damper is a misnomer. It's not a damper for regulating airflow volume but a diverter for directing the airflow.

spot collector hood: (Airflow) A small, localized ventilation hood usually connected by a flexible duct to an exhaust fan. The spot collector hood is not considered a laboratory fume hood.

spread: (Airflow) The divergence of the air stream after it leaves the outlet.

spring range: (Control) The spring range of an actuator restricts movement of the controlled device within set limits. Typical spring ranges for pneumatic actuators are 3 to 7 pounds per square inch gauge pressure (psig), 3 to 8 psig, 8 to 13 psig and 9 to 13 psig.

standard air conditions (Psychrometrics): Standard conditions for air are 70F temperature, 0.075 lb/cu ft density and 0.24 Btu/lb specific heat.

standard cubic feet per minute: (Airflow) The volumetric rate of airflow at standard air conditions.

static discharge head: (Pump) The vertical distance from the centerline of the pump to the free discharge liquid level.

static head: (Pump) The static pressure of a fluid expressed in terms of the height of a column of the fluid.

static pressure: (Airflow) The pressure or force within the fan unit or duct that exerts pressure against all the walls and moves the air through the system. On the discharge (positive) side of the fan it is "bursting pressure." On the suction side of the fan it is "collapsing pressure."

static suction head: (Pump) The vertical distance from the centerline of the pump to the suction liquid free level.

static suction lift: (Pump) The vertical distance from the centerline of the pump down to the suction liquid free level.

steam: (Heat) In an open vessel, at standard atmospheric pressure, water vaporizes or boils into steam at a temperature of 212F. But the boiling temperature of water, or any liquid, is not constant. The boiling temperature can be changed by changing the pressure on the liquid. If the pressure is to be changed, the liquid must be in a closed vessel. In the case of water in a heating system, the vessel is the boiler. Once the water is in the boiler it can be boiled at a temperature of 100F or 300F as easily as at 212F. The only requirement is that the pressure in the boiler be changed to the one corresponding to the desired boiling point. For instance, if the pressure in the boiler is 0.95 psia the boiling temperature of the water will be 100F. If the pressure is raised to 14.7 psia the boiling temperature is raised to 212F. If the pressure is raised again to 67 psia the temperature is correspondingly raised to 300F. A common low pressure HVAC steam heating system will operate around 15 psig (30 psia) or about 250F. The amount of heat required to bring the water to its boiling temperature is its "sensible heat." Additional heat is then required for the change of state from water to steam. This addition of heat is steam's "latent heat" content or "latent heat of vaporization." To vaporize one pound of water at 212F to one pound of steam at 212F requires 970 Btu. The amount of heat required to bring water from any temperature to steam is called "total heat." Total heat is the sum of the sensible heat and latent heat.

step controller: (Control) An electro-mechanical device used with electric or pneumatic systems which may be set to initiate a series of events and to make the events follow in sequence.

strainer: (Water Flow) Strainers are installed before pumps to catch sediment or other foreign material in the water. A strainer contains a fine mesh screen formed into a sleeve or basket that fits inside the strainer body. This sleeve must be removed and cleaned. A strainer with a dirty sleeve or a sleeve with a screen that's too fine means there will be excessive pressure drop across the strainer and lower water flow. Individual fine mesh sleeve strainers may also be installed before automatic control valves or spray nozzles which operate with small clearances and need protection from materials which might pass through the pump strainer.

strainer-dryer: (Refrigeration) A combination device used as a strainer and moisture remover. Removes moisture and solid particles from the refrigerant before entering the metering device. Usually found in the liquid line. Aka filter-dryer.

strainer screen: (Water Flow) A strainer contains a fine mesh screen formed into a sleeve or basket that fits inside the strainer body at the water pump. This sleeve must be removed and cleaned. A strainer with a dirty sleeve, or a sleeve with a screen that's too fine, means there will be excessive pressure drop across the strainer and lower water flow. Automatic control valves or spray nozzles which operate with small clearances and need protection from materials which might pass through the pump strainer may also have fine mesh sleeve strainers installed before the valve or nozzle.

strainer system: (Water Flow) See economizer.

stratification: (Airflow) See stratified air.

stratified air: (Airflow) Layers of air at different temperatures or different velocities flowing through a filter, coil, duct or plenum.

stretched string distance: (Airflow) The shortest distance from an exhaust air opening to an air intake opening over and along the building surface.

strobe light tachometer: (Instrumentation) A strobe light tachometer has an electronically controlled flashing light which is manually adjusted to equal the frequency of the rotating part so the part will appear motionless. To avoid reading harmonics of the actual rpm use nameplate rpm or calculate the approximate rpm and start at that point. See harmonics.

stuffing box: (Pump) A stuffing box seal has a "packing" which has rings made of graphite-impregnated cord, molded lead foil or some other resilient material formed into fitted split rings. These packing rings are compressed into the stuffing box by a packing gland. The tension on the packing gland is critical to the proper operation of the pump. If a packing gland must be replaced, consult the manufacturer's published data for tension recommendations. If there is too much tension the proper water leakage will not occur and this will cause scoring of the shaft and overheating of the packing. Another problem is that as the seal gets older and the packing gland has been tightened over time the packing becomes compressed and loses its resiliency, overheating the stuffing box. When the packing gland is backed off to allow cooler operation there is excessive leakage. When this happens replace the packing.

subcooled liquid: (Refrigeration) A liquid at any temperature below the saturation temperature. A subcooled liquid happens when the temperature of the liquid is decreased below its saturation temperature.

subcooling: (Refrigeration) Cooling the liquid refrigerant out of the condenser below the condensing temperature. Subcooling is sensible heat.

sublimation: (Chemistry) The transition of a substance (e.g., carbon dioxide, CO_2) from the solid phase (state) to the gas (vapor) phase without passing through an intermediate liquid phase. At normal pressures most chemical compounds and elements possess three different states (solid, liquid, gas) at different temperatures. Therefore, the normal transition from solid state to gas state requires an intermediate liquid state.

suction head: (Pump) When the source of supply is above the pump centerline.

suction lift: (Pump) When the source of supply is below the pump centerline.

suction line: (Refrigeration) The refrigerant line (pipe) from the end of the evaporator to the entrance of the compressor. The line suction conveys superheated vapor. See superheated vapor.

suction line accumulator: (Refrigeration) A suction accumulator in the suction line protects the compressor from liquid slugging by ensuring that only gas is returned to the compressor. In large holdover plate refrigerator and freezer systems liquid refrigerant can condense in the plate evaporator lines when the compressor is not running. On startup the liquid is suddenly dumped into the compressor causing liquid "slugging" which can cause damage. Suction line accumulators must be installed with the "inlet" and "outlet" properly connected. Reversing the flow through any of these suction accumulators will result in oil being trapped and severe compressor damage.

suction pipe: (Water System) A water pipe going into the suction side (entrance) of a pump. Example: A return pipe from the water system coil(s).

suction pressure: (Refrigeration) The pressure of the refrigerant vapor in the suction line aka low-side pressure, evaporator pressure or back pressure.

superheat: (Refrigeration) The sensible heat added to a vapor after vaporization. The sensible temperature of the vapor above its boiling temperature.

superheated: (Refrigeration) When the temperature of a vapor has been increased above its saturation temperature.

superheated vapor: (Refrigeration) A vapor at any temperature above the saturation temperature. Superheating is sensible heat.

supply air devices: (Airflow) Devices or openings through which air flows into the conditioned space. See supply air outlet.

supply air outlet: (Airflow) A supply air diffuser, grille, register, or other opening to allow supply air into the conditioned space to mix with the room air to maintain a uniform temperature throughout the occupied zone. Supply air diffusers, grilles, and registers, are chosen and located to control airflow patterns to avoid drafts and air stagnation and to complement the architectural design of the building.

surface effect: (Airflow) The effect caused by entrainment of secondary air when an outlet discharges air directly parallel and against a wall or ceiling. Surface effect is good for cooling applications, especially variable air volume systems because it helps to reduce the dumping of cold air. Surface effect contributes to smudging.

surface tension: (Physics) A property of the surface of a liquid that causes it to behave as an elastic sheet. It allows small objects to float on the surface of water and it is the cause of capillary action.

switching relay: (Control) General-purpose switching relays are used to increase switching capability and isolate electric circuits.

synergism (Definition) Conditions such that the total effect is greater than the sum of the individual parts or efforts. Synergy. Working together.

system curve: (Fan, Pump) Drawn on a fan (pump) curve. Fan: Start with the fan operating point static pressure and cfm. Arbitrarily select several static pressures above and below the operating point and calculate cfm using the 2nd fan law. From these calculated points draw a line to connect the point. This is the system curve along which the fan will operate as rpm is changed. Pump: Start with the pump operating point total dynamic head and gpm. Arbitrarily select several head pressures above and below the operating point and calculate gpm using the 2nd pump law. From these calculated points draw a line between the points. This is the system curve along which the pump will operate as impeller size is changed.

system effect: (Fan) Term used to describe any condition in the fan plenum, inlet, outlet or distribution system that adversely affects the aerodynamic characteristics of the fan and reduces fan performance. Aka system effect factor.

System International Units, SI Units: (Definition) The modern metric system. It is the world's most widely used system of units. SI Units from Système International d'Unités. SI is dynamic as units are created and definitions are modified through international agreement among nations as the technology of measurement progresses.

system level controller: (Control) A microprocessor-based controller that controls centrally located HVAC equipment such as air handlers, chillers, etc. These controllers typically have an expandable input/output device capability and a library of control programs. They may control more than one mechanical system from a single controller.

T is for TEAM

tachometer: (Instrumentation) An instrument for measuring rotational speed. Aka "tach."

TEAM: (Acronym) Together Everyone Accomplishes More. Synergy.

temperature sensor: (Control) A control component to sense a change in temperature. There are various types of temperature sensing elements for pneumatic and electrical/electronic systems. See Chapter 10. Thermal expansion elements such as bimetal or metal rod and tube elements, and vapor or liquid-filled elements such as sealed bellows, remote bulb or capillary fast response or averaging elements. Bimetal elements are commonly used in room thermostats. Rod and tube elements are generally used in insertion and immersion temperature controllers, such as those located in boilers or storage tanks. Sealed bellows elements are commonly used in room thermostats. Remote bulb elements are used where the temperature measuring point is a distance from the controller location, such as in a duct or pipe. Fast response and averaging capillary elements are used instead of the bulb in a remote bulb element. The fast response element is a tightly coiled capillary with a response time many times faster than the standard remote bulb. The averaging element is a capillary evenly strung across a duct to obtain the average temperature in the duct. Change of temperature can be sensed with changes in electrical resistance (thermistor) or voltage (thermocouple).

temperature well: (Instrumentation) Temperature wells are installed at specific points in the piping so a test thermometer can be inserted to measure the temperature of the water in the pipe. Generally, temperature wells are installed on the entering and leaving sides of chillers, condensers, boilers and coils. Thermometer wells must be long enough to extend into the pipe so good contact is made with the water. The well forms a cup to hold a heat-conducting liquid (usually an oil) so good heat transfer is made from the water in the pipe to the liquid in the well to the thermometer. Therefore, wells should be installed vertically or not more than 45 degrees from vertical, so they will hold the liquid. You will not get an accurate reading from inserting a thermometer into a dry well, as air will act as an insulator.

terminal box: (Airflow) A device or unit which regulates supply airflow, temperature and humidity to the conditioned space. Terminal boxes are classified as single duct, dual duct, constant volume, variable volume, medium pressure, high pressure, pressure dependent, pressure independent, system powered, fan powered, induction, terminal reheat and bypass. They may also contain a combination of heating or cooling coils, dampers and sound attenuation. The airflow through the box is normally set at the factory but can also be adjusted in the field. Terminal boxes also reduce the inlet pressures to a level consistent with the low pressure, low velocity duct connected to the discharge of the box. Any noise that's generated within the box in the reduction of the pressure is attenuated. Baffles or other devices are installed which reflect the sound back into the box where it can be absorbed by the box lining. Commonly, the boxes are lined with fiber glass which also provides thermal insulation so the conditioned air within the box won't be heated or cooled by the air in the spaces surrounding the box. Terminal boxes work off static pressure in the duct system. Each box has a minimum inlet static pressure requirement to overcome the pressure losses through the box plus any losses through the discharge duct, volume dampers, and outlets.

terminal velocity: (Airflow) The maximum air velocity of the mixed air stream at the end of the throw.

test plug: (Water System) Test points installed in the piping on the entering and leaving sides of chillers, condensers, boilers and coils to take temperatures or pressures. A hand held thermometer or pressure probe is inserted.

therm: (Energy, Heat) A unit of heat. A therm is 100,000 Btu.

thermal conductivity: (Physics) The property of a material to conduct heat. Heat transfer occurs at a higher rate across materials of high thermal conductivity than across materials of low thermal conductivity. The unit of conductivity is Btu in./hr-sf-F. Symbol k. Conductivity is the amount of heat in Btu flowing through a piece of homogeneous material 1 inch thick in one-hour when the area is 1 ft.^{2} and the difference in temperature is 1F.

thermal overload protection: (Electrical) See overload protection. *ther-*

mals: (Electrical) See overload protection.

thermistor: (Control, Electrical) A sensing element whose resistance varies significantly and predictably with temperature.

thermocouple: (Control, Electrical) A junction between two dissimilar materials that produces a voltage that is a function of the temperature.

thermometer: (Instrumentation) A device that measures temperature or temperature gradient using a variety of different principles. Types include: alcohol, bi-metal mechanical, digital, electrical resistance, Galileo, infrared, liquid crystal, mercury, thermistor and thermocouple.

three-phase motor: (Motors) A motor supplied with three-phase current. For the same size, three-phase motors have a capacity of about 150% greater, are lower in first cost, require less maintenance and generally do better than single-phase motors.

three-pipe main: (Water Flow) A water system which has a three-pipe main arrangement. It has two supply mains and one return main. One supply circulates chilled water from the chiller and the other supply circulates heating water from the boiler. This permits any space to be independently cooled or heated. A three-way valve in each supply branch switches to deliver either chilled water or heating water, but not both, to the terminal. The supply flows are not mixed. The return main, however, receives water from each terminal. This means that frequently the return will be handling a mixture of chilled and hot waters. This results in a waste of energy as both the chiller and the boiler receive warm water and must work harder to supply their proper discharge temperatures. The return connections from the terminals can be made either direct or reverse return.

three-way air valve: (Control, Pneumatic) A solenoid valve. Used in pneumatic controls, e.g., supplies compressed air to open an outside air damper. On a call to close the damper, the valve's supply port closes and the air (pressure) on the damper is exhausted through the open port closing the normally closed OA damper. See solenoid valve.

three-way automatic temperature control valves: (Control, Water Flow) Three-way Automatic Temperature Control Valves are used to mix or divert water flow and are generally classified as mixing or diverting valves. They may be either single seated (mixing valve) or double seated (diverting valve). The single seated, mixing valve is the most common. The terms "mixing" and "diverting" refer to the internal construction of the valve and not the application. The internal difference is necessary so the valve will seat against flow. Also see two-way valves.

threshold limit values: (Indoor Air Quality) The values for airborne toxic materials which are to be used as guides in the control of health hazards. They represent time weighted concentrations to which workers may be exposed 8 hours a day over extended periods of time without adverse effects. Aka TLV.

throttling characteristics: (Control) Throttling characteristics refer to the relationship of the position of a valve disc or damper blade and its percent of flow. A valve or damper has a linear throttling characteristic when the disc or blade open percentage is the same as the flow percentage. An example of a valve with a linear or straight line throttling characteristic would be a disc that's 50% open and the flow is measured at 50%. Then if the disc is closed to 30% open, the flow would be reduced to 30%.

throttling range: (Control) The change in the controlled condition necessary for the controller output to change over a certain range. For example, if a pneumatic thermostat has a 4 throttling range, the thermostat's branch line output will vary from 3 to 15 psi over a 4 degree change in temperature.

throw: (Airflow) The horizontal and vertical distance an air stream travels after leaving the outlet before it is reduced to its specified terminal velocity.

throw characteristics: (Definition) Characteristics of airflow performance from the manufacturer of an outlet device (diffuser, etc.)

time based scheduling: (Energy Management) The process of scheduling electrical loads on and off based on the time of day, the day of the week, month and the date.

time clock: (Energy Management) A mechanical, electrical or electronic timekeeping device connected to electrical equipment for the purpose of turning the equipment on and off at selected times.

timed two-position control: (Control) Timed two-position control is a variation of straight electrical two-position action. It is typically used in room electric/electronic thermostats to reduce operating differential. An anticipator control is used in the electric thermostat. The electronic thermostat uses a programmable clock to set the minimum on and off times.

time value of money: (Energy Management, Financial) The value of money based on a given amount of interest earned over a given amount of time. Money is worth less in the future (if there is inflation).

tip speed: (Fan) The velocity in feet per minute at the tip of the fan blade.

titanium tetrachloride: (Airflow) A chemical that generates white smoke. It is used to test the air pattern in laboratory fume hoods. Titanium tetrachloride is corrosive and irritating. It can stain the hood and will produce a residue that must be cleaned up. Care must be taken to minimize the effects on the hood. Skin contact or inhalation should be avoided.

top dead center: (Refrigeration) Reciprocating compressors. When the piston is at the top of its stroke.

torque: (Physics) The force which produces or tends to produce rotation. Measured in pound-foot, alternatively: foot-pound or inch-pound.

total discharge head: (Pump) The static discharge head plus friction losses plus velocity head.

total dynamic head: (Pump) The total discharge head minus the total suction head or the total discharge head plus suction lift. Suction head is when the water source is above the pump centerline. Suction lift is when the water source is below the pump centerline. For test and balance purposes. The total dynamic head is the difference between the gauge pressure at the pump discharge and the gauge pressure at the pump suction.

total energy: (Energy) Total energy equals kinetic energy plus potential energy.

total enthalpy: (Heat) The enthalpy of the entire mass of substance.

total head: (Pump) In a flowing fluid (e.g, water), the sum of the static and velocity heads at the point of measurement. Total head is called total pressure in air systems.

total heat: (Psychrometrics) The sum of latent heat and sensible heat. The units of total heat are Btu per pound.

total heat rejected: (Refrigeration) The total heat rejected at the condenser includes both the heat absorbed in the evaporator and the heat of compression plus any superheat.

total pressure: (Airflow) The sum of the static pressure and the velocity pressure taken at a given point of measurement in a fan or duct. Total pressure is called total head in water systems.

total static head: (Pump) The vertical distance in feet from the suction liquid level to the discharge liquid level. The sum of static suction lift and static discharge head. The difference between static suction head and static discharge head.

total static pressure: (Fan) The static pressure rise across the fan calculated from static pressure measurements at the fan inlet and outlet.

tower approach: (Refrigeration) The difference between the temperature of the water leaving the cooling tower and the wet bulb temperature of the air entering the tower.

tower range: (Refrigeration) The difference between the temperature of the water leaving the cooling tower and the temperature of the water entering the cooling tower.

transducer: (Control) A device which converts signals from one physical form to another, such as mechanical to electronic.

troffer: (Lighting) A metal ceiling recess (inverted trough) with its bottom next to the ceiling. Typically used to enclose fluorescent lamps. See light troffer.

tube axial fan: (Airflow) An axial fan that is a heavy duty propeller fan used in such HVAC applications as fume hood exhaust systems, paint spray booths and drying ovens. The wheel of the tube axial fan is enclosed in a cylindrical tube and is similar to the propeller type wheel. The main exception is that the wheel has more blades, usually 4 to 8, and the blades are of much heavier construction. The static pressure range is medium pressure, typically up to 3 inches water gauge. The airflow out of the fan is a circular or spiral pattern. The tube axial fan is more efficient than the propeller fan and is most efficient when it is operating at its highest air volume. Tube-axial fan is used in ducted systems requiring medium pressure and medium to high air volume applications such as fume exhausts, paint spray booth exhaust and drying ovens.

tubes: (Equipment) Tubes can refer to evaporator coil tubes, fire or water tubes in a boiler, or tubes in a heating water or cooling water coil, etc.

turbine pump: (Pump) See diffuser vane pump.

turbulent flow: (Fluid Flow) Fluid flow in which the velocity varies in magnitude and direction in an irregular manner.

turbulator: (Definition) A helical device (metal or other material) inserted into the tubes of fire-tube boilers, shell & tube heat exchangers and other types of heat transfer equipment that helps to increase heat transfer efficiency by swirling the fluid.

two-pipe system—direct return: (Water Flow) Two-pipe arrangements have two mains, one for supply and one for return. Each terminal is connected by a supply and return branch to its main. This design not only allows separate control and servicing of each terminal, but because the supply water temperature is the same at each terminal, two-pipe systems can be used for any size application. Two-pipe systems are further distinguished by their return piping. In a two-pipe direct return system the return is routed to bring the water back to the pump by the shortest possible path. The terminals are piped, "first in, first back; last in, last back." The direct return arrangement is popular because generally, less main pipe is needed. However, since water will follow the path of least resistance the terminals closest to the pump will tend to receive too much water while the terminals farthest from the pump will starve. To compensate for this balancing valves are required.

two-pipe system—reverse return: (Water Flow) In a two-pipe reverse return system the return is routed so the length of the circuit to each terminal and back to the pump is essentially equal. The terminals are piped, "first in, last back; last in, first back." Because all the circuits are essentially the same length, reverse return systems generally need more piping than direct return systems but are considered more easily balanced.

two-position control: (Control) The controlled device can only be positioned to either a minimum or maximum condition or an on or off condition.

two-way automatic temperature control valves: (Control, Water Flow) Two-way Automatic Temperature Control Valves control flow rate. They may be either single seated or double seated (balanced valve). The single seated valve is the most common. The valve must be installed with the direction of flow opposing the closing action of the valve plug. The water pressure tends to push the valve plug open. If the valve is installed the opposite way, it will cause chattering. To understand why this is so, it is important to note that as the valve plug modulates to the closed position, the velocity of the water around the plug becomes very high. Therefore, if the flow and pressure were with the closing of the plug, then at some point near closing the velocity pressure would overcome the spring resistance and force the plug closed. Then, when flow is stopped the velocity pressure goes to zero and the spring takes over and opens the plug. The cycle is repeated and chattering is the result. The double seated or balanced valve is generally recommended when high differential pressures are encountered and tight shutoff is not required. The flow through this valve tends to close one port while opening the other port. This design creates a balanced thrust condition which enables the valve to close off smoothly without water hammer despite the high differential pressure.

U is for ULPA

ultra low penetration air filters: (Airflow) Filters used in Environmentally Controlled Areas. They are 99.9995% efficient in removing particles 0.12 micron and larger. Like HEPA filters, ULPA filters are made of spun-hooked glass fiber rolled into a paper-like material. See High Efficiency Particulate Air filters.

uniform flow: (Fluid Flow) The smooth, straight line motion of a fluid across the area of flow.

unitary system: (Definition) A unitary system combines heating, cooling and fan components, all in one package.

unloaders: (Refrigeration) Cylinder unloaders are used for capacity control in reciprocating compressors. See cylinder unloaders.

U-tube manometer: (Instrumentation) An instrument used to measure pressures in air and water systems.

U-tube manometer, Air: Air U-tube manometers have a U-shaped glass or plastic tube partly filled with tinted water, or oil. They are made in various sizes and are recommended for measuring pressures of several inches of water gauge or more. They are not recommended for readings of less than 1.0 inches water gauge. To take a reading, open both tubing connectors. The liquid will be at the same height in each leg. A Pitot tube, impact tube or static pressure tip is connected to the manometer. When the sensing device is inserted into the duct or fan compartment the liquid is forced down in one leg and up in the other. The difference between the heights of the two legs is the pressure reading.

U-tube manometer, Water: Water U-tube manometers are primarily used for measuring pressure drops across terminals, heat exchangers, and flow meters. Fundamentally, the principles that apply to U-tube manometers that measure air pressures also apply to U-tube manometers that measure water pressures. The differences are (1) the pressures measured in a hydronic system are usually much greater than in air systems, (2) the U-tube manometer contains mercury instead of oil or water and must have over-pressure traps or use a manifold to prevent the mercury from entering the piping system (mercury causes rapid deterioration of copper and copper alloy pipes), (3) all air must be purged from the manometer and hoses, (4) water from the system must completely fill the hoses and both manometer legs and rest on top of the mercury. U-tube manometers may also be used with a water Pitot tube to take a velocity head traverse of the pipe. Caution: The high pressures associated with some water systems can drive the Pitot tube back into the user.

V is for Variable

V-belt: (Drive) The two types of V-belts generally used on HVAC equipment are light duty, fractional horsepower belts, sizes 2L through 5 L and industrial belts, sizes "A" through "E." Fractional horsepower belts are generally used on smaller diameter sheaves because they're more flexible than the industrial belt for the same equivalent cross-sectional size. For example, a 5L belt and a "B" belt have the same cross-sectional dimension, but because of its greater flexibility, the 5L belt would be used on light duty fans that have smaller sheaves. The general practice in HVAC design is to use belts of smaller cross-sectional size with smaller sheaves instead of large belts and large sheaves for the drive components. Multiple belts are used to avoid excessive belt stress. V-belts are rated by horsepower per belt, by length and minimum recommended pitch diameter.

vacuum gauge: (Instrumentation) Vacuum gauges measure pressures below atmospheric and read in inches of mercury. Vacuum gauges are calibrated to read zero at atmospheric pressure. Aka vacuum pressure gauge.

valve: (Control, Water Flow) Valves are used in hydronic systems for regulating water flow and isolating part or all the system.

vane axial fan: (Airflow) A vane axial fan is basically a tube axial fan with straightening vanes. They are used in HVAC ducted systems in office buildings or other commercial applications to provide airflow to conditioned spaces where good downstream air distribution is needed; in medium to high pressure and medium to high air volume applications. The housing is a cylindrical tube similar to the tube axial fan with the addition of air straightening vanes. The straightening vanes straighten out the spiral motion of the air and improve the efficiency of the fan. The vane axial fan has the highest efficiency of all the axial type fans. The wheel of the vane axial fans has shorter blades and a larger hub than the tube axial fan

vapor: (Fluid Flow) A state of matter in which the molecules move freely, thereby causing the matter to expand indefinitely occupying the total volume of any vessel in which it is contained. Term for gas.

vapor barrier insulation: (Energy Management, Duct, Building) External insulation on air conditioning (cooling) ducts to keep ducts from excessive heat transfer and having condensation (sweating) on the duct. Rule of thumb: a 10 degree temperature difference between the duct (55F) and the ambient air (65F) will cause condensation. A duct wrap with the fiberglass side touching the duct and the vapor barrier material facing away from the duct. An insulation to reduce heat transfer and moisture in building walls, floors, and ceilings. Typically, the vapor barrier faces into the structure preventing moisture on the walls, etc. In very humid climates the vapor barrier may be placed on the outside so that moist air from the outside cannot penetrate walls, etc.

vapor-compression cycle: (Refrigeration) A four-step process which includes expansion, vaporization, compression, and condensation.
vaporization: (Physics) The conversion of a solid or a liquid into a vapor. Changing from a liquid to a vapor. Occurs by boiling or evaporation. Evaporation takes place only at the surface of a liquid and at any temperature below its saturation temperature. Boiling takes place throughout the liquid, but only occurs at the saturation temperature. Vaporization is accomplished by adding heat to a liquid, decreasing the pressure on the liquid, or a combination of the two.

vapor pressure: (Water Flow) The vapor pressure of a liquid at any given temperature is that pressure necessary to keep the liquid from boiling (flashing) into a vapor. For example, the vapor pressure of 60 degree Fahrenheit water is 0.59 feet absolute (0.25 psia) while water at 180F has a vapor pressure of 17.85 feet absolute (7.72 psia).

variable air volume system: (Airflow, Energy Management) An airflow system (fans, ductwork, terminal boxes, dampers, etc.) which varies air volume to maintain space temperature at setpoint. Basic system is VAV with constant temperature. As air volume varies the horsepower also varies. Energy and energy cost are significantly reduced when system airflow is reduced. See constant air volume system.

variable air volume box: (Airflow) See the various variable air volume terminal boxes listed below.

variable air volume bypass terminal box: (Airflow) A bypass box uses a constant volume supply fan but provides variable air volume to the conditioned space. The supply air comes into the box and can exit either into the conditioned space through the discharge ductwork or back to the return system through a bypass damper. The conditioned space receives either all the supply air or only a part of it depending on what the room thermostat is calling for. Since there is no reduction in the main supply air volume feeding the box this type of system has no savings of fan energy.

variable air volume ceiling induction terminal box: (Airflow) A ceiling variable air volume induction box has a primary damper at the box inlet and an induction damper in the box which allows air in from the ceiling plenum. On a call for cooling the primary damper is full open and the induction damper is closed. As the conditioned space cools down the primary damper throttles back and the induction damper opens to maintain a constant mixed airflow to the conditioned space. At some point the induction damper is wide open and the primary damper is throttled to allow for the maximum induction ratio. Another type of induction box has a constant pressure nozzle inducing either primary air from the main supply system or return air from the ceiling plenum. The room thermostat opens or closes a primary air bypass damper to allow for the induction of primary or return air. This box uses volume regulators to reduce the airflow to the conditioned space. A reheat coil may be installed in the box or in the attached discharge duct.

variable air volume dual duct terminal box: (Airflow) A variable air volume dual duct terminal box is supplied by separate hot and cold ducts through two inlets. A variety of control schemes vary the air volume and discharge air temperature. One type uses a temperature deadband which supplies a varying quantity of either warm or cool air, but not mixed air, to the conditioned space.

variable air volume fan powered terminal box: (Airflow) A variable air volume fan powered box has the advantage of the energy savings of a conventional, single duct variable air volume system with the addition of several methods of heating and a relatively constant

airflow to the conditioned space. The box contains a secondary
fan and a return air opening from the ceiling space. When the
room thermostat is calling for cooling the box operates as would
the standard variable air volume box. However, on a call for heat
the secondary fan draws warm (secondary) air from the ceiling
plenum and recirculates it into the rooms. Varying amounts of cool
(primary) air from the main system are introduced into the box
and mix with the secondary air. A system of dampers, backdraft or
motorized, control airflow and mixing of air streams. As the room
thermostat continues to call for heat the primary air damper closes
off and more secondary air is drawn into the box and it alone is
recirculated. Therefore, the airflow to the conditioned space stays
constant. If more heat is needed reheat coils may be installed in the
boxes. The fan may operate continuously or it may shut off. A common application of fan powered boxes is around the perimeter or
other areas of a building where (1) air stagnation is a problem when
the primary air throttles back, (2) zones have seasonal heating and
cooling requirements, (3) heat is needed during unoccupied hours
when the primary fan is off, or (4) heating loads can be offset mainly with recirculated return air. See Chapter 9.

variable air volume fan powered bypass terminal box: (Airflow) This
box acts the same as the conventional bypass box with the addition
of a secondary fan in the box. The bypass box uses a constant volume supply primary fan but provides variable air volume to the
conditioned space. The supply air comes into the box and can exit
either into the conditioned space through the secondary fan and
the discharge ductwork or back to the return system through a bypass damper. The fan in the box circulates the primary air or return
air into the room. The conditioned space receives all primary air,
all return air or a mixture of the two, depending on what the room
thermostat is calling for. Since there's no reduction in the main supply air volume feeding the box this type of system has no savings
of primary fan energy.

variable air volume pressure dependent terminal box: (Airflow) A pressure dependent variable air volume box is essentially a pressure reducing and sound attenuation box with a motorized damper that's
controlled by a room thermostat. These boxes don't regulate the
airflow, but simply position the damper in response to the signal
from the thermostat. Because the airflow to these boxes is in direct

relation to the box inlet static pressure it is possible for the boxes closest to the supply fan, where the static pressure is the greatest, to get more air than is needed. Therefore, the boxes farther down the line may be getting little or no air. Therefore, pressure dependent boxes should only be installed in systems where there is no need for limit control and the system static pressure is stable enough not to require pressure independence. Pressure dependent maximum regulated volume boxes may be used where pressure independence is required only at maximum volume and the system static pressure variations are only minor. These boxes regulate the maximum volume but the flow rate at any point below maximum varies with the inlet static pressure. This may cause "hunting."

variable air volume pressure independent terminal box: (Airflow) Pressure independent variable air volume boxes can maintain airflow at any point between maximum and minimum regardless of box inlet static pressure, as long as the pressure is within the design operating range. Flow sensing devices regulate the flow rate through the box in response to the room thermostat's call for cooling or heating.

variable air volume single duct terminal box: (Airflow) To maintain the correct airflow in a pressure independent box over the entire potential range of varying inlet static pressure a sensor reads the differential pressure at the inlet of the box and transmits it to the controller. The room thermostat responding to the load conditions in the space also sends a signal to the controller. The controller responds by actuating the volume damper and regulating the airflow within the preset maximum and minimum range. For example, as the temperature rises in the space the damper opens for more cooling. As the temperature in the space drops the damper closes. If the box also has a reheat coil the volume damper on a call for heating would close to its minimum position and the reheat coil would be activated. Because of its pressure independence the airflow through the boxes is unaffected as other VAV boxes in the system modulate and change the inlet pressures throughout the system.

variable air volume system powered terminal box: (Airflow) System powered variable air volume boxes use the static pressure from the supply duct to power the VAV controls. The minimum inlet static pressure with this type of box is usually higher than other VAV systems in order to (1) operate the controls and (2) provide the proper airflow quantity.

variable air volume terminal box: (Airflow) Variable air volume boxes are available in many combinations that include pressure dependent, pressure independent, single duct, dual duct, cooling only, cooling with reheat, induction, bypass and fan powered. Variable air volume boxes can also be classified by (1) volume control throttling, bypass, or fan powered, (2) intake controls and sensors pneumatic, electric, electronic, or system powered, (3) thermostat action direct acting or reverse acting and (4) the condition of the box at rest normally open or normally closed. The basic variable air volume box has a single inlet duct. The quantity of air through the box is controlled by throttling an internal damper. If the box is pressure dependent, the damper will be controlled just by a room thermostat, whereas, the pressure independent version will also have a regulator to limit the air volume between a preset maximum and minimum. Inside the pressure independent box is a sensor. Mounted on the outside is a controller with connections to the sensor, volume damper and room thermostat. The quantity of air will vary from a design maximum cubic feet per minute down to minimum cubic feet per minute. The main feature of the variable air volume box is its ability to vary the air delivered to the conditioned space as the cooling/heating load varies. Then, as the total required volume of air is reduced throughout the system the supply fan will reduce its cubic feet per minute output. This means a savings of energy and cost to operate the fan. The exception to this is the variable air volume bypass box. The types of controls used to regulate the flow of air through variable air volume boxes are as varied as the types of boxes. Many boxes are designed to use external sources of power pneumatic, electric or electronic. These boxes are sometimes called non-system powered. Other boxes are system powered which means that the operating controls are powered by the static pressure from the main duct system. System powered boxes don't need a separate pneumatic or electric control system. This reduces first costs, however, they usually have a higher required minimum inlet static pressure which means that the supply fan will be required to produce higher static pressures resulting in increased operating costs.

variable pitch sheave: (Mechanical, Drive) A sheave which has adjustable belt grooves. Aka variable speed sheave, or adjustable sheave.

variable volume refrigerant system: (Refrigerant Flow) A variable volume refrigerant system is basically a traditional vapor-compression DX split system with electronic metering devices and electronic inverters to vary motor and compressor speed and therefore directly vary refrigerant flow in response to load variations as rooms require more or less cooling or heating. See Chapter 9.

variable water volume: (Water Flow) A variable water volume system is used to achieve either full- or part-load heating or cooling conditions while reducing the energy consumed by the pump. Variable water volume systems typically use two-way automatic control valves with either a constant speed or a variable speed system pump.

vector: (Physics) A quantity that has magnitude and direction. *velocity:* Speed or rate of motion in feet per second or feet per minute. Velocity equals distance divided by time.

velocity head: (Pump) The head required to create flow. The height of the fluid equivalent to its velocity pressure.

velocity pressure: (Fan, Duct) Velocity pressure is the pressure caused by the air being in motion and has a direct mathematical relation to the velocity of the air. Velocity pressure cannot be measured directly as can static and total pressure. However, since total pressure is the sum of static and velocity pressure then velocity can be determined by subtracting static pressure from total pressure.

vena contracta: (Airflow) The smallest area of an air stream leaving an orifice.

ventilation: (Airflow) Supplying air to or removing air from a space by natural or mechanical means.

ventilation hood: (Air Systems) A device (typically metal: iron, stainless steel, galvanized, copper, etc.) to capture contaminated air. Examples: paint spray hood, kitchen hood, welding hood, fireplace hood, etc.

Venturi: (Instrumentation) A Venturi operates on the same principle as the orifice plate but its shape allows gradual changes in velocity and the "permanent" pressure loss is less than is created by an orifice plate. Calibration data which show flow rate in cubic feet per minute (air) or gallons per minute (water) versus measured pressure drop are furnished with the Venturi. The pressure drop is measured with a differential gauge.

verification of system performance: (Energy Management) To test and evaluate the performance condition and efficiency HVAC systems.

VOSP may be done as part of the commissioning process, energy management verification and retrofit and test and balancing.

viscosity: (Physics) A measure of a fluid's resistance to flow. Viscosity is the "thinness or thickness" of a fluid. The less viscous a fluid is, the greater its ease of movement (fluidity). Water is "thin" having a lower viscosity or is less viscous and therefore has more fluidity, while honey is "thick," having a higher viscosity, and therefore is more viscous than water and has less fluidity.

voltage: (Electrical) A measure of electric force or potential. Aka electro-motive force.

voltage range: (Control) The change in the voltage necessary for the controller output to change over the throttling range. For example, a 6 to 9 volt system will have a voltage range of 3. See throttling range.

volt-ammeter: (Instrumentation) A portable multi-meter for measuring voltage and amperage. The volt-ammeter may have an analog or digital scale. To prevent pegging the movable pointer when using an instrument with an analog scale start at the highest scale and work down until the measured voltage or amperage is read in the upper half of the scale. The ammeter part of the instrument permits taking amperage readings without interrupting electrical service. On a three-phase motor measure the voltage between each phase. The standard practice is to read the phase-to-phase voltage from left to right. Read the voltage between the left and center line (L) terminals—L1 to L2, the left and right line terminals—L1 to L3, and then the center and right line terminals—L2 to L3. To get voltage readings on a single-phase motor measure voltage between phase (L1 or L2) and ground. To get amperage readings on a three-phase motor read the motor terminal (T) wires from left to right, T1, T2 and T3. Only one amperage reading is needed when measuring a single-phase circuit and that reading can be on either the hot wire or the neutral wire.

volt-ampere: (Electrical) A unit of apparent power. Volts times amperes.

voltmeter: (Instrumentation) An instrument for measuring voltage.

volume: (Physics) Space taken up by a body. Volume equals Mass divided by Density. Volume equals Mass times Specific Volume. Expressed in cubic feet or cubic inches.

volume damper: (Control) Manual volume dampers are used to control the quantity of airflow in the system by introducing a resistance to flow. If not properly selected, located, installed and adjusted, they

(1) don't control the air as intended, (2) they add unnecessary resistance to the system and (3) they can create noise problems.

volume flow rate: (Fluid Flow) Flow volume per unit of time. Typical units are cubic feet per minute (air) or gallons per minute (gpm).

volute pump: (Pump) A pump having a casing made in the shape of a volute (spiral or scroll-like). The volute casing starts with a small cross-sectional area near the impeller and increases gradually to the pump discharge.

vortex: (Definition) A mass of fluid, such as a liquid, with a whirling or circular motion that tends to form a cavity (or vacuum) in the center of the circle and to draw toward this cavity (or vacuum) bodies subject to its action. A region within a mass of fluid in which the fluid elements have an angular velocity. Plural: Vortices (or vortexes).

W is for Water Hammer

wall reflectance: (Lighting) See reflectance.

water balancing: (Water Flow) Testing, adjusting and balancing the water flow to pumps, boilers, coolers, coils, condensers, and various other heat exchangers to provide the correct water volume to achieve desired heat transfer.

water cooler: (Water Flow) Name for the evaporator portion of a chiller. Aka water chiller.

water differential pressure gauge: (Instrumentation) Analog or digital water differential pressure gauges are used to measure pressure drop across flow meters, terminals, and heat exchangers. An analog differential pressure gauge should be selected so the pressures measured don't exceed the upper limits of the scale. Purge all the air from the gauge before reading.

water hammer: (Steam System) Water hammer can occur in a steam distribution system when condensate is allowed to accumulate on the bottom of horizontal pipes and is pushed along by the velocity of the steam passing over it. As the velocity increases the condensate can form into a non-compressible slug of water. If this slug of water is suddenly stopped by a pipe fitting, bend or valve the result is a shock wave which can, and often does, cause damage to the system (such as blowing strainers and valves apart).

water horsepower: (Power) The theoretical horsepower required to drive a pump if the pump were 100% efficient. Also, gallons of water times pressure times specific gravity divided by 3960.

waterlogged compression tank: (Water Flow) A compression tank (aka expansion tank) becomes waterlogged when the air in the compression tank leaks out and is replaced by water. When this occurs the compression tank cannot maintain the proper pressure to accommodate the fluctuations in water volume and control pressure change in the system. A waterlogged tank must be drained and the air leaks found and sealed.

water tube boiler: (Boiler) A boiler which has water in tubes with fire and combustion gases around the tubes heating the water.

water vapor: (Psychrometrics, Chemistry) In HVAC systems the amount of water vapor or moisture present in the air (outside air, return air, mixed air, or supply air) is measured in pounds of moisture or grains of moisture per pound of air. There are 7,000 grains of moisture in one pound of moisture. For example, 70 grains is equal to 0.01 pounds. Water vapor can be produced from the evaporation or boiling of liquid water or from the sublimation of ice.

watt: (Electrical) A unit of actual power. For resistance loads (strip heaters) Watts equal volts times amps. For induction loads such as (alternating current motors) Watts equal volts times amps divided by power factor.

watt-hour: (Electrical) A measure of electrical energy. Watts times hours.

watt transducer: (Electrical) A device that converts a current signal into a proportional millivolt signal. Used to interface between current transformers and a load management panel.

wet bulb depression: (Psychrometrics)) For each given condition point at dry-bulb temperature (x) and wet-bulb temperature (y) it is the temperature difference between dry bulb temperature and wet bulb temperature. For example, the temperature of the air is 75F DB and 65F WB. The wet bulb depression is 10F.

wet bulb temperature: (Psychrometrics) The temperature obtained by an ordinary thermometer whose sensing bulb is covered with a wet wick and exposed to rapidly moving air. Wet bulb temperatures below 32 degrees Fahrenheit are obtained by an ordinary thermometer with a frozen wick.

wiredrawing: (Water Flow) Occurs when a high velocity water stream through a valve causes the erosion of the valve seat. Eventually, this erosion will cause leakage when the valve is fully closed. See gate valve.

work: (Physics) Force through a distance. Expressed in foot-pounds.

work station: (Environmentally Controlled Area) An open or closed work surface with direct air supply. A work station may be classified as a laminar (air) flow work station or a non-laminar (air) flow work station.

X is for X-axis

X-axis: (Math) Horizontal axis or line on a graph.

Xena: (Trivia) Warrior Princess.

Xenia: (Definition) The ancient Greek concept of guest-host friendship and hospitality. It is just as important for today's "road warriors" and will continue to be for future time-travelers.

Xenia, OH: A city located in southwestern Ohio, about 21 miles from Dayton, Ohio where I started working in HVAC for a design-build MEP company.

Y is for Y-axis

Y-axis: (Math) Vertical axis or line on a graph. Perpendicular to and through the x-axis.

Z is for Zone

Zone: (HVAC) A designated area having a dedicated thermostat. HVAC systems can be single zone or multizone.

Zone Damper: (Airflow, Control) Automatic or manual volume damper in each zone of a multizone HVAC unit. The volume damper in any designed zone.

Zone Pressurization: (Environmentally Controlled Area) Zone pressurization is a means of isolating spaces that generate harmful contaminants. The air distribution system is designed so the hazardous areas have negative pressure and any airborne contaminants are contained in the negatively pressurized areas. Zone pressurization is also for areas that want to be positive to adjacent areas so that there is no infiltration of contaminated or unwanted air into the controlled space.

Chapter 13

Abbreviations, Acronyms, Symbols

A is for Acceleration

A: acceleration, amperage, ampere, amps, area

AABC: Associated Air Balance Council

AB: as built

AC: air conditioning, alternating current

ACCA: Air Conditioning Contractors of America

ACG: Associated Air Balance Council Commissioning Group

ACGIH: American Conference of Governmental Industrial Hygienists

ACS: air conditioning system

A/D: analog to digital

ADF: Air Diffusion Council.

ADP: apparatus dew point

AEE: Association of Energy Engineers

AEV: automatic expansion valve

AFLV: automatic flow limiting valve (aka self-limiting valve)

Ag: silver

AH: air handler

AHJ: authority having jurisdiction

AHP: air horsepower

AHU: air handling unit

AI: analog input, artificial intelligence, as installed

AIA: American Institute of Architects

AKA: also known as

Al: aluminum (US), aluminium (UK)

AM: amplitude modulation, as manufactured

AMCA: Air Movement and Control Association.

AO: analog output

AOR: architect of record

ASHRAE: American Society of Heating, Refrigerating and Air Conditioning Engineers

ASME: American Society of Mechanical Engineers

ATC: automatic temperature control

ATCF: after tax cash flow
AU: as used
Au: gold
AXV: automatic expansion valve

B is for Boiler
B: boiler
BA: bypass air
BACnet: Building Automation and Controls Network
BAS: building automation system
BAT: bypass air temperature
BCA: Building Commissioning Association
BHP: brake horsepower, boiler horsepower
BP: boiling point
BMS: building management system, boiler management system
BOD: basis of design, bottom of duct
Br: bromide
BWG: Birmingham Wire Gauge
Btu: British thermal unit
Btuh: British thermal unit per hour
Btu/hr: British thermal unit per hour
Btuh$_L$: British thermal unit per hour, latent heat
Btuh$_S$: British thermal unit per hour, sensible heat
Btuh$_T$: British thermal unit per hour, total heat
Btu/lbF: British thermal unit per pound, degree Fahrenheit
Btum: British thermal unit per minute
Btu/min: British thermal unit per minute

C is for Carbon
C: carbon, circumference, coil, condenser, conductance
°C: degree(s) Celsius (centigrade)
c: specific heat (capacity)
CA: commissioning authority, commissioning agent (agency)
CAD: computer aided design
CAV: constant air volume
CBF: coil bypass factor
CD: ceiling diffuser
CEA: controlled environment area
CEM: Certified Energy Manager

CER: controlled environment room
CF: ceiling fan, cubic feet
CFC: chlorofluorocarbon
CF/lb: cubic feet per pound
CFL: compact fluorescent light
CFM: cubic feet per minute
cGMP: current good manufacturing practices
CH: chiller
CHR: chilled water return
CHS: chilled water supply
CHW: chilled water
CHWP: chilled water pump
CHWR: chilled water return
CHWS: chilled water supply
Cl: chlorine
cm: centimeter
CMMS: computerized maintenance management system
CO: carbon monoxide, change order
CO$_2$: carbon dioxide
COMP: compressor
COND: condenser
COP: coefficient of performance
CP: commissioning plan, commissioning professional
CPM: commissioning project manager
CRI: color rendition index
CS condenser water return
CT: compression tank, cooling tower, current transformer
CTAB: Certified Testing Adjusting Balancing Professional
CTI: Cooling Tower Institute
Cu: copper
CD: ceiling diffuser
Cube Root: 3x, 32xCu. ft.: cubic feet
Cu. in.: cubic inch
CV: constant volume
Cv: valve coefficient, coefficient of flow
CWP: condenser water pump
CWR: condenser water return
CWS: condenser water supply
Cx: commissioning

D is for Delta

Δ: delta, difference
ΔP: delta pressure difference
ΔT: delta temperature difference
d: density
D: diameter
D/A: digital to analog, direct acting
DB: design build, dry bulb
dB: decibel
DC: direct current, document control
DD: double duct, dual duct
DDC: direct digital control
DH: duct heater
DI: digital input
DIA: diameter
DID: design intent document
DIDW: double inlet double wide
DIFF: diffuser
DO: digital output
DOAS: dedicated outside air system
DOE: Department of Energy
DOP: dioctyl phthalate oil
DP: delta pressure, dew point
DT: delta temperature
DTW: dual temperature water
Dx: direct expansion, direct exchange

E is for Energy

E: energy, electric motive force (voltage)
EA: exhaust air, exhaust air duct or inlet
EAT: entering air temperature, exhaust air temperature
ECT: effective coil temperature (aka ADP)
EDR: energy design resources
EF: exhaust fan
EFF: efficiency
η: efficiency (eta)
EG: exhaust grille
EM: energy management, energy manager
EMCS: energy management computer system or control system

EMF: electromotive force, electromagnetic field
EMS: energy management system
EOR: engineer of record
E-P: electric to pneumatic
EPA: Environmental Protection Agency
ET: expansion tank, extraterrestrial
EVAP: evaporator
EWT: entering water temperature

F is for Fan
F: fan, fluorine
°F: degree(s) Fahrenheit
FC: fan coil, forward curved (fan)
FCL: fluorescent compact light
FCU: fan coil unit
FM: frequency modulation
FPM: feet per minute
FPS: feet per second
FPT: fan powered terminal, functional performance test
FSP: fan static pressure
Ft: foot, feet
Ft. hd: feet of head
Ft. H_2O: feet of water
Ft-lb: foot-pound
Ft/sec^2: feet per second squared
Ft. wc: feet of water column
Ft. wg: feet of water gauge
FTE: fan total efficiency
FTP: fan total pressure
FURN: furnace
FV: fan velocity, future value
FVAV: fan powered variable air volume terminal
FVP: fan velocity pressure

G is for Giga
G: giga, grains, grille
g: acceleration due to gravity, gram
GIGO: garbage in, garbage out
GAAP: generally accepted accounting practices

GMT: Greenwich Mean Time
GPM: gallons per minute
Gr: grains, grille
gr: grain
GRD: grilles, registers, diffusers

H is for Heat
H: heat, humidity, hydrogen, unit of inductance (Henry)
h: enthalpy, hecto, hour(s),
H$_2$O: water
HEPA: High Efficiency Particulate Air (filter)
HF: humidity factor
Hg: mercury
HID: high intensity discharge (lamp)
H$_L$: latent heat
HOPEC: hand operated positive energy control (laboratory fume hood)
HP: horsepower
HR: humidity ratio
hr: hour(s)
H$_S$: sensible heat
H$_T$: total heat
HTR: high temperature return
HTS: high temperature supply
HTW: high temperature water
HVAC: heating, ventilating, and air conditioning
HWP: hot water pump
HWR: heated water return, hot water return
HWS: heated water supply, hot water supply
HX: heat exchanger
Hz: Hertz

I is for Current
I: current (intensity)
IAQ: indoor air quality
IC: integrated circuit, installing contractor
IEEE: Institute of Electrical and Electronics Engineers (Former name, now just the letters.)
IES: Illuminating Engineering Society
IEQ: indoor environmental quality

in: inch(es)
in-lb: inch-pound
in Hg: inches of mercury
in wc: inches of water column
in wg: inches of water gauge
IRCTAB: International Registry of CTAB Professionals
IS: information systems
ISPE: International Society of Pharmaceutical Engineers
IT: information technology

J is for Joule
J: Joule
jpeg: Joint Photographic Experts Group

K is for Kelvin
K: Kelvin
k: kilo (1000), thermal conductivity
kg: kilogram(s)
km: kilometer(s)
kW, kw: kilowatt(s)
kWh, kwh: kilowatt-hour(s)

L is for Inductance
L: inductance
LAD: linear air diffuser
LAT: leaving air temperature
LCD: liquid crystal display
LED: light emitting diode
LF: load factor
lb: pound(s)
lb/cf: pound(s) per cubic foot
lb-ft: pound-foot
lb/hr: pound(s) per hour
lb/min: pound(s) per minute
LEED: Leadership in Energy and Environmental Design
Li: lithium
LMTD: log mean temperature difference
L/s, lps: liters (litres) per second
LT: light troffer

LTR: low temperature return
LTS: low temperature supply
LTW: low temperature water
LWT: leaving water temperature

M is for Mega
M: mega
m: meter, milli, minute
MA: mixed air
MARR: minimum attractive rate of return
MAT: mixed air temperature
MBH: thousand Btu per hour
MD: manual damper, metering device
MEP: mechanical, electrical and plumbing
MERV: minimum efficiency rating value
MKS: meter-kilogram-second
mJ: megaJoule
μ: micro, (mu)
μm: micrometer
mm: millimeter
MMBH: million Btu per hour
MTR: medium temperature return
MTS: medium temperature supply
MTW: medium temperature water
MVD: manual volume damper
MWP: maximum working pressure
MZ: multizone

N is for Nano
n: nano
N: nitrogen
NBS: National Bureau of Standards.
NC: noise criteria, normally closed
NEBB: National Environment Balancing Bureau
NFPA: National Fire Protection Association.
NO: normally open
NPV: net present value

O is for Oxygen
O: oxygen

OA: outside air
O&M: operation and maintenance
OAT: outside air temperature
OBD: opposed blade damper
OEM: original equipment manufacturer
OPM: other people's money
OPR: owner's project recommendations
OSA: outside air
OSHA: Occupational Safety and Health Administration
OT: outlet temperature, outlet total
OV: outlet velocity
Ω: Ohm (omega)

P is for Pico
p: pico
π: pi
P: pressure, proportional
Pa: absolute pressure
PBD: parallel blade damper
PD: pressure differential
pd: pitch diameter
P.E.: professional engineer (licensed)
P-E pneumatic to electric
PF: power factor
PFC: power factor correction
PH, Ø: phase of a motor, 3 ph or 3 Ø
P-I: proportional-integral
P-I-D: proportional-integral-derivative
PM: preventative maintenance
PPM: predictive and preventative maintenance
ppm: parts per million
PR: pressure range
PRAM: parameter random access memory
PROM: programmable read only memory
PRV: pressure reducing valve, pressure relief valve
Ps: static Pressure
psf: pounds per square foot
psia: pounds per square inch absolute
psig: pounds per square inch gauge

Pt: total pressure
PV: present value
Pv: velocity pressure
PVC: polyvinyl chloride
Pw: Vapor Pressure

Q is for Quantity
Q: Quantity
Quad: Quadrillion

R is for Resistance
R: electrical resistance (ohms), radius, register
R: resistance (building materials and insulation, R value)
R: Rankine
RA: return air, return air duct or inlet
R/A reverse acting
RAF: relief air fan, return air fan
RAG: return air grille
RAM: random access memory
RAT: return air temperature
RC: room cavity, room criteria
Rd, Ø: round
RF: return fan
RFI: request for information
RFP: request for proposal
RFQ: request for qualifications
RFS: request for submittal
RG: return grille
RH: relative humidity
RO: return opening, reverse osmosis
ROI: return on investment
ROM: read only memory
ROR: rate of return
Rn: radon
RPM: revolution(s) per minute
RR: return register
RMS: root-mean-square
RTD: resistance temperature detector, aka "resistive thermal devices"
RTF: resolution tracking form

RTU: roof-top unit
RVA: rotating vane anemometer

S is for Entropy

S: electrical conductance
s: entropy, seconds
SA: supply air
SAD: supply air diffuser
SAF: supply air fan
SAG: supply air grille
SAR: supply air register
SAT: supply air temperature
sec: seconds
SCFM: standard cubic feet per minute
SCR: silicon controlled rectifier
SF: service factor, square feet, supply fan
SG: specific gravity
SHF: sensible heat factor
SHR: sensible heat ratio
SI: Systeme International d'Unites (metric system)
SISW: single inlet single wide
SMACNA: Sheet Metal & Air Conditioning Contractors' Nat'l Assoc
SOP: standard operating procedure
SP: static pressure
SpGr: specific gravity
SpHt: specific heat (the symbol c)
SPL: sound pressure level
SpV: specific volume
SpVol: specific volume
SR: supply register
sq. ft.: square foot (feet)
sq. in.: square inch(es)
Square Root: $\sqrt{}$ or $\wedge$ or $\sqrt{}$
STP: standard temperature and pressure
SWG: British Standard Wire Gauge, sidewall grille
SVC: system verification checklist
SZ: single zone

T is for Tera

T: temperature, tera
TAB: test and balance, or testing, adjusting and balancing
TD: temperature difference
TDH: total dynamic head
TES: thermal energy storage
TEV: thermostatic expansion valve
TF: transfer fan
TFP: total fan pressure
TG: transfer grille
THD: total head
Ti: titanium
TLV: threshold limit value
Tn: ton(s)
Tn-hr: ton-hours
Tn/hr: tons per hour
TOD: top of duct
TP: total pressure
TS: tip speed
TSP: total static pressure
TR: throttling range
TXV: thermostatic expansion valve

U is for Unit
U: unit
UH: unit heater
u: heat transfer coefficient
UL: Underwriters Laboratories
ULPA: Ultra Low Penetration Air filters
USGBC: United States Green Building Council
UV: ultraviolet, unit ventilator

V is for Voltage
V: voltage, volts, also velocity, volume, specific volume
VA: volt-ampere
VAC: volts alternating current
VAV: variable air volume
VDC: volts direct current
VFD: variable frequency drive
VFR: volumetric flow rate

VOC: volatile organic compound
VOSP: verification of system performance
VP: velocity pressure
VRF: variable refrigerant flow
VRV: variable refrigerant volume
VSD: variable speed drive
VV: variable volume, variable water volume
VWV: variable water volume

W is for Watt
W: Watt
WAN: wide area network
WB: wet bulb
WC: water column
WG: water gauge
WHP: water horsepower
Whr: Watt-hour(s)
Ws: Watt-second(s)

X is for Unknown
X: x-axis, reactance, unknown (name, quantity, etc.)

Y is for Y- axis
Y: y-axis, unknown quantity

Z is for impedance
Z: impedance, zone

Chapter 14

Tables

TABLE OF TABLES

Table 1. Air Changes per Hour

Air Changes Per Hour	Ceiling Height, feet	Cubic Feet per Minute per square foot
6	8 ft.	0.8
7.5	8 ft.	1
10	8 ft.	1.33
6	9 ft.	0.9
7.5	9 ft.	1.13
10	9 ft.	1.5

Table 2. Air Velocities (Feet per Minute), Recommended

Component	FPM
Coil, Chilled water	500-600
Coil, Hot water	400-700
Duct, Branch, Office	800-1,600
Duct, Fume Exhaust	1,500-2,000
Duct, Main, Office	1,200-2,400
Filter, Electronic	500
Filter, Fiber, Dry	750
Filter, Fiber, HEPA	250
Filter, Fiber, Viscous	700
Filter, Renewable, Dry	200
Filter, Renewable, Viscous	500
Louvers, Exhaust	500
Louvers, Intake	400
Outlets	400-800
Stack, Fume Exhaust	2,500-3,000

Table 3. Altitude and Pressure

Altitude in feet	Inches of mercury	Feet of water (specific gravity 1)
0	29.92	33.9
1,000	28.86	32.8
2,000	27.72	31.6
3,000	26.81	30.5
4,000	25.84	29.4
5,000	24.89	28.2
6,000	23.98	27.3
7,000	23.09	26.2
8,000	22.22	25.2
9,000	21.38	24.3
10,000	20.58	23.4

Table 4. Compressor Horsepower and Kilowatts Required per Ton of Cooling versus COP

Compressor HP/ton = 4.71 divided by COP
3.5 COP = 1.34 hp/ton = 1 kW/ton

Table 5. Condenser Tons and Cooling Tons

Condenser water tons = (chiller gpm x chiller water ΔT) x 1.25 ÷ 24
Condenser water tons = (condenser gpm x condenser water ΔT) ÷ 24
Chilled water (evaporator) tons = (chiller gpm x chiller water ΔT) ÷ 24
Chilled water (evap) tons = (condenser gpm x condenser water ΔT) ÷ 30

Table 6. Duct Friction Loss Correction for Various Materials

Duct Material	Correction Factor
Galvanized Duct	1.00
Fiberglass Duct	1.35
Fiberglass Lined Duct	1.42
Flex Duct, Fully Extended	1.85
Flex Duct, Compressed	3.65
Note: Multiply correction factor times calculated friction loss per 100 feet	

Table 7. Duct Pressure Classes

Class	Static Pressure, inches wg	Velocity
Low	To 2 inches	To 2,500 fpm
Medium	2 to 6 inches	2,000 to 4,000 fpm
High	Above 6 inches	Above 2,000 fpm

Table 8. Math Conversion

1 atmosphere	33.9 ft. wg 407 in. wg 14.7 psi 29.92 in. Hg
1 boiler horsepower	33,475 Btuh (34.5 lb of water evaporated at 970 Btu/lb)
1 Btu	778 foot-pounds 0.000393 horsepower-hours
1 Btu	0.000293 kilowatt-hours
1 Btuh	0.000393 horsepower 0.000293 kilowatts
1 cubic foot	1728 cubic inches
1 cu. ft. of water	7.5 gallons 62.4 pounds
1 ft. wg	0.883 in. Hg 12 in. wg 0.433 psi
1 gallon of water	8.33 pounds
1 horsepower	2545 Btuh 42.42 Btu per minute
1 horsepower	550 foot-pounds per second 33,000 foot-pounds per minute
1 horsepower	0.746 kilowatts 746 Watts
1 in. Hg	1.13 ft. wg 13.6 in. wg 0.491 psi
1 in. wg	0.036 psi 5.2 pounds per square foot (psf)
1 kilowatt	3413 Btuh 1.34 horsepower 56.9 Btu per minute
1 mile per hour	88 feet per minute
1 pound	7000 grains
1 psi	2.04 in. Hg 2.31 ft. wg 27.7 in. wg
1 Ton Refrigeration	12,000 Btuh 200 Btum
1 Watt	3.41 Btuh 0.00134 horsepower 44.26 foot-pounds per min
1 year	8760 hours (24 hrs x 365 days)

Table 9. Motor Amperage Rating

Motor	3 Ø 230V	3Ø 460V	1Ø 115V	1Ø 230V
NPHP	FLA	FLA	FLA	FLA
1/2	2.0	1.0	9.8	4.9
3/4	2.8	1.4	13.8	6.9
1	3.6	1.8	16	8.0
1.5	5.2	2.6	20	10
2	6.8	3.4	24	12
3	9.6	4.8	34	17
5	15.2	7.6	56	28
7.5	22	11		
10	28	14		
15	42	21		
20	54	27		
25	68	34		
30	80	40		
40	104	52		
50	130	65		
60	154	77		
75	192	96		
100	248	124		
125	312	156		
150	360	180		
200	480	240		

NPHP nameplate horsepower, FLA full load amps

Table 10. Motor Efficiency, Power Factor, 3-Phase, Approximate

HP	Efficiency	Power Factor	
1/2	70	69.2	
3/4	72	72.0	
1	79	76.5	
1.5	80	80.5	
2	80	85.3	
3	81	82.6	
5	83	84.2	
7.5	85	85.5	
10	85	88.8	
15	86	87.0	
20	87	87.2	
25	88	86.8	
30	89	87.2	
40	89	88.2	
50	89	89.2	
60	89	89.5	
75	90	89.5	
100	90	90.3	
125	90	90.5	
150	91	90.5	
200	91	90.5	

Note: For most motors, efficiency and power factor curves remain fairly flat until the motor load falls below 50%. Motor efficiency between 82% and 92%   Power factor between 80% and 90%

Table 11. Motor Synchronous Speed @ 60Hz

Number of Poles	Speed
2	3600
4	1800
6	1200
8	900

Table 12. Motor Wire Size

Motor	3 Ø	3 Ø	3 Ø	1Ø	1Ø
HP	230 V	460V	575V	115 V	230 V
1/2	14	14	14	14	14
3/4	14	14	14	12	14
1	14	14	14	12	14
1.5	14	14	14	10	14
2	14	14	14	10	14
3	14	14	14	8	10
5	12	14	14	4	8
7.5	10	14	14		6
10	8	12	14		6
15	6	10	10		
20	4	8	10		
25	4	8	8		
30	3	6	8		
40	1	6	6		
50	2/0	4	6		
60	3/0	3	4		
75	250	1	3		
100	350	2/0	1		
125	2-3/0	3/0	2/0		
150	2-4/0	4/0	3/0		
200	2-350	350	250		

Table 13. Pressure, Gauge and Absolute

	Gauge Pressure psig	Absolute Pressure psia		in. Hg	ft. wg	in. wg
Above Atmospheric	50	64.7				
	40	54.7				
	30	44.7				
	20	34.7				
	10	24.7				
Atmosphere Pressure	0	14.7		29.92	33.9	407
	In. Hg	psig	psia			
Below Atmospheric	10	-4.9	9.8	19.92	22.6	271
	20	-9.8	4.9	9.92	11.3	136
	29.92	-14.7	0	0	0	0
Above Atmospheric Pressure: pounds per square inch gauge and absolute At Atmospheric Pressure: pounds per square inch, inches of mercury, feet wg, inches wg Below Atmospheric Pressure: inches of mercury, feet water gauge and inches water gauge						

Table 14. Roman Numerals

I	one
II	two
IV	four
V	five
VI	six
IX	nine
X	ten
XI	eleven
XX	twenty
XL	forty
L	fifty
XC	ninety
C	One hundred
CC	Two hundred
CD	Four hundred
D	Five hundred
CM	Nine hundred
M	One thousand
*MM	Two thousand
*MM	**One million (thousand thousand) when used in HVAC. For example, MMBH is 1,000,000 Btu per hour, Typically used when describing boiler heating capacity. MBH is 1000 Btu per hour. Examples: 10MMBH is 10 million (10,000,000) Btu/hr. 10 MBH is 10 thousand (10,000) Btu/hr.**

Table 15. Rule of Thumb

Cooling	400 cfm per ton 30 Btuh per cfm
Cooling	1 cfm per sf
Cooling	20-30 cfm per seat Auditorium or Theater
Office	143 Square Feet per Person
Office	143 CFM per person
1 in. Hg is approx	1 ft of water
1 in. Hg is approx	0.5 psi
1 ft. wg is approx	0.5 psi
Centrifugal fans	If fan efficiency is unknown use 80%
Centrifugal pumps	If pump efficiency is unknown use 70%
Determining barometric pressure	0.1 inch Hg reduction from 30" (29.92 rounded off) for each 100' above sea level or subtract 1" for every 1000'.
To find approximate round duct size to carry cfm at 0.1"/100': Duct Diameter $= 0.9(\text{cfm})^{0.40}$	New belt length with sheave change is $1.57(\Delta d)$ Δd is difference between old sheave pitch diameter and new sheave pitch diameter.
Calculating the velocity correction factor for airflow for a change in air density resulting from changes in temperature and altitude. +2% correction for each 1,000' above sea level + or – 1% correction for each 10 degrees above or below 70 degrees Fahrenheit	Heat gain from moderately active office worker 450 Btuh per person total heat 225 Btuh per person sensible heat 225 Btuh per person latent heat

Table 16. Scientific Notation, Power of 10

10^n	Symbol	Name	Decimal
-24	y	yocto	0.000,000,000,000,000,000,000,001
-21	z	zepto	0.000,000,000,000,000,000,001
-18	a	atto	0.000,000,000,000,000,001
-15	f	femto	0.000,000,000,000,001
-12	p	pico	0.000,000,000,001
-9	n	nano	0.000,000,001
-6	u	micro	0.000,001
-3	m	milli	0.001
-2	c	centi	0.01
-1	d	deci	0.1
1	da	deka	10
2	h	hecto	100
3	k	kilo	1,000
6	M	mega	1,000,000
9	G	giga	1,000,000,000
12	T	tera	1,000,000,000,000
15	P	peta	1,000,000,000,000,000
18	E	exa	1,000,000,000,000,000.000
21	Z	zetta	1,000,000,000,000,000,000,000
24	Y	yotta	1,000,000,000,000,000,000,000,000

| 10^0 | | one | 1 |

M: million (thousand thousand)

G: billion   (thousand million)

T: trillion   (thousand billion)

P: quadrillion (quad) (thousand trillion)

E: quintillion, Z: sextillion, Y septillion

Scientific Notation Examples:			
10^3 3 decimal places to the right.	$1 \times 10^3 = 1,000$	$1.23 \times 10^3 = 1,230$
10^6 6 decimal places to the right.	$1 \times 10^6 = 1,000,000$	
10^{-3} 3 decimal places to the left.	$1 \times 10^{-3} = 0.001$	$1.23 \times 10^{-3} = 0.00123$
10^{-6} 6 decimal places to the left.	$1 \times 10^{-6} = 0.000001$	

Table 17. Temperature Conversion

C = (F-32) ÷ 1.8	F = 1.8C +32			
C = (F-32) ÷ 9/5	F = 9/5C +32			
K = C + 273	R = F + 460			
K = R ÷ 1.8	R = 1.8K			
Absolute zero	-273C (273.15)	-460F (459.67)	0K	0R
Freezing point of pure water	0C	32F	273K	492R
Boiling point of pure water	100C	212F	373K	672R
C Celsius K Kelvin F Fahrenheit R Rankin				

Table 18. V-Belt Sizes

Size	Width (inches)	Designation
A	1/2	Standard Industrial
B	21/32	Standard Industrial
C	7/8	Standard Industrial
D	1 1/4	Standard Industrial
E	1 1/2	Standard Industrial
2L	9/32	Fractional Horsepower
3L	3/8	Fractional Horsepower
4L	1/2	Fractional Horsepower
5L	21/32	Fractional Horsepower

Table 19. Water Properties

Temperature	Density	Weight	Vapor Pres	Sp Gravity
50	62.38	8.34	0.41	1.002
60	62.35	8.33	0.59	1.001
70	62.27	8.32	0.84	1.000
80	62.19	8.31	1.17	0.998
90	61.11	8.30	1.62	0,997
100	62.00	8.29	2.20	0.995
110	61.84	8.27	2.96	0.993
120	61.73	8.25	3.95	0.990
130	61.54	8.23	5.20	0.988
140	61.40	8,21	6.78	0.985
150	61.20	8.18	8.74	0.982
160	61.01	8.16	11.20	0.979
170	60.00	8.12	14.20	0.975
180	60.57	8.10	17.85	0.972
190	60.35	8.07	22.30	0.968
200	60.13	8.04	27.60	0.965
210	59.88	8.00	34.00	0.961
Temperature, degree Fahrenheit Density, pounds per cubic foot Weight, pounds per gallon Vapor Pressure, feet of water				

Table 20. Water Temperature Difference and Water Flow in Gallons per Minute per Ton of Cooling

TD degrees F entering and leaving water cooler	gpm/ton
8	3.0
10	2.4
12	2.0
20	1.2

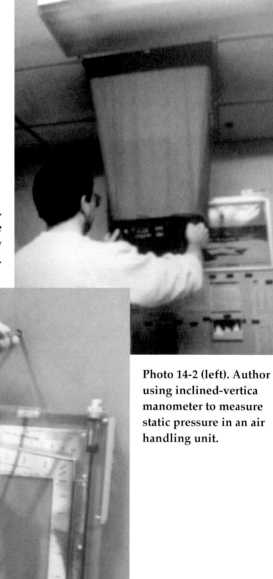

Photo 14-1 (right).
Author using capture
hood to measure airflow
from ceiling diffuser.

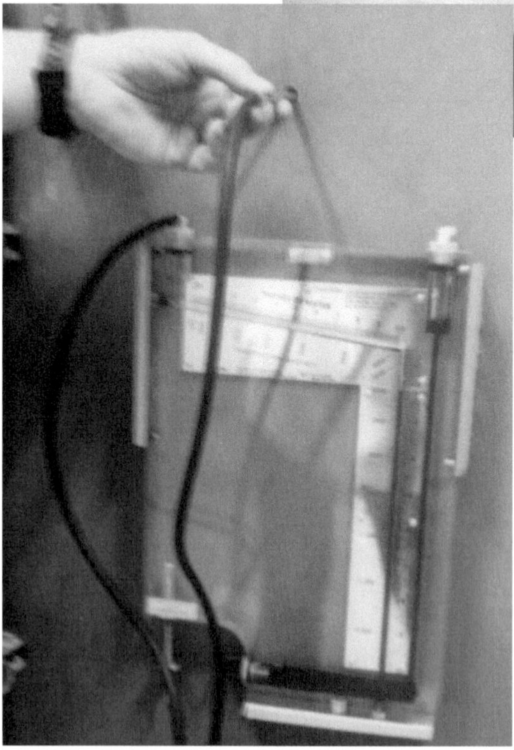

Photo 14-2 (left). Author
using inclined-vertica
manometer to measure
static pressure in an air
handling unit.

Photo 14-3. Using electronic micro-manometer to measure static pressure in an air handling unit. Reading: 1.82 in wg.

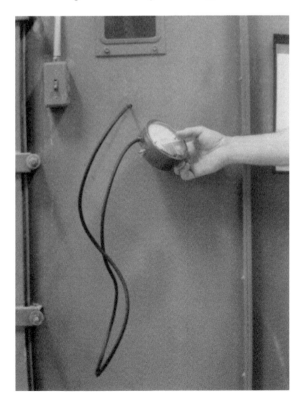

Photo 14-4. Using air differential gauge to measure static pressure in an air handling unit.

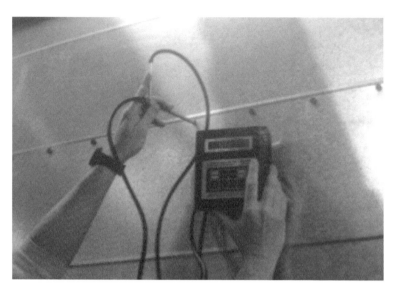

Photo 14-5. Using a electronic micro-manometer to measure velocity pressure for a Pitot tube traverse in the duct of an air handling unit to determine cubic feet per minute airflow.

Photo 14-6. Using an ammeter to measure amperes.

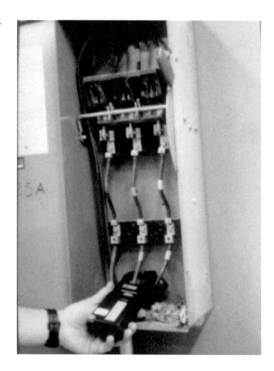

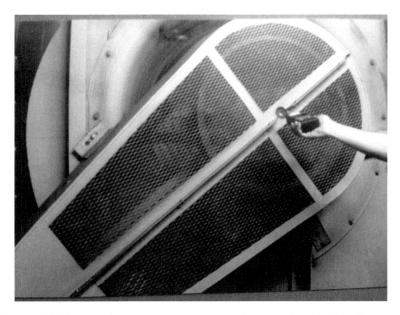

Photo 14-7. Using a tachometer to measure rotating speed on the fan sheave.

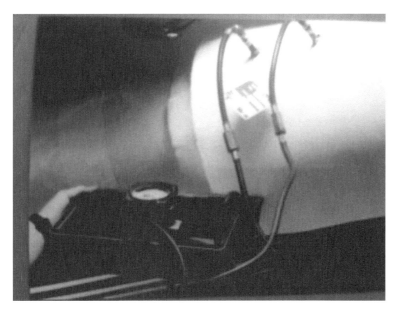

Photo 14-8. Using water differential gauge to measure pressure drop to determine gallons per minute in water pipe.

Chapter 15
HVAC Math

Air Changes per hour
AC/hr = (cfm x 60) ÷ Vol
cfm = (AC/hr x Vol) ÷ 60
Terminology:
AC/hr   Air Changes per hour
cfm   cubic feet per minute, quantity of airflow
60   constant, minutes per hour
Vol   room volume, length x width x height, in cubic feet
Air Density
d = 1.325 (Pb ÷ T)
Terminology:
d   air density in pounds per cubic foot
1.325   constant, 0.075 divided by 29.92/530
Pb   barometric pressure, inches of mercury
T   absolute temperature (indicated temperature in degrees Fahrenheit plus 460)
Air Density, Correction Factor
$cf = \dfrac{\sqrt{0.075}}{d}$
Terminology:
cf   correction factor
0.075   constant, density of air in pounds per cubic foot @ standard air
d   actual density of air, pounds per cubic foot
Example: The density of air is 0.12. Find correction factor.
cf = square root of 0.075 divided by 0.12
cf = 0.79
Air Velocity
V = 4005 √VP
VP = (V/4005)²
Terminology:
V   velocity of airflow in feet per minute
4005   constant
√VP   square root of the velocity pressure, inches of water gauge
VP   velocity pressure, inches of water gauge
Air Velocity, Corrected for density

Vc = Vm x cf	
Terminology:	
Vc	corrected velocity in feet per minute
Vm	measured velocity in feet per minute
cf	correction factor for density
Air Volume	
Q = A x V	
A = Q ÷ V	
V = Q ÷ A	
Terminology:	
Q	quantity (volume) of airflow in cubic feet per minute (cfm)
A	cross sectional area of the duct in square feet (sf)
V	velocity of airflow in feet per minute (fpm)
Air Volume, Corrected for density	
Q = A x Vc	
Terminology:	
Q	quantity of airflow in cubic feet per minute
A	area in square feet
Vc	corrected velocity
Air Volume, Pounds per Hour	
Lb/hr = cfm x 4.5	
Terminology:	
Lb/hr	airflow, pounds per hour
cfm	cubic feet per minute, quantity of airflow
4.5	constant, 60 minutes per hour x 0.075 pounds per cubic foot
Amperage **Single-phase (1Ø) electric circuit**	
A = (bhp x 746) ÷ (V x PF x eff) or alternate equation, I = (bhp x 746) ÷ (E x PF x eff)	
Amperage **Three-phase (3Ø) electric circuit**	
A = (bhp x 746) ÷ (V x PF x eff x 1.73) or alternate equation, I = (bhp x 746) ÷ (E x PF x eff x 1.73)	
Terminology:	
bhp	brake horsepower
V or E	Volts (for three-phase circuits, this is average volts)
A or I	Amps (for three-phase circuits, this is average amps)
eff	motor efficiency

PF	Power Factor
746	constant, watts per horsepower
1.73	constant, square root of 3, for three-phase circuits
Example: Find amps for a three-phase motor operating at 80 bhp, 480 volts, 0.90 power factor, and an efficiency of 0.85	
Solution:	
A = (bhp x 746) ÷ (V x PF x eff x 1.73)	
A = (80 x 746) ÷ (480 x 0.90 x 0.85 x 1.73)	
A = (59680) ÷ (635.26)	
A = 94	

Area, Rectangular and Round Duct

A = ab ÷ 144	
A = πr^2 ÷ 144	
Terminology:	
A	Area of the rectangular duct, square feet
a	length of one side of rectangular duct, inches
b	length of adjacent side of rectangular duct, inches
144	constant, square inches per square foot
A	Area of the duct, square feet
π	3.14
r^2	radius in inches, squared
144	constant, square inches per square foot

Area, Triangle

A = bh ÷ 2	
Terminology:	
A	Area
b	base of the triangle
h	height of the triangle

Belt Length

BL = 2C + 1.57 (D + d) + (D − d)^2 ÷ 4C	
Terminology:	
BL	Belt Length (belt pitch diameter or outside diameter)
C	Center-to-Center distance of the shafts
D	pitch or outside Diameter of the large sheave
d	pitch or outside diameter of the small sheave
1.57	constant, π ÷ 2

Brake Horsepower, Electrical

bhp = (V x A x PF x eff) ÷ 746 Single-phase (1O) electric circuit
bhp = (V x A x PF x eff x 1.73) ÷ 746 Three-phase (3O) electric circuit
Terminology:
bhp   Brake horsepower
V   Volts   (for three-phase circuits, this is average volts)   (E can be substituted for V)
A   Amps   (for three-phase circuits, this is average amps)   (I can be substituted for A)
eff   motor efficiency
PF   Power Factor
746   constant, watts per horsepower
1.73   constant, square root of 3, for three-phase circuits
Calculation:
To find bhp for a single-phase circuit multiply V times A times PF times eff and divide results by 746
To find bhp for a three-phase circuit multiply V times A times PF times eff times 1.73 and divide results by 746
Example: A three-phase alternating current motor is operating at 240 volts, 30 amps, 0.90 power factor and 83% eff. Find bhp.
Solution:
bhp = (V x A x PF x eff x 1.73) ÷ 746
bhp = (240 x 30 x 0.90 x 0.83 x 1.73) ÷ 746
bhp = (9304.63) ÷ 746
bhp = 12.47
Brake Horsepower, Electrical, No Load Amps (motors 10 horsepower and larger)
bhp = (RLA - 0.5 NLA) ÷ (FLAc - 0.5 NLA) x HPn (First calculate FLAc = (Vn x FLAn) ÷ Vm)
Terminology:
bhp   brake horsepower
RLA   Running Load Amps, field measured
NLA   No Load Amps (motor sheave in place, belts removed)
FLAc   Full Load Amps, field corrected
HPn   nameplate Horsepower
Vn   nameplate Volts
FLAn   nameplate Full Load Amps
Vm   Volts, field measured
Brake Horsepower, From Percent Load

bhp = HPn x %load
Terminology:
bhp   brake horsepower
HPn   Horsepower, nameplate
%load   percent load on motor
Example: A 100 HPn is operating at 75% load. Find bhp.
Solution:
bhp = HPn x %load
bhp = 100 x 0.75
bhp = 75
Brake Horsepower, Mechanical, Fan
bhp = (cfm x TSP) ÷ (6356 x eff)
Terminology:
bhp   brake horsepower
cfm   cubic feet per minute, air volume
TSP   Total Static Pressure
6356   constant, 33,000 ft-lb/min divided by 5.19
eff   efficiency, as a percent or decimal
Example: Find bhp if a fan operates at 50,000 cfm , 4.6" TSP. Fan efficiency is 70%.
Solution:
bhp = (cfm x TSP) ÷ (6356 x eff)
bhp = (50,000 x 4.6") ÷ (6356 x 0.70)
bhp = (230,000) ÷ (4449.2)
bhp = 51.7
Brake Horsepower, Mechanical, Pump
bhp = (gpm x TDH) ÷ (3960 x eff)
Terminology:
bhp   brake horsepower
gpm   gallons per minute, water volume
TDH   Total Dynamic Head in feet of water
3960   constant, 33,000 ft-lb/min divided by 8.33 lb/gal
eff   efficiency, as a percent or decimal
Example: Find bhp if a chiller pump operates at 100 ft. hd., and 500 gpm. Pump efficiency is 70%.
Solution:
bhp = (gpm x TDH) ÷ (3960 x eff)
bhp = (500 x 100) ÷ (3960 x 0.70)
bhp = (50,000) ÷ (2772)
bhp = 18

Bypass Air

MSAT = (%BA x MAT) + (%SA x CSAT)

Terminology:

MSAT   Mixed Supply Air Temperature (bypass air temperature and coil supply air temperature)

%BA   percent of Bypass Air

MAT   Mixed Air Temperature

%MA   percent of Mixed Air

CSAT   Coil Supply Air Temperature

Example: Find MSAT if bypass air quantity is 10%, mixed air temperature is 78F, mixed air quantity is 90%, and supply air temperature is 55F.

Solution:

MSAT = (%BA x MAT) + (%SA x CSAT)

MSAT = (10% x 78) + (90% x 55)

MSAT = (7.8) + (49.5)

MSAT = 57.3F

Carbon Dioxide in Airflow

MAC = (%OA x OAC) + (%RA x RAC)

%OA = (MAC - RAC) x 100 ÷ (RAC- OAC)

Terminology:

MAC   Mixed Air, Carbon dioxide amount

%OA   percent Outside Air

OAC   Outside Air, Carbon dioxide amount

%RA   percent Return Air

RAC   return air, carbon dioxide amount

Coefficient of Performance

COP = (ton x Btuh/ton) ÷ (kW x Btuh/kW)

COP = (1 kW/ton ÷ kW/ton) x 3.5

Terminology:

COP   Coefficient of Performance

ton   ton of refrigeration (one ton of refrigeration is 12000 Btuh)

Btuh   Btu per hour

kW   kiloWatt (an electrical unit of power and is equal to 3413 Btuh)

3.5   the COP equal to 1 kW/ton

Example: A 200 ton chiller has an electrical load of 100 kW. Find COP.

Calculation:

To find COP first multiply tons times the constant 12000 Btuh per ton.
Next, multiply kW times the constant 3413 Btuh per kW. Finally, divide the mechanical units by the electrical units
Solution:
COP = ton x Btuh/ton ÷ kW x Btuh/kW
COP = 200 ton x 12000 Btuh/ton ÷ 100 kW x 3413 Btuh/kW
COP = 2,400,000 Btuh ÷ 341,300Btuh
COP = 7.03
Drives
RPMm x Dm = RPMf x Df   (The motor speed times motor sheave pitch diameter is equal to the fan speed times the fan sheave pitch diameter.)
RPMm = (RPMf x Df) ÷ Dm   (To find motor speed, multiply fan speed times fan sheave pitch diameter and divide result by motor sheave pitch diameter.)
Dm = (RPMf x Df) ÷ RPMm   (To find motor sheave pitch diameter, multiply fan speed times fan sheave pitch diameter and divide result by motor speed.)
RPMf = (RPMm x Dm) ÷ Df   (To find fan speed, multiply motor speed times motor sheave pitch diameter and divide result by fan sheave pitch diameter.)
Df = (RPMm x Dm) ÷ RPMf   (To find fan sheave pitch diameter, multiply motor speed times motor sheave pitch diameter and divide result by fan speed.)
Terminology:
RPMm   speed of motor shaft
Dm   pitch diameter of motor sheave
RPMf   speed of fan shaft
Df   pitch diameter of fan sheave
Example: A motor is operating at 1750 rpm. The fan speed is 875 rpm. The fan sheave has a pitch diameter of 16". Find motor sheave pitch diameter.
Solution:
Dm   = (RPMf x Df) ÷ RPMm
Dm   = (875 x 16) ÷ 1750
Dm   = 8
Fan, Air Horsepower
ahp = cfm x P ÷ 6356
Fan, Brake Horsepower

bhp = cfm x P ÷ 6356 x eff
bhp = cfm x FSP ÷ 6356 x FSE
bhp = cfm x FTP ÷ 6356 x FTE
Fan, Efficiency
eff = cfm x P ÷ 6356 x bhp
FSE = cfm x FSP ÷ 6356 x bhp
FTE = cfm x FTP ÷ 6356 x bhp
Terminology:
ahp   air horsepower
cfm   cubic feet per minute, airflow volume
P   fan pressure, in. wg.
6356   constant, 33,000 ft-lb/min divided by 5.19
bhp   brake horsepower
eff   fan efficiency, percent
FSP   fan static pressure, in. wg
FSE   fan static efficiency, percent
FTP   fan total pressure, in. wg
FTE   fan total efficiency, percent
Fan, Air Temperature Rise w/motor in (TR$_I$) or out (TR$_O$) of air stream
TR$_I$ = TSP x 0.371 ÷ (eff$_f$ x eff$_m$)
TR$_O$ = TSP x (0.371 ÷ eff$_f$)
Terminology:
TR   Temperature Rise through the fan, degrees Fahrenheit
TSP   Total Static Pressure rise through the fan
0.371   constant, 2545 ÷ (6356 x 1.08)
eff$_f$ fan efficiency, percent
eff$_m$ motor efficiency, percent
Fan, kiloWatt
kW = cfm x P ÷ 8520 x eff$_m$ x eff$_f$
Terminology:
kW   kiloWatts
cfm   cubic feet per minute, airflow volume
P   fan pressure, in. wg
8520   constant, 6356/.746
eff$_m$ motor efficiency, percent
eff$_f$ fan efficiency, percent
Fan Laws
Terminology:

cfm1	original volume of airflow in cubic feet per minute
cfm2	new volume of airflow in cubic feet per minute
rpm1	original fan speed in revolutions per minute
rpm2	new fan speed in revolutions per minute
Pd1	original pitch diameter of the motor sheave
Pd2	new pitch diameter of the motor sheave
SP1	original static pressure in inches of water column
SP2	new static pressure in inches of water column
bhp1	original brake horsepower
bhp2	new brake horsepower
d1	original density in pounds per cubic foot
d2	new density in pounds per cubic foot
Hz1	original fan motor frequency in Hertz
Hz2	new fan motor frequency in Hertz

Air volume varies in direct proportion to fan speed and pitch diameter of the motor sheave.

$$cfm2 = cfm1 \times (rpm2 \div rpm1) \qquad Pd2 = Pd1 \times (rpm2 \div rpm1)$$

Fan Law #2

Static pressure varies as the square of the fan speed, air volume and pitch diameter.

$$SP2 = SP1 \times (rpm2 \div rpm1)^2 \qquad SP2 = SP1 \times (cfm2 \div cfm1)^2$$
$$SP2 = SP1 \times (Pd2 \div Pd1)^2$$

Fan Law #3

Brake horsepower varies as the cube of the fan speed, air volume and pitch diameter.

$$bhp2 = bhp1 \times (rpm2 \div rpm1)^3 \qquad bhp2 = bhp1 \times (cfm2 \div cfm1)^3$$
$$bhp2 = bhp1 \times (Pd2 \div Pd1)^3$$

Brake horsepower varies as the square root of the static pressures cubed. $\qquad bhp2 = bhp1 \sqrt{(SP2 \div SP1)^3}$

Fan Laws and Density

Air volume remains constant with changes in air density. A fan is a constant volume machine and will handle the same airflow regardless of air density. It must be remembered, however, that many instruments are calibrated for standard air density (70 degrees at 29.92in. Hg) and any change in air density will require a correction factor for the instrument. Static pressure and brake horsepower vary in direct proportion to density.

$$SP2 = SP1 \times (d2 \div d1) \qquad bhp2 = bhp1 \times (d2 \div d1)$$

Fan Tip Speed

$$TS = \pi \times D \times RPM \div 12$$

Terminology:

TS	tip speed in feet per minute
D	fan wheel diameter in inches
RPM	revolutions per minute of the fan
π	3.14
12	constant, inches per foot

Heat Transfer, Air, Sensible, Latent, and Total Heat

$Btuh_s = cfm \times 1.08 \times \Delta T \qquad cfm = Btuh \div (1.08 \times \Delta T)$

$\Delta T = Btuh \div (1.08 \times cfm)$

$Btuh_L = cfm \times 4.5 \times \Delta H_L \qquad cfm = Btuh \div (4.5 \times \Delta H_L)$

$\Delta H_L = Btuh \div (4.5 \times cfm)$

$Btuh_T = cfm \times 4.5 \times \Delta h \qquad cfm = Btuh \div (4.5 \times \Delta h)$

$\Delta h = Btuh \div (4.5 \times cfm)$

Terminology:

$Btuh_s$ Btu per hour Sensible heat (heating coil load or dry cooling coil load or room load)

cfm   cubic feet per minute volume of airflow

1.08   constant, 60 min/hr x 0.075 lb/cu ft x 0.24 Btu/lb/F

ΔT dry-bulb degrees Fahrenheit (F) Temperature difference of air entering and leaving the coil. In applications where cfm to conditioned space needs to be calculated TD is the difference between the supply air temperature dry bulb and the room temperature dry bulb.

$Btuh_L$ Btu per hour Latent heat

4.5   constant, 60 minutes per hour x 0.075 pounds per cubic foot

ΔH_L change in Latent Heat content of air, Btu/lb

$Btuh_T$ Btu per hour Total heat

Δh change in total heat content of air, Btu/lb

Example: Find tons of cooling (total heat) required for a system producing 40,000 cfm. The air entering the cooling coil is 75F and 50% relative humidity (rh). The air leaving the cooling coil is 55F and 90% relative humidity.

How to do it: From a psychrometric chart find heat content (enthalpy, h) of the air at 75F and 50% rh. Next, find heat content (enthalpy, h) of the air at 55F and 90% rh. Then subtract the leaving air enthalpy from the entering air enthalpy. This is Δh. Multiply cfm times 4.5 x Δh and divide by 12000 Btuh per ton. (90 tons)

Heat Transfer, Water

$Btuh = gpm \times 500 \times TD \qquad gpm = Btuh \div (500 \times TD)$

$TD = Btuh \div (500 \times gpm)$

Terminology:	
Btuh	Btu per hour
gpm	gallons per minute, water volume
500	constant, 60 min/hour x 8.33 lbs/gallon x 1 Btu/lb/F
TD	temp. difference between the entering and leaving water
F	degrees Fahrenheit
Mixed Air Temperature	
MAT = (%OA x OAT) + (%RA x RAT)	
Terminology:	
MAT	Mixed Air Temperature
%OA	percent of Outside Air (Note: %OA and %RA must add to 100%)
OAT	Outside Air Temperature
%RA	percent of Return Air
RAT	Return Air Temperature
Example: OAT is 90F, RAT is 75F, with 80%RA and 20%OA. Find MAT.	
Solution:	
MAT = (%OA x OAT) + (%RA x RAT)	
MAT = (20% x 90) + (80% x 75)	
MAT = (0.20 x 90) + (0.80 x 75)	
MAT = (18) + (60)	
MAT = 78F	
Motor Efficiency and kiloWatt	
Terminology:	
Eff	motor efficiency
bhp	brake horsepower
746	constant, watts per horsepower
VA	volts times amps
PF	Power Factor
kW	kiloWatt (an electrical unit of power)
HPn	Horsepower, nameplate
0.746	constant, kW per horsepower
LF	Load Factor, as a percentage or decimal
eff	motor efficiency, as a percent or decimal
kVA	1000 VA
PF	Power Factor
V	Volts (for three-phase circuits, this is average volts)
A	Amps (for three-phase circuits, this is average amps)
1.73	constant, square root of 3, for three-phase circuits

Single-phase (1Ø) electric circuit
Eff = (bhp x 746 watts per brake horsepower) ÷ (VA x PF)
Three-phase (3Ø) electric circuit
Eff = (bhp x 746) ÷ (V x A x PF x 1.73)
Example 1: A three-phase motor operating at 86.4 bhp, 460 volts, 100 amps, 0.90 PF. Find efficiency.
Eff = (bhp x 746) ÷ (V x A x PF x 1.73)
Eff = (86.4 x 746) ÷ (460 x 100 x 0.90 x 1.73)
Eff = (64454.4 ÷ (71622)
Eff = 0.899 or 90%
Single-phase (1Ø) electric circuit
kW = kVA x PF   or   VA x PF ÷ 1000
kW = (bhp x 0.746) ÷ eff
kW = HPn x 0.746 x LF ÷ eff
Three-phase (3Ø) electric circuit
kW = kVA x (PF x 1.73) or (VA x PF x 1.73) ÷ 1000
kW = HPn x 0.746 x LF ÷ eff
kW = (bhp x 0.746) ÷ eff
Example 2: A three-phase 100 hp motor is 90% efficient and has an 85% LF. Find kW.
kW = (HPn x 0.746 x LF) ÷ eff
kW = (100 HPn x 0.746 kw/hp x 0.85) ÷ 0.90
kW = (63.41) ÷ 0.90
kW = 70.45
Example 3: A three-phase 86.4 bhp motor operates at 460 volts and 100 amps. The PF is 0.90. Find kW.
kW = kVA x PF
kW = (460 x 100 ÷ 1000) x 0.90 x 1.73
kW = 71.6
Percent of Outside Air
%OA = [(MAT – RAT) ÷ (RAT – OAT)] x 100
Terminology:
MAT   Mixed Air Temperature
%OA   percent of Outside Air
OAT   Outside Air Temperature
%RA   percent of Return Air (Note: %OA and %RA must add to 100%)
RAT   Return Air Temperature
Calculation: To find %OA first find difference between MAT and RAT. Next, find difference between RAT and OAT (Note: These two calculations are delta temperatures, there are no negative or positive

numbers). Divide the result of (MAT – RAT) by the result of (RAT – OAT) and multiply the result times 100. Example: OAT is 90F, RAT is 75F, MAT is 78F. Find %OA.
Solution:
%OA = [(MAT – RAT) ÷ (RAT – OAT)] x 100
%OA = [(78 – 75) ÷ (75 – 90)] x 100
%OA = [(3) ÷ (15)] x 100
%OA = [0.20] x 100%
%OA = 20%

Power Factor
PF = W ÷ (VA)
PF = kW ÷ (kVA)
PF = (bhp x 746) ÷ (V x A x eff) 1Ø
PF = (bhp x 746) ÷ (V x A x eff x 1.73) 3Ø
Terminology:

PF	Power Factor
W	Watts
VA	Volt-Amps
kVA	kilo Volt-Amps
V	Volts (for three-phase circuits, this is average volts)
A	Amps (for three-phase circuits, this is average amps)
bhp	brake horsepower
746	constant, watts per horsepower
eff	motor efficiency
1.73	constant, square root of 3, for three-phase circuits (3Ø)

Calculation:
To find PF divide watts by volt times amps or divide kilowatts by kilovolt-amps. To find PF for a single-phase circuit multiply bhp times 746. Next, multiply V times A times eff. Finally, divide the result of bhp times 746 by the result of V times A times eff. To find PF for a three-phase circuit multiply bhp times 746. Next, multiply V times A times eff times 1.73. Finally, divide the result of bhp times 746 by the result of V times A times eff times 1.73.

Pump Brake Horsepower and Efficiency
bhp = (gpm x hd) ÷ (3960 x eff)
eff = (gpm x hd) ÷ (3960 x bhp)
Terminology:

bhp	brake horsepower
gpm	gallons per minute, water flow rate
hd	pressure (head) against which the pump operates, feet of water

3960	constant, 33,000 ft-lb/min divided by 8.33 lb/gal
eff	pump efficiency, percent
Pump Head (developed by a centrifugal pump, approximate)	
$hd = (D \times rpm \div 1840)^2$	
Pump impeller diameter (approximate)	
$D = (1840 \sqrt{hd}) \div rpm$	
Terminology:	
hd	head in feet of water
D	diameter of the impeller in inches
rpm	revolutions per minute of the impeller
1840	constant
Pump, kiloWatt	
$kW = gpm \times hd \div 5308 \times eff_m \times eff_p$	
Terminology:	
kW	kilowatts
gpm	gallons per minute, water flow volume
hd	pressure, feet of water
5308	constant, 3960/0.746
eff_m	motor efficiency, percent
eff_p	pump efficiency, percent
Pump Laws	
Terminology:	
gpm1 = original volume of water flow in gallons per minute	
gpm2 = new volume of water flow in gallons per minute	
rpm1 = original pump speed in revolutions per minute	
rpm2 = new pump speed in revolutions per minute	
D1 = original diameter of the impeller in inches	
D2 = new diameter of the impeller in inches	
hd1 = original head in feet of water	
hd2 = new head in feet of water	
bhp1 = original brake horsepower	
bhp2 = new brake horsepower	
Hz1 = original pump motor frequency in Hertz	
Hz2 = new pump motor frequency in Hertz	
Pump Law #1	
Water volume varies in direct proportion to pump speed	
$(gpm2 \div gpm1) = (rpm2 \div rpm1)$	
Water volume varies in direct proportion to impeller diameter.	

Most HVAC pumps are directly connected (direct drive) so pump impeller diameter is substituted for rpm. (gpm2 ÷ gpm1) = (D2 ÷ D1)
Pump Law #2
Head varies as the square of the pump speed, gpm, or impeller diameter
(hd2 ÷ hd1) = (rpm2 ÷ rpm1)² (hd2 ÷ hd1) = (gpm2 ÷ gpm1)² (hd2 ÷ hd1) = (D2 ÷ D1)²
Pump Law #3
Brake horsepower, or amperage varies as the cube of the pump speed, gpm or impeller diameter
(bhp2 ÷ bhp1) = (rpm2 ÷ rpm1)³ (bhp2 ÷ bhp1) = (gpm2 ÷ gpm1)³ (bhp2 ÷ bhp1) = (D2 ÷ D1)³
Rectangular Duct and Round Duct
Terminology:
d   duct diameter, inches
a   length of one side of rectangular duct, inches
b   length of adjacent side of rectangular duct, inches
π 3.14
Round Duct Equivalent for Rectangular Duct)
$d = \sqrt{4ab} \div \pi$
Example: When a is 30" and b is 12"
$d = \sqrt{4\,(30 \times 12)} \div \pi$
d = 21.4
Rectangular Duct Equivalent for Round Duct
a = (d² x 3.14 ÷ 4) ÷ b
Example: When b is 12"
a = [(21.4² x 3.14) ÷ 4] ÷ 12
a = 30"
Standard Air
air temperature   70F
density   (ρ) or (d) 0.075 lb/cf
d = 1/v   1 ÷ v   (1 ÷ 13.34 = 0.075)
specific volume   (v)   13.34 cf/lb
v = 1/d   1 ÷ d   (1 ÷ 0.075 = 13.34)
Specific Heat   0.24 Btu/lb-F (British thermal units per pound per degree Fahrenheit)
Barometric Pressure   29.92" Hg (inches of mercury)
Barometric Pressure   14.7 psia
Terminology:

F	degrees Fahrenheit
d	air density in pounds per cubic foot
ρ(rho)	math symbol for density
lb/cf	pounds per cubic foot
v	specific volume in cubic feet per pound
c	math symbol for specific heat
cf/lb	cubic feet per pound
Btu/lb-F Fahrenheit	British thermal units per pound per degree
psig	pounds per square inch gauge
psia (psia = psig + 14.7)	pounds per square inch absolute

Volt-Amps and kiloVolt-Amps

VA = V x A	
VA = (bhp x 746) ÷ (PF x eff) Single-phase (1Ο) electric circuit	
VA = (bhp x 746) ÷ (PF x eff x 1.73) Three-phase (3Ο) electric circuit	
kVA = kW ÷ PF	

Terminology:

V	Volts	(for three-phase circuits, this is average volts)
A	Amps	(for three-phase circuits, this is average amps)
bhp	brake horsepower	
746	constant, watts per horsepower	
PF	Power Factor	
eff	motor efficiency	
VA	Volt-Amps (volts times amps)	
kVA	1000 VA	
1.73	constant, square root of 3, for three-phase circuits	

Calculation:

To find VA multiply volt times amps

To find kVA multiply volt times amps and divide by 1000

To find VA for a single-phase circuit multiply BHP times 746. Next, multiply PF times eff. Finally, divide the result of BHP times 746 by the result of PF times eff.

To find PF for a three-phase circuit multiply Bhp times 746. Next, multiply PF times eff times 1.73. Finally, divide the result of BHP times 746 by the result of PF times eff times 1.73.

Voltage Imbalance and Current Imbalance

%V = (ΔDmax ÷ Vavg) x 100

%C = (ΔDmax ÷ Cavg) x 100
Terminology:
%V   percent Voltage imbalance (should not exceed 2%)
ΔDmax   maximum Deviation from average voltage
Vavg   average Voltage
%C   percent Current imbalance (should not exceed 10%)
ΔDmax   maximum Deviation from average amps
Cavg   average Current (amps)
Water Flow Through Coil (using valve as a flow meter)
gpm = Cv x square root of ΔP
Terminology:
gpm   gallons per minute, water flow
Cv   valve Coefficient
ΔP   Pressure drop, psi
Calculation: To find gpm first determine the square root of ΔP. Then multiply gpm times the result of square root of ΔP
Example: A chilled water control valve has a flow coefficient (Cv) of 10. Find gpm if the pressure drop through the valve is 3 psi.
Solution:
gpm = Cv x   square root of ΔP
gpm = 10 x   square root of 3
gpm = 10 x   1.73
gpm = 17.3
Water Flow Through Coil (using coil pressure drop)
(gpm2 ÷ gpm1) = $\sqrt{\Delta P2 \div \Delta P1}$
Terminology:
gpm1   gallons per minute, rated water flow
gpm2   gallons per minute, measured or calculated water flow
ΔP1   rated Pressure drop, psi or feet of water
ΔP2   measured or calculated Pressure drop, psi or feet of water
(This equation can only be used when the rated pressure drop is a tested value by the manufacturer)
Water Flow, using Heat Transfer
Wet Cooling Coil)
gpm = cfm x 4.5 x Δh ÷ 500 x ΔTw
Heating Coil

gpm = cfm x 1.08 x ΔT_A ÷ 500 x ΔTw
Terminology:
gpm   gallons per minute, water volume
500   constant, 60 min/hour x 8.33 lbs/gallon x 1 Btu/lb/F
Δh   change in total heat content of the supply air, Btu/lb (from wet-bulb temperatures and psychrometric chart)
ΔT_A Temperature difference between the entering and leaving air
ΔTw   Temperature difference between the entering and leaving water
Water Horsepower
whp = (gpm x hd x SG) ÷ 3960
Terminology:
whp   water horsepower
gpm   gallons per minute, water flow
hd   pressure, feet of water
SG   Specific Gravity, pure water
Note: For temperatures between freezing and boiling, the specific gravity is taken as 1.0 and is therefore dropped from the equations for bhp and efficiency.

ENERGY CALCULATIONS
After Tax Savings
ATS = (1-T)S + (TD)
Equation: Find ATS, given T, S, and D
Calculation: To find ATS, subtract T from 1 and multiply times S. Then multiply T times D. Finally add the two results.
Terminology:
ATS   After Tax Savings, annual
T   Tax Rate, as a decimal
D   Depreciation, annual
S   Savings, annual
Example:
A project has an annual savings of $30,000. The annual depreciation is $10,000. The tax rate is 25%. What is the after tax savings?
Solution:
ATS = (1-T)S + (TD)
ATS = (1 - 0.25)$30,000 + (0.25 x $10,000)
ATS = $22,500 + $2500
ATS = $25,000

Annual Value or Cash Flow
A = P (A/P, I, N)
Equation: Find A, given P, and (A/P, I, N)
(A/P, I, N) is a factor from interest tables and is based on interest rate (I) and number of years (N)
Calculation: To find A, multiply P times (A/P, I, N)
Terminology:
P   Present value or cash flow either positive or negative, savings or cost
A   Annual value or cash flow either positive or negative, savings or cost
I   Interest rate*
N   Number of years
*Interest rate is a percentage and is expressed in various ways such as:
Discount Rate
Interest Rate
Internal Rate of Return (IRR)
Minimum Attractive Rate of Return (MARR)
Rate of return (ROR)
Return on Investment (ROI)
Example: A project requires a $20,000 investment. The project life is 5 years. The rate of return is 12%. What must be the annual savings to make this project viable?
How to do it:
Go to 12% Interest Table See Financial Table 1 at end of this chapter
Come down the "n" column to 5
Go right to (A/P, I, N) column to find 0.2774
Multiply $20,000 times 0.2774
Solution:
A = P (A/P, I, N)
A = $20,000 (0.2774)
A = $5548 per year

Boiler Systems, Steam and Water
Boiler Efficiency
% eff = (Heat out ÷ Heat in) x 100
% eff from flue gas composition and temperature chart
Steam Boiler Efficiency
% eff = [(Evaporation ratio x Heat content of steam) ÷ CF] x 100
Water Boiler Efficiency
%eff = [(Water flow rate from boiler x Heat output of water) ÷ CF x Fuel rate] x 100
Savings in Fuel = [(EFFn – EFFo) ÷ EFFn] x FC
% Blowdown = A ÷ (B - A)
Equations:
Boiler
Find %eff given heat in and out
Find %eff given net stack temperature and oxygen or carbon dioxide
Steam Boiler
Find %eff given evaporation ratio, heat content, and CF
Water Boiler
Find %eff given Water flow rate, Heat output, CF and fuel rate
Savings
Find %eff given new efficiency, old efficiency, and fuel consumption
%Blowdown
Find % Blowdown given A and B
Terminology:
eff   efficiency
EFFn   New Efficiency
EFFo   Old Efficiency
FC   Fuel Consumption
NST   Net Stack Temperature (Stack Temp – Room Temp)
CF   Caloric value of fuel
A   (100 - % condensate return) x (PPM (TDS) makeup water)
B   PPM (TDS) allowed in boiler
Example:
Use flue gas composition and temperature chart at the end of this chapter to determine efficiency of a natural gas boiler that has 8% carbon dioxide and a net stack temperature of 550F
How to do it:
At the bottom of the flue gas chart find 8% carbon dioxide (CO_2). Now, go up to intersect with 550F. Finally, go left to find boiler efficiency for natural gas at 77%.

Capacitor Size
Capacitor size = kW Load x Multiplier
Equation: Find capacitor (capacitance required) given kW and a Multiplier (from Power Factor Correction Table)
Example:
Find capacitor size (leading reactive kvars or kilovolt-amps reactive) to increase power factor from 60% to 90% for a 100 kW load.
How to do it:
Use the Power Factor Correction Table at the end of this chapter: kW Multipliers for determining capacitor kvars. Go down left side to find 60% (Original Power Factor in Percentage). Next, go right across the top to find 90% (Desired Power Factor in Percentage). Now, go right from 60 and down from 90. The intersection is the multiplier (0.850). Finally, multiply load times Multiplier.
Solution:
Capacitor size = kW Load x Multiplier
Capacitor size = 100 x 0.850
Capacitor size = 85 kvars

Cogeneration
Cost of NG per year to operate
$/yr = (Mcf/yr x Btu/cf x $/therm) ÷ Btu/therm
Electrical energy generated $/yr = kWh/yr x $/kWh
Natural Gas energy savings per year
$/yr = (Btu/yr x $/Therm) ÷ Btu/Therm
Terminology:
$/yr	dollar cost or savings per year
$/kWh	dollar cost or savings per kwh
$/Therm	dollar cost or savings per Therm
Btu	British thermal unit
Btu/cf	Btu per cubic foot
kW	kilowatt
kWh	kilowatt-hour
kW- mo	kilowatt per month
kWh/yr	kilowatt-hour per year
Mcf/yr	1000 cubic feet per year
Therm	100,000 Btu
Example:

A 550 kW natural gas cogen system generates 4,000,000 kWh/yr of electricity. It uses 45,000 Mcf/yr of natural gas. The system also generates 5,700 million Btu of heat. The heating value of the natural gas is 1000 Btu/cf. Electricity is $0.13/kWh and $8/kW-mo. Natural gas is $0.75/Therm. Find costs and savings.
Natural gas cost to operate cogeneration system
$/yr = (Mcf/yr x heating value x $/Therm) ÷ Btu/Therm
$/yr = (45,000 Mcf/yr x 1000 Btu/cf x $0.75/Therm) ÷ 100,000 Btu/Therm
$/yr = (45,000,000 x 1000 x 0.75) ÷ 100,000
$/yr = 337,500
Cost saving from electrical energy and power, and heat generated:
Electrical energy $/yr = kWh/yr x $/kWh
Electrical energy $/yr = 4,000,000 kWh/yr x $0.13/kWh
Electrical energy $/yr = 520,000
Electrical power $/yr = kW x $/kW-mo x 12mo/yr
Electrical power $/yr = 550kW x 8/kW-mo x 12mo/yr
Electrical power $/yr = 52,800
Natural gas $/yr = (Btu/yr x $/Therm) ÷ Btu/Therm
Natural gas $/yr = (5,700,000,000 Btu/yr x $0.75/Therm) ÷ 100,000 Btu/Therm
Natural gas $/yr = (5,700,000,000 x 0.75) ÷ 100,000
Natural gas $/yr = (4275000000) ÷ 100,000
Natural gas $/yr = 42,750
Total electrical and heat generated $/yr = 615,550 ($520,000 + $52,800 + $42,750)
$/yr = energy generated - cost
$/yr = $615,550 - $337,500
$/yr = $278,000

Energy Content, Steam
Example 1:
Find energy required to raise 1 pound of water from 180F to steam at 30 psia, saturated vapor.
How to do it:
Subtract temperature of water from 212F. The difference is in degrees and Btu/lb. A Btu is the amount of heat to raise 1 lb of water 1 degree Fahrenheit. Next, go to steam table and find 30 psia. Now, find *h* (total heat of steam in Btu/lb) of saturated steam. Finally, add the two Btu/lb together

Solution:
212F - 180F = 32F
32F x 1 Btu/lb-F = 32 Btu/lb
From table *h* at 30 psia: 1164 Btu/lb
32 Btu/lb + 1164 Btu/lb
Solution: 1196 Btu/lb
Example 2:
Find heat (energy) content (Btu/lb) of steam at 30 psia and 800F.
Terminology:
Btu/lb   British thermal unit per pound
psia   pounds per square inch absolute
F   degrees Fahrenheit
h   enthalpy (heat content)
How to do it:
Go to superheated steam table (212F steam is saturated steam. Any temperature over 212F will be superheated steam). Find 30 psia. Next find enthalpy (*h*) at 800F
Solution: 1432.7 Btu/lb

Energy Cost – Electrical, Fan or Pump **Energy savings with change in brake horsepower from reduction in size of fan drive or pump impeller.**		
\$/yr = (bhp x 0.746 kW/bhp ÷ eff$_m$) x hr/yr x \$/kWh		
\$/yr = (bhp1 - bhp2) x (0.746 kW/bhp ÷ eff$_m$) x hr/yr x \$/kWh		
Terminology:		
\$/yr	dollar cost or savings per year	
bhp	brake horsepower	
0.746	constant	
kW/bhp	kilowatt per brake horsepower	
eff$_m$	motor efficiency	
kWh	kilowatt-hour	
hr/yr	hours per year	
\$/kWh	dollar cost or savings per kWh	
Example 1: A fan operates 10 hours a day, 6 days a week, 52 weeks a year. Calculate cost to operate. The fan motor works at 45 bhp at 87% efficiency. The cost of electricity is \$0.12 per kWh.		
Solution:		
\$/yr = (bhp x 0.746 kW/bhp ÷ eff$_m$) x hr/yr x \$/kWh		
\$/yr = (45 x 0.746 kW/bhp ÷ 0.87 x 10 x 6 x 52 hr/yr x 0.12/kWh		
\$/yr =\$14,447		
Example 2: The fan in example 1 has a reduction in airflow and the bhp is reduced to 33. Calculate energy savings. Efficiency of the motor at both		

bhp is 87%.
Solution:
$/yr = (bhp1 – bhp2) x (0.746 kW/bhp ÷ eff$_m$) x hr/yr x $/kWh
$/yr = (45 - 33) x 0.746 kW/bhp ÷ 0.87 x 10 x 6 x 52 hr/yr x 0.12/kWh
$/yr = $3852

Energy Costs and Savings – Electrical or Fossil Fuel, Boiler
Example 1:
A boiler can be fueled by natural gas or fuel oil. Which is the least expensive fuel to burn if the boiler is 82% efficient with natural gas and 77% efficient with fuel oil?
Fuel units: fuel oil is 140,000 Btu/gal and natural gas is 100,000 Btu/Therm
Fuel cost: fuel oil is $0.90 per gallon and natural gas is $0.63/therm.
How to do it:
Convert all fuels to Therm units.
First, determine how many gallons of fuel oil there are in a them by dividing a therm by number of Btu per gallon of fuel oil:
100,000 Btu/ Therm ÷ 140,000 Btu/gal = 0.714 gal/Therm
Next, determine the cost of oil in $/Therm:
0.714 gal/Therm x $0.90 gal = $0.64/Therm
Then determine the cost of oil based on efficiency:
$0.64/Therm ÷ 0.77 = $0.83/Therm
Now do the same for natural gas.
Determine the cost of natural gas in $/Therm:
$0.63/Therm
Determine the cost of gas based on efficiency:
$0.63/Therm ÷ 0.82 = $0.77/Therm
Solution:
Natural gas is the least expensive fuel to burn, $0.77/Therm versus $0.83/Therm
Example 2:
The natural gas for a boiler costs $7.50 per million Btu. The boiler water is heated from 80F to 180F. If the boiler is 82% efficient how much does it cost per hour to heat a gpm of water.
Equations to use:
Btuh = gpm x 500 x TD
Fuel cost = $/10^6 Btu ÷ efficiency
Fuel cost/hr = (Btuh ÷ million Btu) x fuel cost
Equations:
Find Btuh given gpm, the constant for water heat transfer, and TD.
Find fuel cost given cost per million Btu and boiler efficiency.

Find fuel cost per hour given Btuh, and cost per million Btu based on boiler efficiency.
Calculation:
To find Btuh multiply gpm times constant times TD.
To find fuel cost divide cost per million Btu by boiler efficiency.
To find fuel cost per hour divide Btuh by a million Btu, and then multiply the result by fuel cost per based on boiler efficiency.
Terminology:
Btuh   Btu per hour, water
gpm   gallons per minute, water volume
500   constant, 60 min/hour x 8.33 lbs/gallon x 1 Btu/lb/F
TD   temperature difference between the entering and leaving water
F   degrees Fahrenheit
Mcf   1000 cubic feet
10^6 Btu is a million Btu
How to do it:
Find Btuh
Find fuel cost based on efficiency
Find fuel cost per hour
Solution:
Btuh = gpm x 500 x TD
Btuh = 1 gpm x 500 x 100 (180F - 80F)
Btuh = 50,000
Fuel cost = \$/10^6 Btu ÷ efficiency
Fuel cost = \$7.50/10^6 Btu ÷ 0.82
Fuel cost = \$9.15/10^6 Btu
Fuel cost/hr = Btu/hr ÷ million Btu x fuel cost
Fuel cost/hr = (50,000 Btu/hr ÷ 1 million Btu) x \$9.15
Fuel cost/hr = (0.05 hr) x \$9.15
Fuel cost/hr = \$0.4575/hr
Example 3:
Find lowest operating cost between a fuel oil boiler and an electrical boiler. The fuel oil boiler operates at 75% efficiency. The electrical boiler operates at 3.5 COP. Fuel oil costs \$1.35 per gal and is rated at 138,000Btu/gal (calorific value or heat content). Electricity is \$0.06/kWh.
How to do it:
Convert all fuels to Therms.
First, determine how many gallons of fuel oil there are in a Therm by dividing a Therm by number of Btu per gallon of fuel oil:
100,000 Btu/Therm ÷ 138,000 Btu/gal = 0.725gal/Therm
Next, determine the cost of oil in \$/Therm:
0.725 gal/Therm x \$1.35 gal = \$0.98/Therm
Then determine the cost of oil based on efficiency:

$0.98/Therm ÷ 0.75 = $1.31/Therm
Now, determine how many kWh there are in a Therm by dividing a therm by number of Btu per kWh:
100,000 Btu/Therm ÷ 3413 Btu/kWh = 29.3 kWh/Therm
Determine the cost of electricity by :
$0.06/kWh x 29.3 kWh/Therm = $1.76/Therm
Solution:
The fuel oil boiler has the lowest operating cost. $1.31/Therm vs. $1.76/Therm
Example 4:
Determine efficiency of a natural gas boiler that has 8% oxygen and a stack temperature of 600F. If oxygen is reduced to 3% find difference in efficiency. Natural gas is $0.93 a therm. Calculate energy savings dollars per therm.
How to do it:
At the lower left side of the flue gas composition and temperature chart at the end of this chapter find 8% oxygen (O_2). Now, go right horizontally to intersect the curved line and then go vertically to intersect with 600F. Next, go left to find boiler efficiency for natural gas. Repeat for 3% oxygen to find new boiler efficiency. Savings = $/Therm ÷ efficiency at 8% O_2 minus $/Therm ÷ efficiency at 3% O_2

Heat Transfer, Materials and Heating/Cooling Degree Days
Btuh = U x A x TD
U = Btuh ÷ (A x TD)
Btu/yr = U x A x DD x 24
R = 1/U or 1/c or in/k
U = 1/R or 1/Rt
in = R x k
c = Btu/hr-sf-F
k = Btu-in/hr-sf-F
HDD = 65F - average daily temperature
CDD = average daily temperature - 65F
Equations:
Find Btuh given U, A, and TD
Find U given Btuh, A, and TD
Find Btu/yr given U, A, and DD
Find inches (in) given R and k
Find HDD given daily high and low outside air temperature
Find CDD given daily high and low outside air temperature
Terminology:

Btuh	Btu per hour
Btu	British thermal unit
U	coefficient of heat flow, Btu/hr-sf-F
A	area in square feet
TD	temperature difference between the entering and leaving air
R	resistance to heat flow, Btu/hr-sf-F
c	thermal conductance
k	thermal conductivity
DD	degree day, either HDD or CDD
HDD	heating degree day
CDD	cooling degree day
F	degrees Fahrenheit
yr	year
in	inches
hr	hour
sf	square feet
24	hours per day
Example 1:	
Find R-value for 2 inches of insulation applied to a boiler. The k-value of the insulation is 0.25 Btu-in/hr-sf-F.	
Solution:	
R = in/k	
R = inches ÷ Btu-in/hr-sf-F	
R = 2 ÷ 0.25	
R = 8 Btu/hr-sf-F	
Example 2:	
A wall is 10' x 20'. It has a 2.5 R-value. 4 inches of glass fiber insulation (k = 0.25) is added. The inside temperature is 72F. The outside temperature is 30F. Find Btu/hr heat flow through the wall.	
How to do it:	
Find area of the wall. Next, find TD. Now, find R-value of the glass fiber. Next, add the R-value of the wall to the R-value of the insulation (this is total R, or Rt). Now, find U-value. Finally, find Btuh by multiplying U times A times TD.	
Solution:	
A = feet x feet	
A = 10 x 20	
A = 200 sf	
TD = inside temperature - outside temperature	
TD = 72 - 30	
TD = 42F	
R = in/k	
R = inches ÷ Btu-in/hr-sf-F	

R = 4 ÷ 0.25
R = 16 Btu/hr-sf-F
Rt = R1 + R2
Rt = 2.5 (wall) + 16 (insulation)
Rt = 18.5 Btu/hr-sf-F
U = 1/Rt
U = 1 ÷ 18.5
U = 0.054 Btu/hr-sf-F
Btuh = U x A x TD
Btuh = 0.054 Btu/hr-sf-F x 200 sf x 42F
Btuh = 453.6
Example 3:
Cellular glass insulation has a thermal conductivity of 0.39 Btu-in/hr-sf-F. Find thickness of cellular glass insulation required to keep a tank's outside temperature at 120F when the surface temperature is 350F. The tank surface is 314 sf. The heat flow is 9389 Btu/hr.
How to do it:
Find TD. Next, find U-value of the insulation by multiplying A times TD and dividing the result into Btuh. Now, find R-value of the glass insulation. Finally, find number of inches of insulation.
Solution:
TD = surface temperature - outside temperature
TD = 350F - 120F
TD = 230F
U = Btuh ÷ (A x TD)
U = 9389 Btu/hr ÷ (314 sf x 230F)
U = 9389 Btu/hr ÷ (72220 sf-F)
U = 0.13 Btu/hr-sf-F
R = 1/U
R = 1 ÷ 0.13 Btu/hr-sf-F
R = 7.69 Btu/hr-sf-F
in = R x k
in = 7.69 Btu/hr-sf-F x 0.39 Btu-in/hr-sf-F
in = 3
Example 4:
Find Btu per year if 3" of insulation (k = 0.25) is added to a wall (1000 sf) with an R-value of 2. The inside temperature is 75F and the outside temperature is 40F. There are 5,000 heating degree-days in this year.
How to do it:
Find R-value of the insulation. Next, add the R-value of the wall to the R-value of the insulation (this is total R, or Rt). Now, find U-value. Finally, find Btuh by multiplying U times A times DD x 24.
Solution:

R = in/k
R = inches ÷ Btu-in/hr-sf-F
R = 3 ÷ 0.25
R = 12 Btu/hr-sf-F
Rt = R1 + R2
Rt = 2 + 12
Rt = 14 Btu/hr-sf-F
U = 1/Rt
U = 1 ÷ 14
U = 0.0714 Btu/hr-sf-F
Btu/yr = U x A x DD x 24
Btu/yr = 0.0714 Btu/hr-sf-F x 1000 sf x 5000F-day x 24hr/day
Btu/yr = 8,568,000
Example 5:
Find c, k, R for each component of a wall and then find total R-value of the wall constructed as follows: inside air, ½" gypsum board, 3" fiber glass insulation, ¾" air space (0F), ¾" plywood, 3" face brick, outside air (winter conditions)
Example 5:
Find c, k, R for each component of a wall and then find total R-value of the wall constructed as follows: inside air, ½" gypsum board, 3" fiber glass insulation, ¾" air space (0F), ¾" plywood, 3" face brick, outside air (winter conditions)
How to do it:
Go to tables for building materials and tables for thermal resistance for surface film and air space.

	c	k	R
inside air		0.68	
½" gypsum board	2.25	0.44	
3" fiber glass insulation		0.25	12.00
¾" air space (0°F)		1.28	
¾" plywood	1.07		0.93
3" face brick		9.0	0.33
outside air (winter conditions)			0.17
		TOTAL	15.83

R = 1/c
R = 1 ÷ 2.25 Btu/hr-sf-F
R = 0.44 Btu/hr-sf-F
R = in/k
R = 3 in ÷ 0.25 Btu-in/hr-sf-F
R = 12.00 Btu-in/hr-sf-F
R = 1/c

| R = 1 ÷ 1.07 Btu/hr-sf-F |
| R = 0.93 Btu/hr-sf-F |
| R = in/k |
| R = 3 in ÷ 9.0 Btu-in/hr-sf-F |
| R = 0.33 Btu-in/hr-sf-F |
| Rt = R1 + R2 + R3 + R4 + R5 + R6 + R7 |
| Rt = 0.68 + 0.44 + 12 + 1.28 + 0.93 + 0.33 + 0.17 |
| Rt = 15.83 |
| Example 6: |
| Find HDD if the high temperature for the day is 45F and the low is 30F. |
| How to do it: |
| Add together the high temperature and the low temperature for the day and divide by 2 to get average daily temperature. Subtract the average daily temperature from 65. |
| Solution: |
| HDD = 65F - [average daily temperature] |
| HDD = 65F - [(45 + 30) ÷ 2] |
| HDD = 65F - [(75) ÷ 2] |
| HDD = 65F - 37.5 |
| HDD = 27.5F |
| If 27.5 is the HDD for each day in January then the HDDs for January would be 852.5 |

| **Interest Rate** |
| I = P ÷ A |
| Equation: Find I, given P and A |
| (P/A, I, N) is a factor from interest tables (see Interest Table at end of chapter) and is based on interest rate (I) and number of years (N) |
| Calculation: |
| To find I, divide P by A |
| Terminology: |
| P   Present value or cash flow either positive or negative, savings or cost |
| A   Annual value or cash flow either positive or negative, savings or cost |
| I   Interest* |
| N   Number of years |
| *Interest is a percentage and is expressed in various ways such as: |
| Discount Rate |
| Interest Rate |
| Internal Rate of Return (IRR) |
| Minimum Attractive Rate of Return (MARR) |

Rate of return (ROR)
Return on Investment (ROI)
Example: A project saves $15,580 per year. The project life is 10 years. The cost of the project is $100,000. Find the internal rate of return (IRR).
Solution:
P divided by A gives you (P/A, I, N)
P/A = (P/A, I, N)
$100,000/$15,550 = (P/A, I, N)
(P/A, I, N) = 6.4185
How to do it:
Look at each interest table. Come down the "n" column in <u>each interest table</u> to 10 years.
Go right to (P/A, I, N) column to find factor. You are looking for 6.4185 (or as close as possible). When you find it look at the top of the page to find the interest rate. See Financial Table 2 at end of this chapter.
In this example the interest rate is 9% because at 10 years the (P/A, I, N) factor is 6.4177. For 8% interest at 10 years the factor is 6.7101 and at 10% interest the factor is 6.1446.

Lighting
N = (F x A) ÷ (Lu x LLF x Cu)
N = (F x A) ÷ (Lu x L1 x L2 x Cu)
RCR = 5 x h x (1 + w) ÷ (l x w)
$E = e \, (d^2 \div D^2)$
Equations:
Find N given F, A, Lu, LLF, Cu
Find N given F, A, Lu, L1, L2, Cu
Find RCR given h, l, and w
Find E, given e, d and D
Calculation:
To find N first find A by multiplying room length by width. Next, multiply F times A. Then, multiply Lu times LLF times Cu. Finally, divide result of F times A by the result of Lu times LLF times Cu.
To find N first find A by multiplying room length by width. Next, multiply F times A. Then, multiply Lu times L1 times L2 times Cu. Finally, divide result of F times A by the result of Lu times L1 times L2 times Cu.
To find RCR first add room length (l) to width (w). Now, multiply 5 times height (h) times sum of (1 + w). Next, multiply room length (l) by width (w) to get room area. Finally, divide results of 5 times h times (1 + w) by room area.
To find E first find the square of d (d x d). Next, square D (D x D). Now,

| divide d squared by D squared and multiply the result by e. |
| Terminology: |
| N   number of lamps required |
| F   required footcandle level |
| FC   footcandle |
| A   area of the room in square feet |
| Lu   lumen output per lamp |
| LLF   light loss factor (L1 x L2) |
| L1   lamp depreciation factor |
| L2   dirt depreciation factor |
| Cu   coefficient of utilization |
| sf   square feet |
| RCR   room cavity ratio |
| h   height from top of working space to lamp |
| l   length of room |
| w   width of room |
| Rc   ceiling reflectance |
| Rw   wall reflectance |
| E   final illumination in footcandles |
| e   initial illumination in footcandles |
| d   initial perpendicular distance from light source to work surface |
| D   final perpendicular distance from light source to work surface |
| Example 1: A room is 50' x 60'. The light loss factor is 0.68 and the coefficient of utilization is 0.75. The room needs to have an even light at 50 footcandles. Find number of fixtures required if each fixture has two lamps with a rating of 2500 lumens per lamp. |
| Solution: |
| A = 50' x 60' |
| A = 3000 sf |
| N = (F x A) ÷ (Lu x L x Cu) |
| N = (50 x 3000) ÷ (2500 x 0.68 x 0.75) |
| N = (150,0300) ÷ (1275) |
| N = 118 |
| Number of fixtures = N ÷ 2 |
| Number of fixtures = 59 |
| Example 2: A room is 50' x 60'. The lamps are in the ceiling and the floor to ceiling height is 9'. The work space is 36" from the floor. Find RCR. |
| How to do it: |
| Find h by converting 36" to feet and subtracting from floor to ceiling height. |
| Add (50' + 60') |
| Multiply (50' x 60') |
| Multiply (5 x 6 x 110) and divide by (3000) |

Solution:
RCR = 5 x h x (1 + w) ÷ (l x w)
RCR = 5 x (9' - 3') x (50' + 60') ÷ (50' x 60')
RCR = 5 x (6) x (110) ÷ (3000)
RCR = 3300 ÷ 3000
RCR = 1.1
Example 3: Lights are 9 feet above the workspace. The light level is 50 FC. The lights are lowered to 6' above the workspace. Find new light level.
Solution:
$E = e\,(d^2 \div D^2)$
$E = 50\,(9^2 \div 6^2)$
$E = 50\,(81 \div 36)$
$E = 50\,(2.25)$
E = 112.5 footcandles
Example 4: Find coefficient of utilization (Cu) for lamps a room where the room cavity ratio (RCR) is 1.0. The ceiling reflectance (Rc) is 70% and the wall reflectance (Rw) is 50%.
How to do it:
Go to a reflectance chart for the type and maker of lamps in the room. Find RCR on left side of chart. Next, go across top to find Rc at 70. Below Rc 70 find Rw at 50. In Rc 50 column go down while coming across from RCR 1.0. The intersection is Cu for that particular lamp.

Load Factor
LF = kWh ÷ (kW x hr)
Equation: Find LF, given kWh, kW and hours
Calculation: To find LF first multiply kW times number of hours
The number of hours may be known or may have to be calculated for some finite period of time, such as a month, some months, year round or some portion of the year. Finally, divide the kWh by the result of (kW x hours)
Terminology:
LF   Load Factor (as a percentage or decimal)
kWh   kilowatt-hours (an electrical unit of energy)
kW   kilowatt (an electrical unit of power)
hr   hour(s)
Example:
In January, a business used 100,000 kWh with a demand of 150 kW in January. January has 31 days and the system in question operated 24 hours a day. Find load factor.
Solution:

| LF = kWh ÷ (kW x hr) |
| LF = 100,000 kWh ÷ (150 kW x 744 hr in January) |
| LF = 100,000 kWh ÷ (111,600 kWhr) |
| LF = 0.90 or 90% |

| **Motor Costs** |
| Example: |
| A standard TEFC (totally enclosed fan cooled) 10 HPn motor needs to be replaced. A new 10 hp standard motor is rated at 1.15 service factor, 0.82 PF and 83% eff. The cost is $500. A 10 hp EE (energy efficient) TEFC motor is rated at 1.15 service factor, 0.90 PF and 96% eff. The cost is $580. The cost of electricity is $0.12/kWh. Demand is $9.50/kW. The motor operates 4050 hrs/yr. To determine project viability assume motors run at 100% load. Find yearly cost for standard motor and energy efficient motor. Find simple payback for purchasing the energy efficient motor. |
| Solution, Std motor: |
| kW = (HPn x 0.746 x LF) ÷ eff |
| kW = (10 x 0.746 x 1.00) ÷ 0.83 |
| kW = (7.46) ÷ 0.83 |
| kW = 8.99 |
| Cost for energy: |
| $/yr = hr/yr x $/kWh x kW |
| $/yr = 4050 x 0.12 x 8.99 |
| $/yr = 4369 |
| Cost for power: |
| $/yr = $/kW x kW/mo x 12 mo/yr |
| $/yr = 9.50 x 8.99 x 12 |
| $/yr =1025 |
| Total cost to operate motor = cost for energy + cost for power |
| Total cost to operate motor = 4369 + 1025 |
| Total cost to operate motor = $5394 |
| Solution, EE motor: |
| kW = (HPn x 0.746 x LF) ÷ eff |
| kW = (10 x 0.746 x LF) ÷ 0.96 |
| kW = (7.46) ÷ 0.96 |
| kW = 7.77 |
| Cost for energy: |
| $/yr = hr/yr x $/kWh x kW |
| $/yr = 4050 x 0.12 x 7.77 |
| $/yr = 3776 |
| Cost for power: |
| $/yr = $/kW x kW/mo x 12 mo/yr |

$/yr = 9.50 x 7.77 x 12
$/yr = 886
Total cost to operate motor = cost for energy + cost for power
Total cost to operate motor = 3776 + 886
Total cost to operate motor = $4662
Difference in cost of motors $80
SPB = Investment ÷ Net Annual Savings
SPB =   ($580 - $500) ÷ ($5394 − $4662)
SPB = $80 ÷ $732
SPB = 0.11 years (less than 2 months)

Present Value or Cash Flow
P = A (P/A, I, N)
Equation: Find P, given A, and (P/A, I, N)
(P/A, I, N) is a factor from interest tables and is based on interest rate (I) and number of years (N)
Calculation: To find P, multiply A times (P/A, I, N)
Terminology:
P   Present value or cash flow either positive or negative, savings or cost
A   Annual value or cash flow either positive or negative, savings or cost
I   Interest*
N   Number of years
*Interest is a percentage and is expressed in various ways such as:
Discount Rate
Interest Rate
Internal Rate of Return (IRR)
Minimum Attractive Rate of Return (MARR)
Rate of return (ROR)
Return on Investment (ROI)
Example:
A project saves $5,000 per year. The project life is 6 years. The rate of return is 15%. How much can the project cost to be viable?
How to do it:
Go to 15% Interest Financial Table 3 at end of this chapter
Come down the "n" column to 6
Go right to (P/A, I, N) column to find 3.7845
Multiply $5000 times 3.7845
Solution:
P = A (P/A,I,N)
P = $5,000 (3.7845)

P = $18,922

Present Worth
PW = [A (P/A, I, N)] – P
Equation: Find PW, given P, A, and (P/A, I, N)
(P/A, I, N) is a factor from interest tables and is based on interest rate (I) and number of years (N)
Calculation: To find PW, first multiply A times (P/A, I, N) and then subtract P
Terminology:
P   Present value or cash flow either positive or negative, savings or cost
A   Annual value or cash flow either positive or negative, savings or cost
I   Interest*
N   Number of years
PW   Present Worth
*Interest is a percentage and is expressed in various ways such as:
Discount Rate
Interest Rate
Internal Rate of Return (IRR)
Minimum Attractive Rate of Return (MARR)
Rate of return (ROR)
Return on Investment (ROI)
Example:
Project 1 saves $30,000 per year. The project life is 5 years. The cost of the project is $100,000. The rate of return is 15%. Find present worth (PW).
How To Do It:
Go to 15% Interest Financial Table 3 at end of this chapter
Come down the "n" column to 5
Go right to (P/A, I, N) column to find 3.3522
Multiply $30,000 times 3.3522 and then subtract $100,000
Solution:
PW = A (P/A, I, N) – P
PW = $30,000 (3.3522) – $100,000
PW = $566

Simple Payback
Simple Payback = Investment ÷ Net Annual Savings
Equation: Find Simple Payback given Investment and Net Annual Savings

Calculation: To find Simple Payback divide Investment by Net Annual Savings
Terminology:
Simple Payback   The number of years before the savings is paid back.
Simple Payback does not consider the time value of money.
Investment   Initial cost of the project
Net Annual Savings   Gross savings less maintenance costs and other costs
Example:
A project requires a \$200,000 investment. Savings per year are \$22,000. The maintenance cost is \$2,000 per year. What is the Simple Payback in years?
Solution:
SPB = Investment ÷ Net Annual Savings
SPB =   \$200,000 ÷ (\$22,000 - \$2,000)
SPB =   \$200,000 ÷ \$20,000
SPB =   10 years

Utility Bill
Example:
A utility bill states that a business has an electrical PF of 83%. The company consumed 250,000 kWh in January. The demand was 400 kW. Find the January bill if the
The customer charge is \$50/mo.
The demand charge is \$9.50/kW
The energy charge is:
\$0.10 for the 1st 10,000 kWh
\$0.12 for the 10,001 kWh to 200,000 kWh
\$0.15 for over 200,000 kWh
The power factor charge is:
The demand charge is increased 1% for each 1% PF is less than 85%
Taxes: 10%
Terminology:
PF   Power Factor (as a percentage or decimal)
kWh   kilowatt-hours (an electrical unit of energy)
kW   kilowatt (an electrical unit of power)
How to do it:
Calculate all the various charges and subtotal. Note: The power factor charge is the demand charge increased by 2% because the power factor is

2% below the limit of 85%. Finally, multiply subtotal by the tax rate to get taxes and total.		
Solution:		
Customer charge:	$50	
Demand charge:	$3800	($9.50/kW x 400 kW)
Energy charge:	$1000	($.10 for the 1st 10,000 kWh)
	$22800	($.12 for the 10,001 kWh to 200,000 kWh)
	$7500	($.15 for over 200,000 kWh)
Power factor charge:	$76	($9.50/kW x 400 kW x 0.02)
Subtotal	$35226	
Taxes	$3523	($35226 x 0.10)
January bill	$38,749	

Financial Table 1
Time Value of Money
(annual value or cash flow)

12% int.	Find A given P
Years (n)	A/P, I, N
1	1.1200
2	0.5917
3	0.4163
4	0.3292
5	0.2774

Financial Table 2
Time Value of Money
(interest rate)

9% int.	Find P given A
Years (n)	P/A, I, N
1	0.9174
2	1.7591
3	2.5313
4	3.2397
5	3.8897
6	4.4859
7	5.0330
8	5.5348
9	5.9952
10	6.4177

Financial Table 3
Time Value of Money
(present value or cash flow)

15% int.	Find P given A
Years (n)	P/A, I, N
1	0.8696
2	1.6257
3	2.2832
4	2.8551
5	3.3522
6	3.7845
7	4.1604
8	4.4873
9	4.7716
10	5.0188

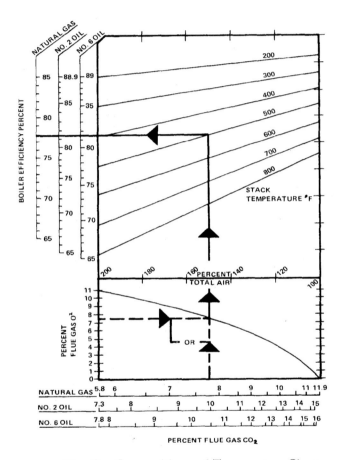

Flue Gas Composition and Temperature Chart

Original Power Factor	Desired Power Factor %												
	80	81	82	83	84	85	86	87	88	89	90	91	92
50	.982					1.112					1.248		1.308
51													
52	.893												
53													
54	.809												
55						.899					1.035		1.090
56	.730												
57													
58	.655												
59													
60	.584	.610	.636	.662	.688	.714	.741	.767	.794	.822	.850	.878	.905
65	.419					.549							.740
70	.270					.400							.591
75	.132					.262							.453
80	.000	.026				.130					.266		.321

Power Factor Correction Table

Index